PRACTICAL MEDICAL PHYSICS

PRACTICAL MEDICAL PHYSICS

A Guide to the Work of Hospital Clinical Scientists

Edited by

Debbie Peet

University Hospitals of Leicester NHS Trust, Leicester, UK

Emma Chung

University Hospitals of Leicester NHS Trust, Leicester, UK and
Department of Cardiovascular Sciences, University of Leicester, UK

CRC Press is an imprint of the
Taylor & Francis Group, an **informa** business

First edition published 2022
by CRC Press
6000 Broken Sound Parkway NW, Suite 300, Boca Raton, FL 33487-2742

and by CRC Press
2 Park Square, Milton Park, Abingdon, Oxon, OX14 4RN

Library of Congress Cataloging-in-Publication Data
Names: Peet, D. J. (Debbie J.) editor. | Chung, Emma, editor.
Title: Practical medical physics : practical medical physics / edited by Debbie Peet and Emma Chung.
Description: First edition. | Boca Raton : CRC Press, 2021. | Includes
bibliographical references and index.
Identifiers: LCCN 2021020333 | ISBN 9781138307537 (paperback) | ISBN
9781138309821 (hardback) | ISBN 9781315142425 (ebook)
Subjects: LCSH: Medical physics–Textbooks.
Classification: LCC R895 .P72 2021 | DDC 610.1/53–dc23
LC record available at https://lccn.loc.gov/2021020333

ISBN: 978-1-138-30982-1 (hbk)
ISBN: 978-1-138-30753-7 (pbk)
ISBN: 978-1-315-14242-5 (ebk)

DOI:10.1201/9781315142425

Typeset in Times
by MPS Limited, Dehradun

Contents

Preface

Scientists have played a major role in developing cancer therapy using radioactive sources in developing imaging using X-rays in radiography and CT scanners and using electromagnetic fields in MR scanners. More recently they have been involved in the development of proton therapy and the PET/CT scanners of the twenty first century. Scientists have also worked on projects from the first vaccination programmes to the genetics programme in life science and from simple measurements of pulse, temperature and blood pressure to Doppler ultrasound, shear wave elastography and magnetoencephalography in physiological measurement.

Scientists in Universities, those working for equipment manufacturers and those in hospitals across the world and in the NHS in the UK have been involved from the initial design stage through development of equipment that makes a significant contribution to healthcare and to the translation of that technology and the introduction of new techniques into the healthcare setting. Indeed, science is seen as key to the plans for the NHS in the short, medium and long term.

Scientists in healthcare span the life sciences, physiological sciences and the physical sciences, including engineering. This book will concentrate on the role of Physical Scientists in Healthcare. Across the world they are commonly called Medical Physicists. As a career, Medical Physics offers many opportunities for a stimulating and rewarding use of the knowledge and understanding gained from a university level education. Whilst it has been a privilege to observe many of these developments first hand and to have been directly involved in a number of these advancements, the pace of change and opportunities are likely to keep proceeding at speed even in the current economically challenging and post COVID environment and scientists in the NHS can look back with pride and to the future with drive, determination and optimism.

This book was conceived back in 1983. At that time Dr Ray Hudson, the Head of Radiotherapy Physics at Mount Vernon Hospital, pointed out to me – a young and quite green Medical Physicist – that the way I approached the work and recorded how to do new things would make a very good textbook. At the time I scoffed but when he called me ten years later to tell me he was dying I committed to completing this book. As I was then on my second period of maternity leave and absolutely exhausted, I couldn't contemplate starting such a project at the time but here it now is – nearly 40 years in the mental drafting and with a lot of help from my current colleagues. I hope it shows the variety and interesting nature of the work and at the same time areas of commonality between the different specialisms and the innate skills and characteristics of those who choose this wonderfully rewarding career path. One of the commonest questions I have been asked throughout my career is what on earth is it that you actually do? That comes not only from my family and friends but from colleagues in other disciplines and career structures. Here is my answer....

Debbie Peet

Acknowledgements

The Editors would like to thank the busy NHS staff who have taken time out of their duties to share their passion for their work and contribute to this book. We hope we have succeeded in producing a descriptive and informative text to inspire the next generation of Medical Physicists.

Special thanks to:

Sarah Fletcher, Nicole Jones, Natalie Cooper-Raynor, Malula Masonda, Daniel Bullin, Jade Clarke, Roya Jalali and Kevin Martin.

Contributors

Debbie Peet was the Head of Medical Physics in the University Hospitals of Leicester NHS Trust. She has worked in the NHS for nearly 40 years as a Clinical Scientist in all disciplines but has specialised in Radiation Safety and Diagnostic Radiology Physics for over 25 of those years. She has an interest in facility design in Radiotherapy and Nuclear Medicine including PET/CT and has published several book chapters on these topics.

Emma Chung is a Lecturer in Medical Physics within the University of Leicester Department of Cardiovascular Sciences. She is a registered NHS Clinical Scientist specialising in ultrasound physics and is currently External Examiner for Imperial College London's MSc in Medical Ultrasound. Emma has worked jointly within the NHS and academia since 2004 and has been the recipient of numerous research awards. Throughout her career, Emma has been active in supporting the academic advancement of Clinical Scientist and encouraging students from all backgrounds to consider a career in Medical Physics.

Philip Baker works at the University Hospitals of Leicester NHS Trust where he is the Lead Physicist for dosimetry in Radiotherapy. He has been specialising in Radiotherapy for twelve years.

Alimul Chowdhury is the Head of Non-Ionising Radiation at the University Hospitals of Leicester NHS Trust. He has worked as a MR Physicist since 2001, both in NHS Trusts and in academic institutions. He has been registered as a Clinical Scientist since 2008. Ali has an interest in research, and he has co-authored several articles related to functional MRI and Magnetic Resonance Spectroscopy (MRS) measurements.

Joanne Cowe is the Head of the Electrodiagnostics Service in the University Hospitals of Leicester NHS Trust. She began her NHS career in 2005 in Ultrasound; since then she spent some time working in Clinical Engineering prior to moving to Electrodiagnostics where she has worked since 2014.

Elizabeth (Lizzie) Davies is Head of Leicester Radiation Safety Service at University Hospitals of Leicester NHS Trust. She acts as Radiation Protection Adviser, Radioactive Waste Adviser and Medical Physics Expert for Diagnostic Radiology to the Trust. She says that the best thing about her job is the variety of the work; no day is ever the same as the next and she enjoys applying physics to practical situations.

Richard Farley is a Clinical Scientist working at the University Hospitals of Leicester NHS Trust. He has worked within Radiation Safety and Diagnostic Radiology Physics since 2009. Originally starting as a Clinical Technologist, he has performed assessments on a broad spectrum of clinical ionising radiation equipment.

John Gittins works as a Lead Clinical Scientist at the University Hospitals of Leicester NHS Trust. He began his career at the Trust in 1994 working as a Research Associate in Ultrasound, Physiological Measurement and Medical Instrumentation. In 2010, he joined the Radiotherapy department at UHL, having responsibility for Radiotherapy Computing, Imaging and Engineering support.

Justyna Janus works as Non-Ionising Radiation Specialist within the Medical Physics Department at the University Hospitals of Leicester, NHS Trust. She has been involved in ultrasound research for over a decade and provides NHS Medical Physics trainees with training in carrying out ultrasound QA, risk assessments, and audits. Justyna has experience in managing clinical and pre-clinical research using various imaging systems to study human disease.

Greg Jolliffe works as a Senior Clinical Scientist with responsibility for sealed sources in Radiotherapy Physics at the University Hospitals of Leicester NHS Trust.

Jasdip Mangat began his career in the NHS in 2002 as a Clinical Scientist trainee at St Georges Hospital, London. From 2005 to 2011 Jay worked within the Clinical Engineering department at Addenbrookes Hospital in Cambridge, where he became Lead Clinical Scientist for the design and development sub-section in 2008. He joined the team at Leicester in 2011 and went on to become Head of Service in 2015. Jay has significant experience in the field of medical devices relating to all aspects of the medical device lifecycle, including design and development, management, and support. Jay has also held numerous professional roles during his career as a Clinical Engineer.

Anna Mason is a Clinical Scientist with 16 years of experience in Radiotherapy Physics. Prior to this she completed her PhD studying the radiobiology of neutron beams at the University of Birmingham. Anna's main area of expertise is treatment planning and she has a keen interest in training. She has taught FRCR part 1 physics candidates in both Oncology and Radiology and contributed to writing OSFA exams for the NSHCS.

Caroline Reddy works at the University Hospitals of Leicester NHS Trust where she is the Lead Clinical Scientist for Treatment Planning. She began her career as a Clinical Scientist in Radiotherapy Physics in 2002.

Lisa Rowley is Head of Section in Nuclear Medicine. She is a Medical Physics Expert, a certified Radiation Protection Adviser and Radioactive Waste Adviser, and was formerly a Dangerous Goods Safety Adviser. She is currently developing the molecular radiotherapy service at Leicester.

David Towey is Head of Nuclear Medicine Physics at Northampton General Hospital NHS Trust. He has over 20 years of experience in Nuclear Medicine and has a special interest in image processing and image analysis. He completed his PhD in brain SPECT imaging and analysis at Imperial College and has published papers in the fields of SPECT, PET and clinical image processing.

Andrea Wynn-Jones works at University Hospitals of Leicester NHS Trust where she is the Head of Radiotherapy Physics. She began her career as a Clinical Scientist in Nuclear Medicine and Ultrasound before changing to Radiotherapy where she has worked for more than 32 years. Andrea has a special interest in dosimetry audit and has acted as a national assessor for IPEM and contributed to curriculum development for the NSHCS Clinical Scientist Training Programme.

1 Introduction

Debbie Peet, Emma Chung, Jasdip Mangat and Joanne Cowe

CONTENTS

From the early days of hospital care and the inception of the NHS in 1948, the field of Medical Physics has driven advances in technology and aided innovations in healthcare. In 1901, the first Nobel Prize for Physics was awarded to Wilhelm Roentgen for his discovery of X-rays in 1895. Roentgen found that passing X-rays through human tissue revealed the structure of the bones (Figure 1.1). The increasing reliance of doctors on diagnostic X-ray imaging then paved the way for the development of X-ray computed tomography (CT) and further innovations including

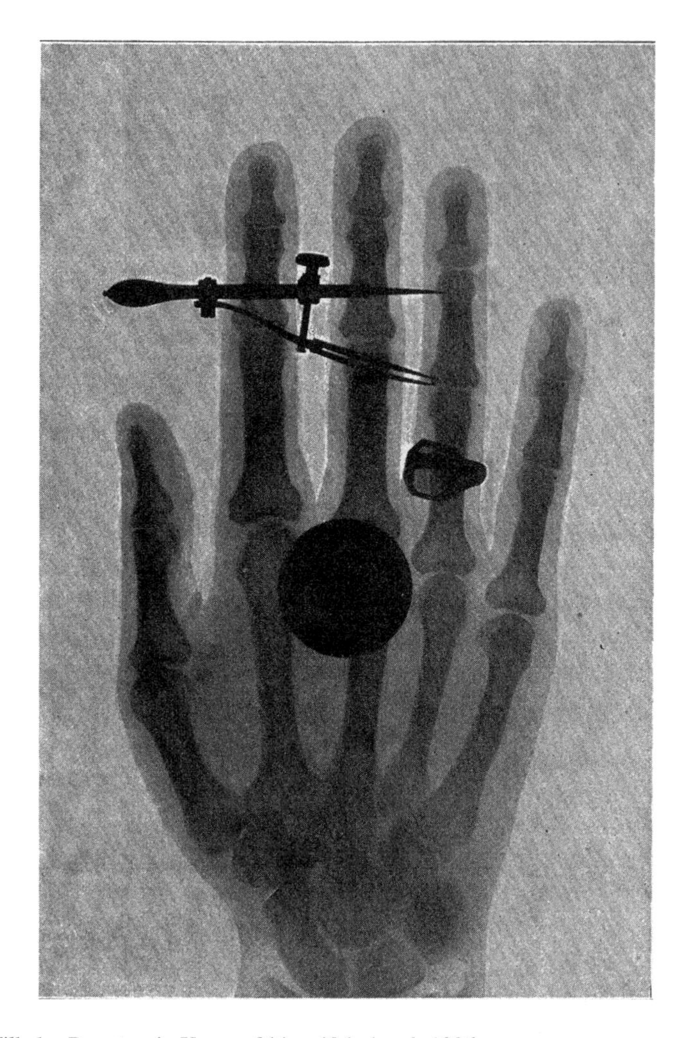

FIGURE 1.1 Wilhelm Roentgen's X-ray of his wife's hand, 1896.

magnetic resonance imaging (MRI) and ultrasound. Today, high-energy X-rays generated by linear accelerators are also applied to radiotherapy treatment of cancer.

In 1911, the Polish-born physicist Marie Sklodowska Curie was awarded the Nobel Prize for Chemistry for her discovery of the radioactive element, radium (Gasinska 2016). Her work demonstrated potential applications of radioactive sources in therapeutic medical procedures, leading to the development of Molecular Imaging and Radiotherapy techniques in the field of Nuclear Medicine and Radiotherapy.

Like many physicists, Marie Curie operated mobile X-ray imaging units during the First World War, training the first Radiologists, which contributed to saving millions of lives. Today, we face a very different set of challenges, but the essential role of Medical Physicists in advancing medical device development, training practitioners, and ensuring patient safety continues.

Although early clinical users of X-rays were unaware of the potential harmful effects of radiation exposure, injuries were quickly observed. The field of Health Physics and Radiation Safety developed further after the bombings of Hiroshima and Nagasaki, and the Chernobyl disaster. We now have a far better understanding of the risks associated with ionising radiation and have developed protocols to reduce unnecessary exposure.

The development of sonar during the First and Second World Wars paved the way for new medical ultrasound technologies, which began to become widely used clinically for obstetrics scanning during the 1960s. Ultrasound avoids the use of ionising radiation, so it is safer for use in pregnancy than X-rays. A further breakthrough in imaging occurred in the 1980s through the introduction of a further non-ionising imaging technique – MRI. Today, the use of medical imaging (diagnostic X-rays, ultrasound and MRI), together with other technologies and medical devices for diagnosis and treatment, forms a large part of the work conducted by Clinical Medical Physicists.

Medical technologies often rely on software, information technology and artificial intelligence to generate diagnostically useful information. From simple measurements of pulse, temperature and blood pressure, to more complicated models and simulations, our reliance on healthcare technology requires specialist scientific knowledge and assessment skills.

All Clinical Scientists working in Medical Physics need to work closely with other NHS staff, to ensure safe and effective diagnostics and therapeutic care. General professional skills expected of Clinical Scientists are introduced later in this chapter. These include key clinical skills, the ability to conduct research and critically evaluate clinical services, development of quality assurance (QA) programmes, leadership and professional development. This book is not intended to cover the underlying theory of Medical Physics. There are many excellent specialist texts referenced in indiviual chapters. This book aims to bridge the disparity between the theory of Medical Physics described within university textbooks, and the "real-life" practice of Medical Physics by hospital Clinical Scientists. We illustrate how Medical Physics is applied within hospitals through the practical skills and knowledge that Clinical Scientists bring to their daily work.

1.1 MEDICAL PHYSICISTS AND HEALTHCARE SCIENTISTS

Medical Physicists can be found in academia, industry and healthcare, leading device development and translation of scientific advances to a clinical setting. They are key workers within the healthcare workforce, supporting day-to-day delivery and improvements in patient care. Across most of the world, the title Medical Physicist is used to describe scientists working at masters/doctoral level in healthcare. In the UK, a protected broader title of Clinical Scientist has also been adopted. This brings Medical Physicists under the governance of the Health and Care Professions Council (HCPC 2020a), which requires Clinical Scientists to be listed on a statutory register.

Throughout this book, the title Clinical Scientist can be interchanged with "Medical Physicist" used elsewhere in the world, although it should be recognised that within healthcare Clinical Scientists can also hold other roles drawn from other specialties including engineering, biological sciences, mathematics, chemistry and informatics.

Clinical Scientists may be laboratory-based, engaged in testing samples, investigating genetics or developing and trialling new drugs. Others work directly with patients, performing scans, administering treatment or taking measurements. Scientists also work behind the scenes, ensuring that medical equipment is working safely, and driving the development and evaluation of advances in medical research. All Clinical Scientists are involved in research and innovation within and across specialist areas.

The title of Clinical Scientist forms part of a wider Healthcare Science career structure, which covers more junior Healthcare Science Practitioners and Healthcare Science Associates and Assistants, and more senior Consultant Clinical Scientists. All Healthcare Scientists (including Clinical Scientists) fall under the umbrella of the National School of Healthcare Science (NSHCS 2020). The place of Clinical Scientists within the NSHCS career structure is summarised in Figure 1.2. For more information, please see the NSHCS "Careers in healthcare science" webpage (NSHCS 2020).

In the UK, there are estimated to be over 50,000 Healthcare Scientists working in more than 50 specialist areas. Healthcare Scientists are estimated to be involved in 80% of all clinical decisions and are behind the introduction of life-saving clinical and technological advancements for preventing, diagnosing and treating a wide range of medical conditions (NHS England 2020a). If you

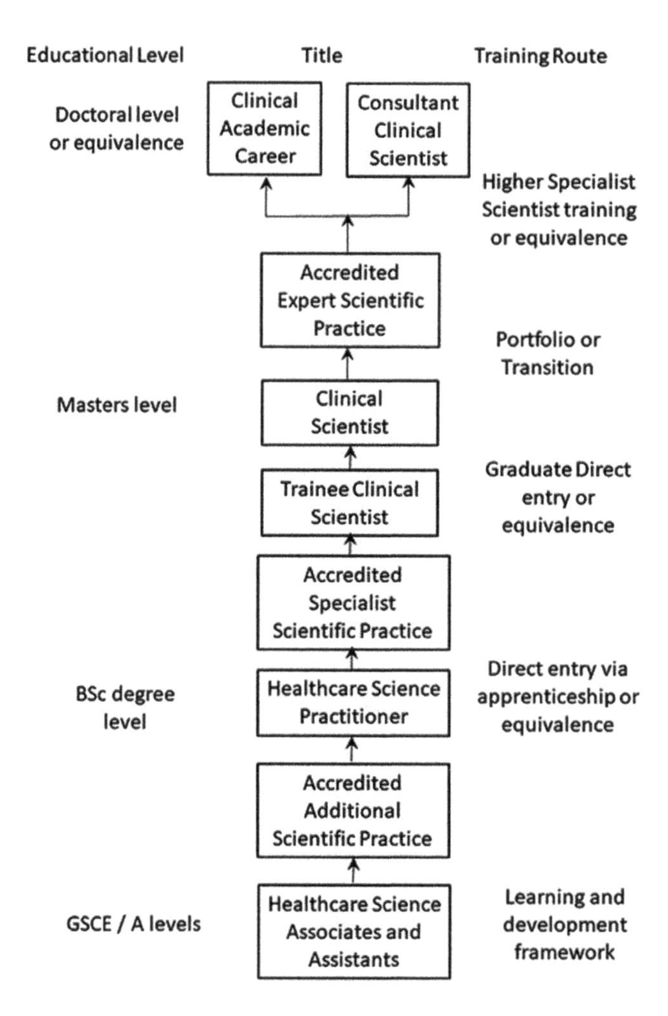

FIGURE 1.2 Career structure for UK Healthcare Sciences.

have ever had a blood test, been given a new treatment, had an X-ray, or undergone a hearing test, it is more than likely that a healthcare scientist was involved. In this book, we focus on describing the role of NHS Clinical Scientists specialising in Medical Physics.

1.2 CLINICAL SCIENTIST TRAINING FOR MEDICAL PHYSICISTS

A career in Medical Physics offers a stimulating and rewarding opportunity to apply the skills and knowledge gained as part of an undergraduate degree for patient benefit. Good communication and the ability to work as part of a team are essential. Clinical Scientists who specialise in Medical Physics usually have a strong background in physics, followed by a specialist master's and/or doctoral degree.

1.2.1 Entry Requirements and Career Path

The NHS offers apprenticeship schemes to enable school leavers with GCSEs to train as Associate or Assistant Healthcare Scientists. School leavers with good A levels are also able to train as Healthcare Science Practitioners, which involves studying for a degree-level qualification through the Practitioner Training Programme (PTP).

If you are already working within the NHS as an Assistant Healthcare Scientist or Healthcare Science Practitioner, progression is possible through demonstrating equivalence to the next level without enrolling on a formal training programme. These equivalence progression routes allow existing NHS staff to transition between career paths and gradually "work their way up".

If you have a good science degree, and are not already working in the NHS, you can apply directly to a "fast track" graduate-level Scientist Training Programme (STP). This provides a full salary during 3 years of hospital-based training, during which students also prepare for an MSc in Clinical Science. Successful completion of the STP training programme leads to registration as a Clinical Scientist working at Band 7 of the NHS pay scale. The STP scheme welcomes Healthcare Science Practitioners, graduates with higher level qualifications (such as doctorates), and scientists who have relevant experience working in industry. The STP scheme is understandably popular and places are limited. The selection process is therefore highly competitive. Successful candidates will typically have at least a 2:1 degree in a relevant discipline, as well as hospital experience, gained either as part of their studies, or through additional shadowing and work experience.

Beyond STP training and equivalence routes, Clinical Scientists are increasingly encouraged to combine a clinical and an academic career. Qualified Clinical Scientists will find opportunities to progress to leadership and Consultant Clinical Scientist roles through further training. The NHS Higher Specialist Scientific Training (HSST) scheme mirrors the progression of clinically qualified doctors to consultant status. For Clinical Scientists, this may involve studying for a doctoral-level qualification in the form of a Doctoral of Clinical Science (DClinSci) or demonstrating equivalent experience through equivalence routes.

In the UK, the regulatory framework governing the registration of Clinical Scientists mirrors that of registration of doctors with the General Medical Council (GMC). Clinical Scientists specialising in Medical Physics are registered by the HCPC (2020a). This means that they have been assessed for their leadership, research, scientific, technical, clinical and communication skills, and have been judged fit to practice. Although Clinical Scientists may be based in healthcare, industry, or academia, practical hospital experience is essential. Careers can be flexible, with many scientists choosing to work part time or take career breaks.

1.3 THE ACADEMY OF HEALTHCARE SCIENCE

The body responsible for assessing whether UK Healthcare Scientists meet key HCPC criteria is currently the Academy of Healthcare Science (AHCS 2020). The AHCS works alongside professional bodies, such as the Institute of Physics in Medicine (IPEM), to review initial applications for a Certificate of Attainment obtained through the graduate-level NHS STP. Or, for existing NHS staff, a Certificate of STP Equivalence can be applied for through submission of a portfolio of evidence. The AHCS also assesses individuals working at the level below Clinical Scientists through the PTP.

Regardless of whether applying for a Certificate of Attainment, or Certificate of Equivalence, Clinical Scientist registration requires all candidates to demonstrate "competencies" across the following scientific, technical, clinical, research and professional domains appropriate to each specialist area.

1.4 THE HCPC STANDARDS OF PROFICIENCY

Clinical Scientists operate under the HCPC Standards of Conduct as outlined on the HCPC website (HCPC 2020b), as well as adhering to Good Scientific Practice (GSP) guidelines as laid down by the AHCS (AHCS 2012). The application of basic scientific principles to medicine continues to evolve, so Clinical Scientists need to be at the forefront of advances in patient care, proficient within their chosen field, and demonstrate adherence to the HCPC "Standards of Proficiency" (HCPC 2020a).

The aims of the HCPC standards of proficiency are to:

- Outline what service users and the public should expect from their health and care professional
- Set standards to protect the public from harm
- Set clear expectations of knowledge and abilities for each registered profession
- Ensure registered professionals continue to meet the standards of proficiency that apply to the scope of practice

It is important that all healthcare professionals registered under HCPC protected titles, such as Paramedics, Clinical Scientists and Radiographers, meet these standards to be able to practise lawfully, safely and effectively. HCPC approved training programmes, such as the STP programme, are tailored to enable staff to meet these standards. For registered Clinical Scientists, the HCPC may intervene if someone raises a concern about a registrant's practice.

1.4.1 Good Scientific Practice

Standards of GSP are split into five domains (Figure 1.3). These are broadly discussed in the following sections, although how competency in each area is demonstrated is specific to each role.

1. **Professional Practice**
   Members of the workforce observing GSP will put the patient first, ensuring that patients are treated as individuals, with respect and dignity, and without discrimination; not everyone will work in a patient-facing role, but as a Clinical Scientist, your work will directly contribute to patient care. Patients and other professionals may ask for your professional opinion, so it is imperative that you understand the limits of your knowledge and experience, and only work within your scope of practice. Most healthcare professionals will be working as part of a multi-disciplinary team. Good teamwork and communication, and an understanding of others' roles and expertise, facilitates the reduction of delays to patient care and effective delivery of services. Healthcare Scientists work in a position of trust and should have good ethical principles, demonstrating honesty and integrity, and a willingness to take responsibility for their decisions and actions.

2. **Scientific Practice**
   GSP requires that individuals think and act scientifically; this may seem obvious, but it is worth pointing out that this applies to multiple aspects of the role, not just to research studies. Critical thinking and problem solving are key to improving practices and services.

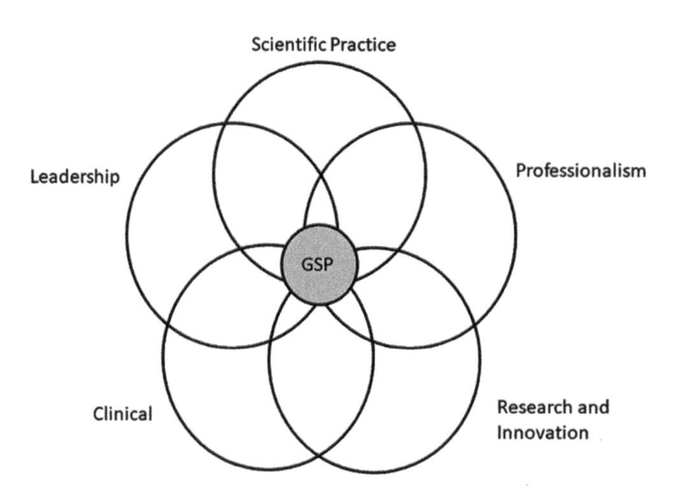

FIGURE 1.3 Elements of Good Scientific Practice (GSP).

Clinical Scientists need to be competent in undertaking a range of scientific investigations. When reporting on scientific investigations or measurements, scientists need to be aware of who the expected audience is; it may be that information needs to be pitched differently for patients, staff or other scientists. Reports should be clear and concise and ensure that appropriate methods for analysing and displaying the information are used. Again, good critical thinking skills are essential – what do these results show? Is the question/issue originally raised answered? Would this report enable others to repeat the investigation – does it accurately describe what has been done? Have appropriate references been cited? Have others involved had the opportunity to contribute and provide feedback? Scientists should critically evaluate their own work, as they would if they were auditing or assessing the work of others.

Scientists provide scientific and technical advice to ensure safe and effective delivery of services. They will be expected to take an active role in ensuring a high quality of services, including participation in audits and quality control (QC) checks (described in more detail in future chapters). Clinical Scientists are expected to identify and manage sources of risk by performing risk assessments; this may include risks related to radiation, electricity, equipment or clinical waste (among others). Setting, applying and maintaining quality standards, and keeping an effective audit trail are imperative. The ability to maintain clear, accurate and complete records are essential.

3. **Clinical Practice**
   Regardless of whether Clinical Scientists have a patient-facing role, there are aspects of GSP to which all Clinical Scientists must adhere. This includes the need to inform patients, carers and staff of any procedures being undertaken, and processes for receiving and evidencing informed consent. Scientists also have a responsibility to ensure that any staff they supervise understand this requirement. It is imperative that scientists and their teams, including administration and clerical staff, maintain the confidentiality of patient information and records accessed during their work.

   For scientists working in patient-facing services, it is vital that protocols and services are evidence-based and align with best practice for their field. Where necessary, the delivery of services should be prioritised based on the clinical needs of the patient. To this end, it is important that scientists keep their scientific knowledge up-to-date and that they understand the wider clinical implications and consequences of their decisions; scientists should only ever act within their scope of practice and competency. Scientific knowledge will increase with time, and delivery of services involves building good working relationships staff and patients, attending conferences and training courses, and representing the service at multidisciplinary team meetings.

   When carrying out clinical assessments, it is vital that clear and accurate records are kept; these may need to be reviewed by yourself and others when analysing and interpreting the results. Records can be recalled at any time and used, together with other findings, to provide a better understanding of the complete clinical picture and its evolution over time – it is therefore important that records are legible and free from any ambiguity; what may seem obvious and unnecessary to record at the time of testing can be vital at a later date.

4. **Research, Development and Innovation**
   Participating in research, development and innovation is a key part of any role in healthcare science. This can take several forms, such as reviewing and critically appraising scientific evidence, to informing your own practice, and ensuring your service stays at the leading edge of evidence-based practice. When implementing new service developments, you may need to design a research study to evaluate the impact of your changes; to assess whether a proposed treatment, diagnostic procedure or analysis

method, provides benefits for patients or improves NHS workflows. Where evidence indicates it may be beneficial to implement a new procedure, you may also be involved in embedding this into routine practice. Scientists often work as part of a wider collaborative research team, and with other professions, to support data gathering and analysis for research studies, to perform laboratory experiments, and to analyse, model, and interpret data that others have collected.

5. **Clinical Leadership**

   Not all Clinical Scientists will be involved in managing other staff, but effective clinical leadership skills need to be demonstrated. All colleagues should be treated fairly and with respect, regardless of status, and it is important that the skills and contributions that staff make to your team and service are recognised and valued. While working in a clinical environment, care should be taken to see each patient as an individual, and to understand the roles and responsibilities of different team members. Imagine if you were a patient and no-one in the clinic introduced themselves to you or explained what they would be doing; putting yourself in the shoes of your patients, or colleagues, is a good exercise to help critically assess the service you are providing.

   Safety is key; we all have a duty to ensure our practices, and the environments we work in, are safe for all. This may include reporting shortfalls in the performance of others that may be putting patients or staff at risk and developing actions for improvement. This may involve working with other professions and departments (e.g. estates, infection prevention, nursing, pharmacy, IT and finance) to ensure risks are minimised. You also have a duty to ensure your role will be covered in the event of your absence; don't assume others will pick up where you left off, good communication and handover are essential.

   Problem solving, critical thinking and the ability to understand how your service fits into the bigger picture, both within your local hospital, and as a profession, are key to good clinical leadership. Finally, remember that as a Clinical Scientist you will be a role model to other staff groups, so adhering to GSP will help perpetuate good practices throughout your service and organisation.

1.5 CONTINUOUS PROFESSIONAL DEVELOPMENT AND PROGRESSION

It is important that Healthcare Scientists are proactive in their own personal development, ensuring their skills and knowledge are kept up to date. Specialist skills and knowledge need to be continually refreshed and maintained through adopting the principles of Continuous Professional Development (CPD). Some staff remain in the same institution, whereas others move to progress. As a Clinical Scientist's career develops, they may choose to specialise in a particular area, or seek to gain managerial and leadership skills towards registration as a Consultant Clinical Scientist. Consultant Clinical Scientist status is also awarded by the AHCS by undertaking Higher Specialist Scientist Training (HSST), or by demonstrating Higher Specialist Scientist (HSS) equivalence.

Many scientists are also involved in the training and development of other staff; students and trainees should be properly supervised, and it is important that they adhere to GSP in addition to being proficient in the theoretical and practical aspects of their role.

1.6 LINKS TO OTHER PROFESSIONS

Nearly all hospital Clinical Scientists work closely with other professional groups responsible for patient care. Some hospital scientists work behind the scenes, whereas others have more patient contact. Some work mainly in research, others are split between research and clinical roles. Senior scientists are responsible for managing hospital services and might provide consultancy advice to external providers; to support private clinics, dentists and GPs in adhering to best practice and regulatory guidelines. A further important role of Clinical Scientists is in the training of other hospital staff. Many also have academic roles, teaching undergraduate and postgraduate courses and performing scientific and clinical research.

Many services require multidisciplinary teams for daily operation; therefore, Clinical Scientists need to be able to work closely with other healthcare staff. Links to other professions differ between specialties and are described in more detail in individual chapters.

With over 350 different NHS careers available, roles and job titles of NHS staff can initially be confusing. Job titles provide important information that enable staff to immediately establish a staff member's area of work, scope of practice and level of responsibility. It is good etiquette to refer to staff by their proper job titles, to avoid inadvertently causing offence; you will soon be corrected if you accidentally refer to a Radiologist as a Radiographer, or a surgeon as "Doctor" rather than "Mr". If in doubt, check their ID badge. Examples of staff groups are summarised in Table 1.1. A full list of roles is provided on the NHS health careers website (NHS 2020).

Technologies and treatments change rapidly, so it is important to attend conferences and keep in contact with colleagues and equipment manufacturers to keep abreast of new developments. Many physicists also need to liaise with external service engineers and equipment maintenance staff to ensure devices and scanners are working properly. For some Clinical Scientists, especially those based at research institutes and university teaching hospitals, there are opportunities to be involved with research, which encourages links with other hospitals and universities. You may also be called upon to act as an ambassador for your role in outreach and communication activities. This includes

TABLE 1.1
Hospital Staff: Individual Titles and Staff Groups

Doctors	There are a wide variety of clinically qualified specialist doctors, including Radiologists (who specialise in medical imaging), and various specialist Surgeons and Physicians. For example, Oncologists specialise in cancer treatment. Some clinicians specialise in treating specific organs, such as the heart (Cardiologists) or Brain (Neurologists). A Specialist Registrar (SpR) is a doctor who is receiving advanced training in a specialist field of medicine.
Allied Health Professionals	There are 14 types of Allied Health Professional. Common examples include: Radiographer, Physiotherapist, Dietician, Operating Theatre Practitioner, Paramedic or Osteopath.
Healthcare Scientists	There are over 50 types of Healthcare Scientist with backgrounds in Clinical Bioinformatics, Life Sciences, Physical Sciences and Biomedical Engineering, and Physiology. Healthcare Scientists with a background in physics usually specialise in clinical measurement, clinical or medical technology in Medical Physics, imaging (ionising), imaging (non-ionising), medical device risk management and governance, Medical Engineering, Nuclear Medicine, Radiation Physics and Radiation Safety Physics or Radiotherapy Physics.
Nurses	Adult and paediatric nurses, critical care nurses.
Medical Associate Professions	Newer roles include physician and anaesthesia associates and critical care and surgical care practitioners.
Wider Healthcare Teams	The NHS relies on staff in other diverse roles, such as health records staff, caterers, cleaners, porters, newborn hearing screeners, healthcare assistants, receptionists, medical secretaries, human resources staff and finance managers. All staff work together and are important for supporting patients and the NHS regardless of their different roles.

FIGURE 1.4 Medical Physics Outreach activities.

communication with patient groups and charities, as well as raising awareness of Medical Physics as a career among school and university students (Figure 1.4).

1.7 WORKING WITH MEDICAL DEVICES

All Clinical Scientists work closely with medical devices. Clinical Scientists often become leading experts in the devices used within their specialism, developing new devices, exploring new clinical applications and training users. Clinical Scientists working in Medical Physics often work closely with those specialising in Clinical Engineering, who tend to be responsible for managing the procurement and QA of hospital equipment, but also play an important role in assessing and developing new devices. Although the Scientist's specialist knowledge of the operation and use of such equipment is vital, Clinical Engineers also play an important role.

Medical devices and health technology are integral to all aspects of patient care. In 2016, the NHS was estimated to spend in the region of £500 million pounds per annum on capital equipment (equipment costing over £5,000). The cost of acquiring and managing NHS medical equipment is estimated to be in the region of £1.2 billion a year. As such, expertise and resources are required to ensure that the purchase and maintenance of medical equipment is of value for money and benefits patients.

Within Medical Physics, much of the work of Clinical Scientists involves equipment that has been classed as a "medical device". Medical devices are more closely regulated than other types of technology. The range of equipment and objects that qualify as medical devices might surprise you, covering simple objects such as walking sticks, to thermometers, imaging equipment, software and linear accelerators. Whether or not an object is formally classified as a medical device depends very much on what you intend to use it for.

MEDICAL DEVICE DEFINITION

In accordance with EU Regulation 2017/745, a medical device is defined as:

"any instrument, apparatus, appliance, software, implant, reagent, material or other article intended by the manufacturer to be used, alone or in combination, for human beings for one or more of the following specific medical purposes:

- *Diagnosis, prevention, monitoring, prediction, prognosis, treatment or alleviation of disease*
- *Diagnosis, monitoring, treatment, alleviation of, or compensation for, an injury or disability*
- *Investigation, replacement or modification of the anatomy or of a physiological or pathological process or state*
- *Providing information by means of in vitro examination of specimens derived from the human body, including organ, blood and tissue donations and which does not achieve its principal intended action by pharmacological, immunological or metabolic means, in or on the human body, but which may be assisted in its function by such means*

The following products are also deemed to be medical devices:

- *Devices for the control or support of conception*
- *Products specifically intended for the cleaning, disinfection or sterilisation of devices"*

This definition covers an enormous number of different products. More recently, the definition of medical devices has been extended to specifically include software, which reflects the increasing use of algorithms to diagnose pathology and predict patient outcomes.

In recent years, Clinical Engineering departments have expanded their procurement and equipment management roles. Clinical Engineering departments take responsibility for performing equipment checks for a wide range of medical technologies. Larger hospitals may also include engineers that are involved with research and innovation, with the main emphasis on supporting research projects involved with the development or assessment of medical devices. Clinical Engineering teams require a range of skill sets and expertise, and it is very common to see such services being managed by qualified Clinical Scientists specialising in clinical measurement, clinical or medical technology in Medical Physics, medical device risk management and governance, or medical engineering.

Staff responsible for compliance of non-certified medical devices need to be familiar with relevant Regulations. Medical devices used in Europe including the UK should carry a CE or UKCA mark. This symbol confirms that products meet relevant safety standards.

In the UK, medical devices are currently regulated via three directives:

Directive 90/385/EEC on active implantable medical devices
Directive 93/42/EEC on medical devices
Directive 98/79/EC on *in vitro* diagnostic medical devices

There are currently two important Medical device regulations laid down in Europe. These are the general medical devices (2017/745) regulations (from 26 May 2021), and the *in vitro* diagnostic medical devices (2017/746) regulations (from 26 May 2022). These directives/regulations exist to assure consumers of the safety and performance of medical devices.

For institutions that undertake medical device research and innovation, there is a need for trained and experienced individuals who are familiar with the requirements of the regulations, and how to apply them. Clinical Scientists and Engineers provide consultancy advice to university researchers, or commercial companies, looking to conduct clinical device trials. Device manufacturers may also wish to obtain clinical data as evidence for CE marking.

Clinicians and researchers occasionally want to use an existing medical device for a purpose that it was not originally designed for, or to custom design a prosthetic device for an individual patient. Clinical Engineers may be asked to generate an "in-house" design or modify a device to fit a research or clinical requirement, in which case additional safety checks and approvals may be required. Scientists that work within Clinical Engineering departments require the training, knowledge and experience to ensure that the use of "off-label" and "in house" devices are compliant with all aspects of the regulations, as well as supporting applications for approval from relevant authorities. In the UK, medical devices are regulated by the Medicines and Healthcare Regulatory Authority (MHRA 2020).

More widespread is the need for expertise in selecting and managing healthcare technologies. In the UK, a large proportion of capital medical equipment is procured through framework providers to provide better value for money by negotiating bulk purchasing with equipment suppliers. This system facilitates rapid procurement, without each hospital having to undergo their own extensive selection processes. The life cycle of medical equipment within any health institution is described in Figure 1.5. Additional information can be found in Healthcare Technology Management, a systematic approach, by Hegarty et al. (2017).

1.8 WORKING ENVIRONMENT AND GENERIC SKILLS

Clinical Scientists typically work in small groups to ensure continuity in service and so that physicists can support each other's training. Depending on the institution, physics teams might be embedded within a clinical area (such as Oncology) or run their own services (e.g. a Nuclear Medicine service) from within a centralised Medical Physics department. Either way, physicists will often find themselves working across a range of clinical settings and hospital sites. Staff working in Nuclear Medicine or radiation protection have even been known to occasionally delve into sewers or climb into a cherry picker to check radiation levels (Figure 1.6).

Where services are under development, a physicist may need to be embedded within the service until it becomes fully established. Clinical Scientists have varying levels of patient contact, and routine duties differ between institutions and specialisms. Different types of role, and level of contact with patients, are indicated in each chapter. Clinical Scientists are skilled in applying research techniques to solving problems, and should be willing to seek advice from other centres to build their knowledge. If no data exist, scientists may need to carry out additional calculations and measurements, communicate their findings and publish recommendations.

Trainee Clinical Scientists learn a range of practical skills, which are assessed through observation and case-based discussions, as well as more conventional forms of assessment (Table 1.2).

Clinical Scientists working in all specialisms will gain experience in the following techniques.

1.8.1 Clinical Skills

Clinical skills will depend on the role but always include basic life support and infection prevention, which is especially important since the advent of COVID. Other skills may include history taking, use of imaging and physiological measurement techniques, cannulation, injection and some drug administration. These depend on the exact role undertaken. The process for demonstrating competency in clinical tasks varies between institutions, but usually requires the task to be performed a minimum number of times under supervision, as well as being "signed-off" by an approved Assessor.

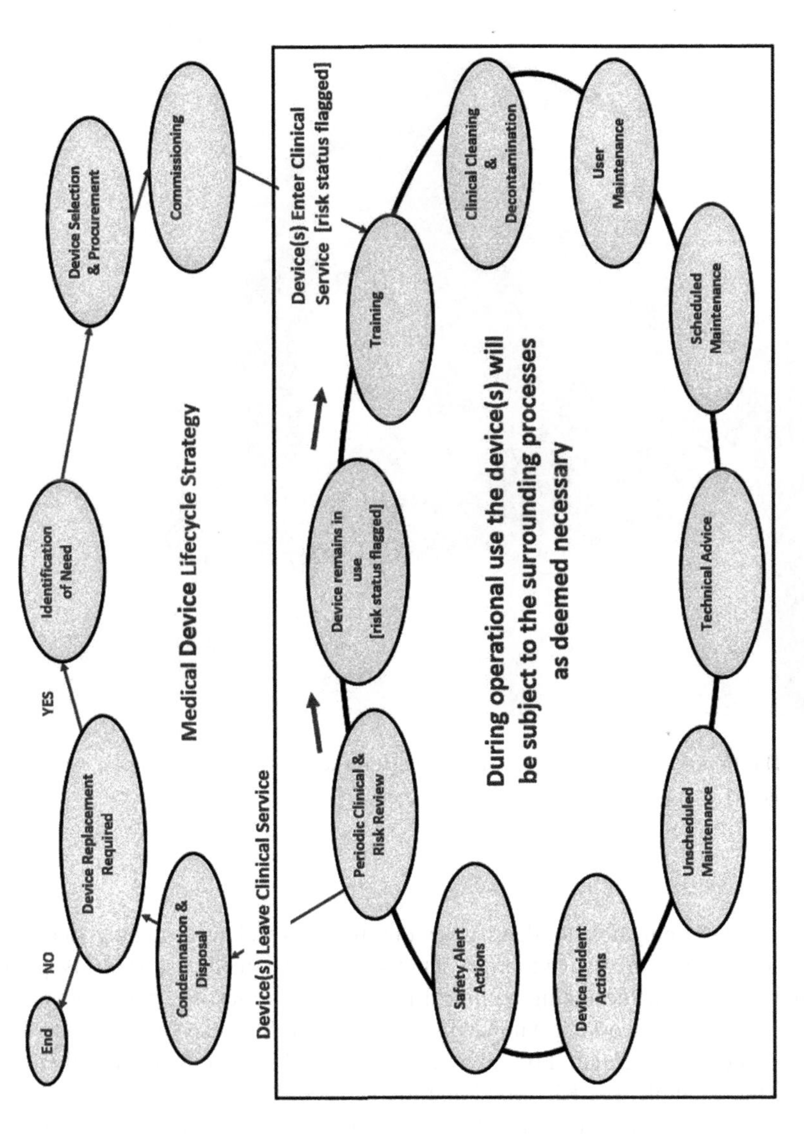

FIGURE 1.5 Life cycle of medical equipment, reproduced with the permission of Dr J McCarthy (Hegarty et al. 2017).

FIGURE 1.6 The Editor, Debbie Peet, measuring radiation levels outside of a linear accelerator bunker in a gale.

1.8.2 Research Skills

Research skills include being able to identify key questions, as well as designing studies to answer them. Investigations may involve adopting standard research methods, such as performing a systematic literature review, designing a clinical trial, laboratory testing, performing simulations, or developing models. A working knowledge of Medical Statistics and Health Technology Assessment methods is essential. For an introduction to Medical Statistics, there are many available texts. For example, see Kirkwood and Sterne (2003).

Good critical thinking is important for designing studies and interpreting the results. Are your conclusions fully supported by data? Have you considered enough samples/examples to be sure of your findings? Was your investigative process robust? What are the implications of your findings? If you don't ask yourself these questions, others will.

It is equally important to be able to describe and present your research, both in written form as reports, papers and posters, and orally as conference presentations or presentations to relevant staff groups. There is little point in conducting a research study or service evaluation if you then keep the results to yourself.

TABLE 1.2
Types of Practical Competency Assessment

Direct Observation of Practical Skills (DOPS)	A practical assessment in which an Assessor observes a trainee carrying out a specific task. The purpose is to test the trainee's ability to complete the practical work they will be undertaking later in their career. The Assessor completes a short report on the trainee's performance and marks it as either satisfactory or unsatisfactory. Practical skills might include measuring output for a radiotherapy treatment, cleaning up a radioactive spill, or completing an ultrasound QA test.
Case-Based Discussion (CBD)	This involves a discussion between the trainee and an Assessor about a piece of work, or case, that the trainee was involved with. The purpose is to test the trainee's understanding of the work they have undertaken.
Objective Structured Final Assessments (OSFA)	OSFAs are held in the final year of STP training. Trainees undertake various timed practical tasks involving clinical scenarios, scientific skills and patient interactions. Separate OSFAs are used to test generic skills, as well as those relating to the trainees chosen speciality. It is necessary to pass both to complete the STP.
Multi-Source Feedback (MSF)	Anonymous questionnaires are completed by colleagues to assess each trainee's professional abilities, such as interpersonal and communication skills. The trainee also completes the same questionnaire to assess their own performance so that the two can be compared. Feedback is obtained at regular intervals throughout the training period.

1.8.3 Service Improvement

Scientists are likely to be involved in the evaluation of services. This could involve evaluating products or equipment used for a specific clinical purpose – requiring an understanding of the clinical purpose and the operation and limitations of the equipment, as well as any methodologies used by staff.

All professions are encouraged to engage in service improvement across the NHS. Clinical Scientists often spend time reviewing hospital processes in order to improve efficiency, safety, use of resources and patient outcome. "First steps towards quality improvement: A simple guide to improving services", NHS England (2014) recommends a five-step approach to service improvement:

1. Preparation
2. Launch
3. Diagnosis
4. Implementation
5. Evaluation

More commonly used is a continuous cycle of "Plan, Do, Study and Act", known as the PDSA cycle. Other approaches adopted in Medical Physics include "Lean" and "Six Sigma" approaches – designed to help eliminate waste and improve efficiency. For more information, see the full list of tools available from the NHS England website (NHS England 2020b).

A major challenge rests in determining whether changes to services have been beneficial. Often, a change might inadvertently influence other factors and cause unforeseen problems. It is important to monitor and review any changes to assess their effects.

1.8.4 Quality Management Systems

Quality Management Systems (QMS) are important in many areas of healthcare. They formalise processes and procedures aimed at ensuring that best practice is consistently delivered across the NHS. A QMS is often also used to align local processes with the strategic direction of a department or organisation. The UK body responsible for accrediting organisations and services is the United Kingdom Accreditation Service (UKAS 2020).

1.8.5 Audit and Service/Product Evaluation

Many scientists will be involved in undertaking audits. Audits can take many forms, including monitoring staff compliance with rules and regulations, evaluating products and services, equipment audits, or monitoring the impact of changes in service or new processes. Clinical audits involving patient treatment are typically led by medical doctors, but scientists may also be involved. Audits can be a useful tool for maintaining QMS accreditation and checking that local working practices are effective.

1.8.6 Risk Assessment

Hazards around the work of Clinical Scientists can have serious consequences for the health and wellbeing of staff and patients. The management of health and safety at work is enshrined in regulations requiring employers to complete full and suitable risk assessments for all work activities. For example, the risks and consequences of radiation exposure or introduction of ferrous metal into an MRI scanner must be carefully considered and mitigated, where possible. Clinical Scientists are often involved in undertaking calculations and offer advice to managers to enable them to complete well-informed risk assessments. They might develop templates and spreadsheets for use in clinical areas. Five steps to risk assessment, written by the Health and Safety Executive (HSE 1998) suggest the following steps:

- All hazards need to be identified.
- Consideration needs to be given to who might be harmed as a result of those hazards.
- The risks need to be evaluated and these sometimes need to be evaluated numerically, which usually involves a scientist.
- Risk reduction measures should be identified and implemented.
- Findings need to be recorded and the whole assessment needs to be regularly reviewed.

Hazards encountered in hospitals include radiation, chemicals and strong magnetic fields. Those that might be harmed include patients, individual health workers, entire groups of staff and members of the public. Evaluating risk usually involves establishing the likelihood that a hazard might cause harm. In healthcare, harm might be physical, psychological or institutional, ranging from minor injuries requiring first aid, to an increased risk of cancer, or death. Risks to the institution might include loss of public reputation or regulator confidence, disruption or closure of hospital services, or financial losses.

The process of risk assessment is often aided by calculating a risk score derived from a "consequence score" (e.g. from 1 to 5), related to the level of potential harm, multiplied by a "likelihood score" describing the likelihood that the risk will occur. The likelihood score can range from a very low chance of an event occurring (score of 1), to risks that are almost certain to happen (score of 5). A risk score between 1 and 25 is then obtained, which indicates whether action is required to reduce the risk, the frequency of review, and how quickly the risk needs to be escalated through the organisation. As part of the risk assessment process, measures are often identified and implemented to reduce the level of risk.

1.8.7 QA Programmes and QC Checks

Most scientists will play a part in delivering QA programmes incorporating processes for ensuring QC. For medical equipment, overarching QA procedures need to cover the entire life cycle of equipment, from product evaluation and procurement, to acceptance testing, commissioning, baseline performance testing, regular QC checks, and disposal. Scientists may be involved in the design of a QA programme, and partially or fully responsible for QC. Equipment management is a whole field of specialisation and is usually led by Clinical Engineering teams. Specific examples of elements of the QA programme and QC checks/performance measurements are provided in the relevant chapters.

1.9 WORKING WITH MEDICAL IMAGES

Medical Images can be produced in many ways but are all stored in a common Digital Imaging and Communications in Medicine (DICOM) format (ISO 2017). Most hospitals also use a Picture Archiving and Communication System (PACS) allowing images to be viewed from anywhere in the hospital. Whether a Scientist works in Diagnostic Radiology (MRI, ultrasound or X-ray imaging), Nuclear Medicine, or Radiotherapy, most Clinical Scientists work with some form of medical imaging data.

1.9.1 Image Properties

A large part of many Clinical Scientists' roles involve assessment of image quality. It is therefore important to understand some fundamental concepts relating to the properties of medical images, such as spatial and contrast resolution and signal to noise ratio (SNR). Image quality is characterised in similar ways across disciplines by obtaining images of test objects and phantoms, and through analysis of image properties using specialist software. Relevant image properties are briefly summarised in the following:

1.9.1.1 Contrast and Greyscale

Most, but not all, medical images are represented as a greyscale image ranging from white to black. It is important that equipment settings used to display medical images show the full range of greyscale; this can be tested by viewing a test card such as that shown in Figure 1.7. Each image also has a certain image resolution related to the number of pixels, or voxels, for a 3D dataset.

1.9.1.2 Signal to Noise

For a perfect imaging system, images would precisely represent the object being imaged. Unfortunately, real medical imaging systems are always imperfect, exhibiting varying degrees of inaccuracy or "noise". For example, an X-ray image of a uniform test object, such as a sheet of metal of constant thickness, does not produce a completely uniform image. An example of noise can be seen in Figure 1.8. A quantity called the SNR is often calculated to quantify the amount of noise in an image.

This noise can be quantified in most imaging modalities by placing a region of interest (ROI) over the image and measuring the standard deviation of pixel values within that ROI. If images are very noisy, small- or low-contrast features may be difficult to visualise. In X-ray imaging, it is important to achieve a balance between reducing image noise and exposure to ionising radiation. In MRI, it may be necessary to reach a compromise between signal to noise and scan duration.

1.9.1.3 Contrast Resolution

Noise can affect the clarity of an image, making it difficult to visualise features generating differing levels of contrast. Contrast is the difference in appearance (or signal) between an object in

FIGURE 1.7 Greyscale test card.

FIGURE 1.8 Example of noise in an image.

the image and its surroundings. The difference in contrast that can be detected by the human eye is assessed by imaging test objects of known contrast. The contrast level at which an object can no longer be distinguished from its background is used to give the contrast resolution.

1.9.1.4 Spatial Resolution

Fine detail, such as small calcifications in breast tissue or tiny emboli in the lung, requires an imaging system with exceptionally high spatial resolution. The extent to which an imaging system "blurs" the image can be assessed by imaging a point-like object, such as a dot or fine wire, and measuring blurring of that object on the resulting image – this is called the point, or line, spread function. By imaging pairs of points (or lines) with varying known separations (Figure 1.9), the minimum distance at which the two points can be separately visualised can be measured. This

FIGURE 1.9 Example line resolution test plate showing parallel lines of different spatial frequencies used to determine the fidelity of an image through estimating a modulation transfer function.

distance defines the spatial resolution. Depending on the imaging modality, spatial resolution may differ in different directions in the image, requiring separate measurements of axial and lateral resolution, depending on the geometry of the scan.

Another way of measuring spatial resolution associated with detector technologies is through determining a modulation transfer function (MTF). The MTF describes the detector's capacity to convert modulations in contrast with a particular "spatial frequency" to intensity levels in the image (Figure 1.9). For more information on assessing medical image properties, see "Webb's Physics of Medical Imaging" (Flower 2012) or "Physics for Diagnostic Radiology" (Dendy and Heaton 2012).

1.10 SUB-SPECIALTIES WITHIN MEDICAL PHYSICS

Clinical Scientists specialising in Medical Physics are typically employed to work within one or more of the following sub-specialties: MRI, ultrasound, diagnostic X-ray imaging, nuclear medicine, radiotherapy treatment of cancer or radiation safety. Hospital Medical Physics departments also employ a range of other staff including Clinical Engineers, technical, and managerial staff. There is often extensive crossover between roles, but this book focuses on the following sub-specialties listed under Medical Physics in the NHS STP training scheme.

1.10.1 Part 1: Non-Ionising Imaging (MRI and Ultrasound)

Each year, millions of medical images are used to aid to patient diagnosis or treatment. Clinical Scientists working as non-ionising radiation imaging specialists are usually experts in the physics of ultrasound (Chapter 2) or MRI (Chapter 3). Staff are responsible for performing regular safety and QC checks of imaging equipment using specially developed test objects.

1.10.2 Part 2: Ionising Radiation, Diagnostic X-rays, Nuclear Medicine and Radiotherapy

Clinical Scientists working with ionising radiation are usually experts in the physics of diagnostic X-ray imaging (Chapter 4), the use of radioactive materials for imaging and therapy (Chapter 5), or Radiotherapy (Chapter 6). A major role of Clinical Scientists working with X-rays is to verify that imaging equipment is producing sufficiently high-quality images for clinical diagnosis while minimising radiation exposure. For more information, see Chapter 4.

In Nuclear Medicine, Clinical Scientists are experts in working with radioactive materials. They usually take the lead in managing nuclear medicine equipment and services, and sometimes also support the preparation, control and administration of radiopharmaceuticals. Physicists working in Nuclear Medicine are also called on to perform calculations and simulations modelling radioactive decay. Nuclear medicine imaging is sometimes also called molecular imaging, as this technique makes it possible to "tag" specific molecules to observe functional changes. Single-photon emission CT (SPECT) and positron emission tomography (PET) are nuclear medicine techniques that provide metabolic and functional information (Chapter 5).

In some centres, Nuclear Medicine staff are involved not only in imaging but also in the therapeutic administration of radioactive material such as iodine 131 for thyroid ablation and other novel radionuclides for other tumours. This is known as molecular radiotherapy.

1.10.3 Radiotherapy

The majority of Clinical Scientists are based within hospital radiotherapy departments. Radiotherapy involves using high-energy beams of X-rays to destroy cancerous tumours within the body. The aim of this form of treatment is to shrink, or eradicate, the tumour, while sparing surrounding tissues and organs. X-rays in the MV energy range are generated using a machine called a linear accelerator. Physicists plan radiotherapy treatment based on images of each patient obtained from X-ray CT, and sometimes MRI imaging, to plan and optimise how radiotherapy treatment should be delivered.

Some radiotherapy physicists also use sealed radioactive sources for cancer treatment, known as brachytherapy. More details about the role of Clinical Scientists in Radiotherapy treatment are provided in Chapter 6.

1.10.4 Radiation Safety

Radiation safety staff are experts in the legislative guidelines governing radiation exposure, and work behind the scenes to ensure staff, patients and the public are not exposed to unnecessary high levels of radiation. Some radiation protection staff specialise in techniques involving "non-ionising" radiation in the form of light, sound or heat generated from ultraviolet (UV) light, lasers, ultrasound or MR equipment. These roles are often combined with those of experts working in non-ionising imaging. Some staff in radiation safety are expert in the safety aspects of the use of sealed and unsealed radioactive material and the equipment used in Radiotherapy and Nuclear Medicine. In some centres, the radiation safety aspects of the service delivered are covered by staff working entirely in those areas. Other centres prefer to run a centralised dedicated radiation safety service. For more details about the role of Clinical Scientists in radiation safety, see Chapter 7.

1.11 SUMMARY

Large hospital Medical Physics departments usually employ Clinical Scientists across a range of specialisms, although the level of training available can vary between institutions. As illustrated in subsequent chapters, Clinical Scientists increasingly need to be flexible and able to cross traditional boundaries. For example, combining radiotherapy treatment with X-ray CT or MRI to improve targeting of tumours, or combining Nuclear Medicine Molecular Imaging methods such as SPECT and PET scanners with anatomical information from X-ray CT scans to highlight areas of dose uptake – showing function as well as anatomy.

This book is intended to provide examples of the important work of Clinical Scientists within hospitals, giving a flavour of the different specialisms, so that the reader can understand more about the role of Medical Physicists in each area. It is not intended to replace traditional academic Medical Physics texts, but to bridge the gap between understanding the physics and the broader skills and knowledge required to work within a hospital environment. This information would be valuable to someone considering a career in Medical Physics, trainee Clinical Scientists, or for established hospital staff who would like to gain an insight into other roles.

REFERENCES

AHCS (2012). "Good Scientific Practice, Version 1". Online Available from https://www.ahcs.ac.uk/2012/12/12/good-scientific-practice/ [accessed 21 August 2020].

AHCS (2020). "Academy for Healthcare Sciences". Online Available from https://www.ahcs.ac.uk/ [accessed 21 August 2020].

Dendy PP and Heaton B (2012). *Physics for Diagnostic Radiology*, 3rd edition. CRC Press, London.

EU (1990). "Council Directive 90/385/EEC Relating to Active Implantable Medical Devices". Online Available from https://eur-lex.europa.eu/legal-content/EN/TXT/PDF/?uri=CELEX:31990L0385&qid=1600252363319&from=EN [accessed 16 September 2020].

EU (1993). "Council Directive 93/42/EEC Concerning Medical Devices". Online Available from https://eur-lex.europa.eu/legal-content/EN/TXT/PDF/?uri=CELEX:31993L0042&qid=1600252703394&from=EN [accessed 16 September 2020].

EU (1998). "Council Directive 98/79/EEC on In Vitro Diagnostic Medical Devices". Online Available from https://eur-lex.europa.eu/legal-content/EN/TXT/PDF/?uri=CELEX:31998L0079&qid=1600252838467&from=EN [accessed 16 September 2020].

EU (2017a). "EU Regulation 2017/745 on Medical Devices". Online Available from https://eur-lex.europa.eu/legal-content/EN/TXT/PDF/?uri=CELEX:32017R0745 [accessed 6 August 2020].

EU (2017b). "EU Regulation 2017/746 on In Vitro Medical Devices (2017)". Online Available from https://eur-lex.europa.eu/eli/reg/2017/746/oj [accessed 6 August 2020].

Flower (2012). *M A. Webb's Physics of Medical Imaging*. CRC Press, London.

Gasinska A. (2016). "The Contribution of Women to Radiobiology: Marie Curie and beyond". *Rep Pract Oncol Radiother* **21**(3): 250–258.

HCPC (2020a). "The Standards of Proficiency for Clinical Scientists". Online Available from http://www.hpc-uk.org/standards/standards-of-proficiency/clinical-scientists/ [accessed 9 September 2020].

HCPC (2020b). "Standards of Conduct, Performance and Ethics". Online Available from https://www.hcpc-uk.org/standards/standards-of-conduct-performance-and-ethics/ [accessed 9 September 2020].

Hegarty F, Ammore J, Blackett P, McCarthy J and Scott R (2017). *Healthcare Technology Management: A Systematic Approach*. CRC Press, London.

HSE Health and Safety Executive (1998). *Five Steps to Risk Assessment*. Leaflet INDG163(rev1). HSE Books.

ISO (2017). "Health Informatics - Digital Imaging and Communication in Medicine (DICOM) Including Workflow and Data Management". International Standards Organisation ISO/IEC 12052:2017. Online Available from https://www.iso.org/standard/72941.html [accessed 4 September 2020].

Kirkwood B and Sterne J (2003). *Essential Medical Statistics*. Blackwell Publishing, UK.

MHRA (2020). "Medical Devices Regulation and Safety". Online Available from https://www.gov.uk/topic/medicines-medical-devices-blood/medical-devices-regulation-safety [accessed 9 September 2020].

NHS (2020). "350 Careers. One NHS. Your Future". Online Available from https://www.healthcareers.nhs.uk/ [accessed 9 September 2020].

NHS England (2014). "First Steps towards Quality Improvement: A Simple Guide to Improving Services". Online Available from https://www.england.nhs.uk/improvement-hub/wp-content/uploads/sites/44/2011/06/service_improvement_guide_2014.pdf [accessed 21 July 2020].

NHS England (2020a). "What Is a Clinical Scientist". Online Available from https://www.england.nhs.uk/healthcare-science/what/ [accessed 9 September 2020].

NHS England (2020b). "Quality Service Improvement and Redesign (QSIR) Tools". Online Available from https://improvement.nhs.uk/resources/quality-service-improvement-and-redesign-qsir-tools/ [accessed 21 July 2020].

NSHCS (2020). "Careers in Healthcare Science". https://nshcs.hee.nhs.uk/careers-in-healthcare-science/ [accessed 21 July 2020].

UKAS (2020). "Management Systems Certification". Online Available from https://www.ukas.com/sectors/management-systems-certification/ [accessed 21 July 2020].

Part I

Non-Ionising Imaging

The NHS performs millions of imaging tests each year, with the most common being plain X-ray radiography, followed by diagnostic ultrasound, X-ray computerised axial tomography (CT scans) and magnetic resonance imaging (MRI). Part 1 of this book covers non-ionising (MRI and ultrasound) imaging. Ionising imaging using X-rays and molecular imaging techniques in Nuclear Medicine are reserved for Part 2.

Ultrasound and MRI represent the second and fourth most performed imaging modalities used in the UK. Of 3.6 million imaging tests taking place in England in November 2019, ultrasound accounted for 860,000 scans and MRI for 310,000.

Both ultrasound and MRI were pioneered in the UK but require fewer hospital physicists than the use of X-rays due to their exemplary safety profiles. Therefore, the roles of physicists specialising in non-ionising imaging tend to focus on research and teaching as an adjunct to hospital support. As the fundamental physics underlying MRI and ultrasound techniques are very different, non-ionising specialists tend not to have studied both in detail.

MRI physicists are often recruited at post-doctoral level and then asked to apply for Clinical Scientist registration through demonstrating equivalence. Larger hospitals will often employ a small "non-ionising" team responsible for QC and safety checks of hospital imaging equipment; however, it is also common for services to be offered regionally or outsourced to external companies.

The smaller footprint of physicists in these areas is partly due to fewer legislative requirements, but it also reflects a trend toward development and testing of new medical technologies within industry rather than by hospital staff and maintenance of scanners through external service agreements.

2 Magnetic Resonance Imaging Physics

Alimul Chowdhury

CONTENTS

The main technological developments in magnetic resonance imaging (MRI) for medical applications began in the 1970s. Raymond Damadian of the State University of New York, Paul C. Lauterbur at the University of Illinois and Peter Mansfield of the University of Nottingham independently showed that imaging could be performed by acquiring signals using the phenomenon of nuclear magnetic resonance (NMR) and by applying magnetic field gradients (Rinck 2018).

MRI involves using the radio-frequency (RF) part of the electromagnetic spectrum. The ability to produce a vast range of different types of image contrast, which exploit differences in both biophysical and chemical properties, makes MRI particularly powerful for investigating both morphological and functional characteristics in clinical and research applications. For example, MRI sequences and methods can be utilised to measure anatomical structure, vascular flow, perfusion and diffusion in a single scan session. Moreover, magnetic resonance spectroscopy (MRS) and functional MRI (fMRI) are increasingly finding novel clinical applications in mapping brain activity and metabolites (Jalali et al. 2018). Clinical Scientists specialising in MRI are often heavily involved in both leading and supporting research and development activities. As MRI increases in popularity, there is likely to be a need for more MR physicists in the future as further scanners are procured across both healthcare and academia.

MRI-related work carried out by Clinical Scientists includes:

- Providing safety advice
- MR Quality Assurance (QA)

- Teaching and training
- Image optimisation
- Image processing and analysis (for both clinical and research applications)
- MRI-related research and development
- Site planning
- Development and review of local rules and operating procedures
- Providing advice related to procurement
- Performing commissioning and acceptance testing for new MRI equipment.

However, this is not an exhaustive list. The Clinical Scientist specialising in MRI has a very broad remit and is often approached for help with solving problems, which requires bespoke solutions to be provided.

Clinical Scientists need to keep up to date with MR issues related to safety, legislation, national guidelines, new technology, conferences, research and development. There is an MRI physics discussion list in "JiscMail" (a national academic mailing list service), which is a useful resource for clinically focused discussions about safety, QA, training, regulations, standards, jobs, meetings and conferences. The Institute of Physics and Engineering in Medicine (IPEM) facilitates meetings related to MR QA and safety updates. International Society for Magnetic Resonance in Medicine (ISMRM) meetings help scientists to keep up to date with cutting-edge research and developments in MRI methods and technology, as well as providing an excellent networking opportunity.

Clinical Scientists specialising in MRI work closely with a broad range of professionals including Radiographers, Radiologists, scanner manufacturer service engineers and applications specialists, clinical consultants, academic research scientists, nurses, healthcare assistants, estates and facilities projects team, and IT specialists. MR physicists are often involved in providing teaching and training for NHS staff, including MRI physics lectures for trainee Radiologists and safety courses for MR Radiographers and other staff that work within the "MR Controlled Access Areas".

Advanced Clinical Scientist trainees specialising in MRI physics will learn to safely operate different models of MRI scanner, running MR sequences to produce images of phantoms, healthy volunteers and patients. Understanding how the MRI hardware and different pulse sequences work to produce different types of image contrast is paramount to being able to optimise MR settings and develop sequences for specific clinical and research applications. There are hundreds of sequence parameters that can be combined in different ways to manipulate image contrast, spatial and temporal resolution, signal-to-noise, and scan acquisition times.

The scanner manufacturer does not provide optimised sequences for all clinical or research applications when they deliver the MR scanner, so the MR physicist will need to test and develop MR sequences to get the best performance from scanner. MRI optimisation can mean developing new methods for improving image quality so that Radiologists can confidently reach a clinical diagnosis within the shortest scan time. In some research settings, a Clinical Scientist with enough MRI physics expertise may work with scanner manufacturers to develop and test bespoke sequences or methods that are not yet commercially available. However, in a clinical setting, the development of advanced sequences by MR physicists is not always feasible due to limited availability of the MR scanner during normal working hours. Clinical Scientists specialising in MRI often need to work outside of normal hours, which can be challenging as clinical scanners are often in use 12 hours a day, 7 days a week. Trainee Clinical Scientists often work evenings to get a reasonable length of time on the MR scanner, in order to complete development projects and annual MRI Quality Control (QC) tests. Figure 2.1 shows an MRI QC test object inside the bore of a modern MRI scanner.

Typical Clinical Scientist training related to MRI would initially provide a basic understanding of MRI theory, hardware and MRI safety aspects. The trainee would then need to learn about MRI

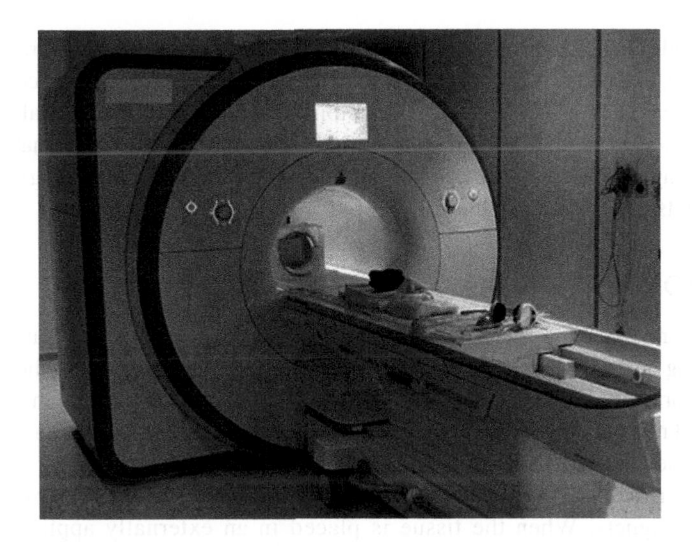

FIGURE 2.1 Modern MRI scanner with a Quality Control test object inside the bore.

scanner performance testing for QC purposes and start to develop the knowledge and skills to operate the scanner. It is important for the trainees to familiarise themselves with basic MRI sequence types and the main sequence parameters, including basic gradient echo, spin echo and inversion recovery sequences. MRI can be used to image all parts of the body and the most common applications include imaging of the brain, spine, breast, heart and blood vessels. Figure 2.2 shows an MRI scan of the entire body.

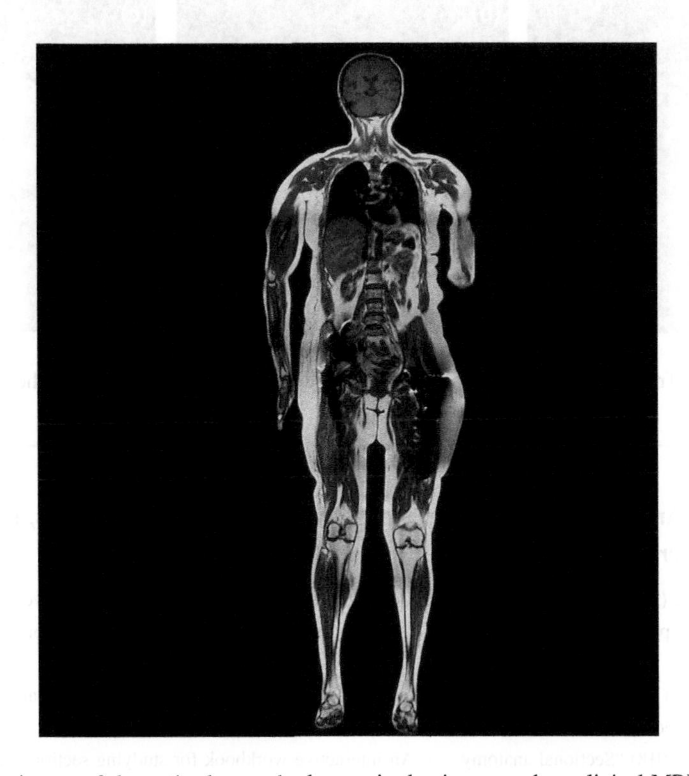

FIGURE 2.2 An image of the entire human body acquired using a modern clinical MRI scanner.

Trainees should have a good understanding of how the main sequence parameters (pulse repetition time (TR), echo time (TE) and flip angle) affect image contrast. To acquire images, it is important to understand the terminology describing the different orthogonal imaging planes, namely transverse, sagittal and coronal (for example, see Figure 2.3) and to have a basic understanding of anatomy. This knowledge can be found in most medical and radiology textbooks. Some recommended texts are listed in Table 2.1.

2.1 HOW DOES MRI WORK?

MRI techniques are based on the phenomenon of NMR, which occurs for atoms possessing a nucleus containing an odd number of protons and/or neutrons. This results in the atom possessing an intrinsic nuclear magnetic moment, causing it to act like a mini bar-magnet in the presence of an externally applied magnetic field. For a detailed description of the physics underlying MRI, see the following recommended textbooks and websites (Table 2.2).

In brief, a macroscopic sample of tissue contains protons, mainly from water and fat, that act as mini-bar magnets. When the tissue is placed in an externally applied static magnetic field, B_0, these align with the field, generating a net magnetization vector, M. The NMR signal can only be measured if there is a component of M in the transverse plane, which is orthogonal to the direction of B_0. For resonance to occur, an additional magnetic field, B_1, is applied in the form of an RF electromagnetic wave with a specific frequency, known as the Larmor frequency. By applying this additional time-dependent magnetic field, B_1, in a direction perpendicular to both M and B_0 under resonance conditions, the net magnetization, M, can be flipped into the transverse plane, enabling signal detection. After B_1 is switched off, the net magnetization, M, returns to the equilibrium state via spin-lattice and spin-spin "relaxation"

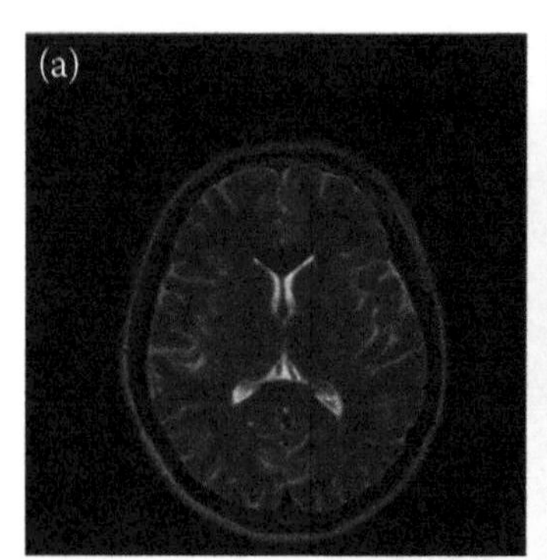

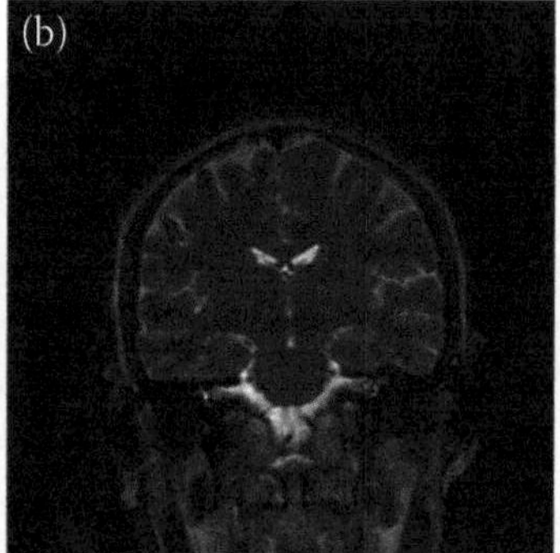

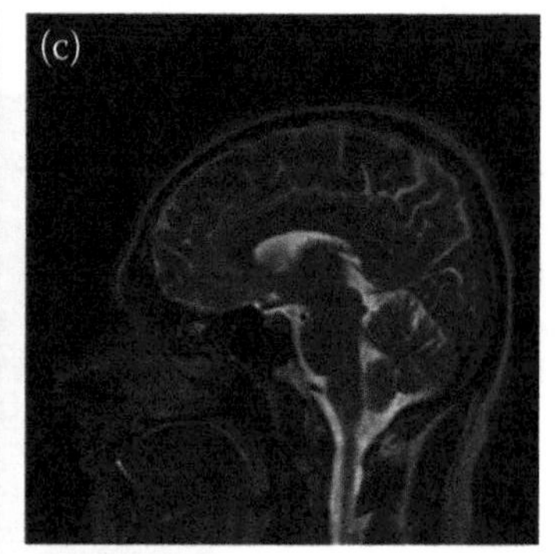

FIGURE 2.3 (a) Transverse plane, (b) coronal plane and (c) sagittal plane scans of the brain.

TABLE 2.1
Recommended Anatomy and Radiology Texts Aimed at Clinical Scientists, Radiologists and MR Radiographers

Paulson and Waschke (2011) "Sobotta Atlas of Human Anatomy", package (15th edition)	Comprehensive atlas of human anatomy, covering general anatomy and the musculoskeletal system, internal organs, and head, neck, and neuroanatomy in several formats.
Haaga and Boll (2016) "CT and MRI of the Whole Body" (6th edition)	A classic reference text that provides an enhanced understanding of advances in CT and MR imaging.
Kelley and Petersen (2018) "Sectional anatomy for imaging professionals" (4th edition)	An interactive workbook for studying sectional anatomy, including illustration labelling, puzzles and fill-in-the-blank questions.

TABLE 2.2
Recommended MRI Physics Texts Aimed at Clinical Scientists, Radiologists, and MR Radiographers

Hornak (2020) "The Basics of MRI", online available at: http://www.cis.rit.edu/htbooks/mri/	Website that contains basic MRI theory including lots of animations.
Elster (2020) "Questions and Answers in MRI", online available at: http://www.MRIquestions.com/	Website that provides explanations for clinical MRI applications.
Ilangovan and Stewart (2020) "MRImaster.com", online available at: https://mrimaster.com/	Website that includes guidance for MRI scanning and image interpretation.
Westbrook (2016) "MRI at a Glance", 3rd edition.	Introductory MRI textbook suitable for beginners.
McRobbie et al. (2017) "MRI from Picture to Proton", 3rd edition.	Intermediate level MRI text, which includes information related to MRI theory, hardware and Quality Assurance.
Bernstein et al. (2004) "Handbook of MRI Pulse Sequences".	Advanced-level MRI textbook. A useful resource for MR physicists involved with pulse programming.
Brown et al. (2014) "Magnetic Resonance Imaging: Physical Principles and Sequence Design", 2nd edition.	Provides detailed explanations of the physical principles involved with magnetic resonance imaging sequence design and implementation.
de Graaf (2019) "*In Vivo* NMR Spectroscopy: Principles and Techniques", 3rd edition. John Wiley and Sons, UK.	Advanced level text that provides detailed information for understanding magnetic resonance spectroscopy.

processes that are characterized by time constants T1 and T2, respectively. The free induction decay (FID) signal detected by an RF coil is dependent on the bio-chemical environment of the tissue, allowing the generation of medical images. Typical B_0 field strengths generated by clinical scanners are either 1.5 Tesla or 3 Tesla. In many research labs, there are 7 Tesla whole-body MRI systems for human imaging and there are also plans to build 14 Tesla human scanners. Ultra-high magnetic field strength MRI scanners offer improved signal-to-noise ratio (SNR) that facilitates higher spatial resolution and faster scans. However, MRI scanners with extremely high field strengths are also associated with additional technical and safety challenges, as well as additional costs, so are limited to research centres.

A handy summary of typical relaxation times for biological materials (Bojorquez et al. 2017; Ethofer et al. 2003; McRobbie et al. 2017) is provided in Table 2.3. These differences in relaxation time are used in MRI to generate contrast in the image. Knowledge of the approximate longitudinal

TABLE 2.3
Relaxation Times for Biological Materials

Tissue	1.5 Tesla		3.0 Tesla	
	T1 (ms)	T2 (ms)	T1 (ms)	T2 (ms)
Cerebral spinal fluid (CSF)	4000	2000	5000	800
Grey matter	1100	90	1300	80
White matter	900	70	1100	110
Fat	250	80	400	70
Heart	1000	50	1500	50
Liver	600	50	800	40
Skeletal muscle	1000	40	1400	40
Kidney	700	60	1200	60
Blood	1400	290	1900	280

(T1) and transverse (T2) relaxation time values of different tissues is usually also useful for sequence optimisation and image interpretation. In general, T1 values are much longer (5–10 times longer) than T2 relaxation times; fluids tend to have very long T1 and T2 values. T1 durations are also longer at a field strength of 3 Tesla compared to 1.5 Tesla.

In some MRI methods, exogenous paramagnetic contrast agents are intravenously injected into the patient to locally change the tissue relaxation times, hence increasing the relaxation time difference and signal contrast intensities between tissues. This enables tissues affected by the contrast agent to be better visualised in the resulting MR image.

2.2 THE MRI SCANNER

The main hardware required for MRI includes (i) a strong magnet (usually a superconducting coil) to produce a static magnetic field, $\mathbf{B_0}$; (ii) a radio frequency tuned coil to produce the $\mathbf{B_1}$ RF field; and (iii) magnetic gradient coils to allow manipulation of the magnetisation vector generating spatial information and differences in image contrast (see, for example, Figure 2.4 and Table 2.4).

2.3 SETTING UP AN MRI SCAN AND CHOICE OF SEQUENCES

MRI scanners usually have specific clinical sequences pre-programmed so that a series of sequences can be run automatically to generate several different types of image within a single scanning session. Each sequence is optimised to highlight different features of anatomy, physiology or pathology. Examples of different types of routine clinical MR sequences and associated image contrast produced are summarised in Table 2.4. It is also useful to be aware that different scanner manufacturers have their own trade name acronyms for similar types of sequences (as shown in Table 2.5).

2.4 PERFORMING A SCAN

Initially, Clinical Scientist trainees learn how to operate the MR scanner by acquiring phantom images for QC purposes using routine clinical sequences. This process may include correctly setting up the test object in the scanner bore and selecting sequence parameters based on current QA protocol guidelines.

The Clinical Scientist specialising in MRI physics might also help to optimise routine clinical sequences. For example, a time-of-flight angiography sequence might be optimised to visualise the arteries supplying the brain (as shown in Table 2.4). With more experience, trainees learn to set up scans for clinical trials and can eventually optimise complicated sequences and scans. For example, an arterial spin labelling (ASL) sequence could be optimised to measure cerebral blood flow, and a

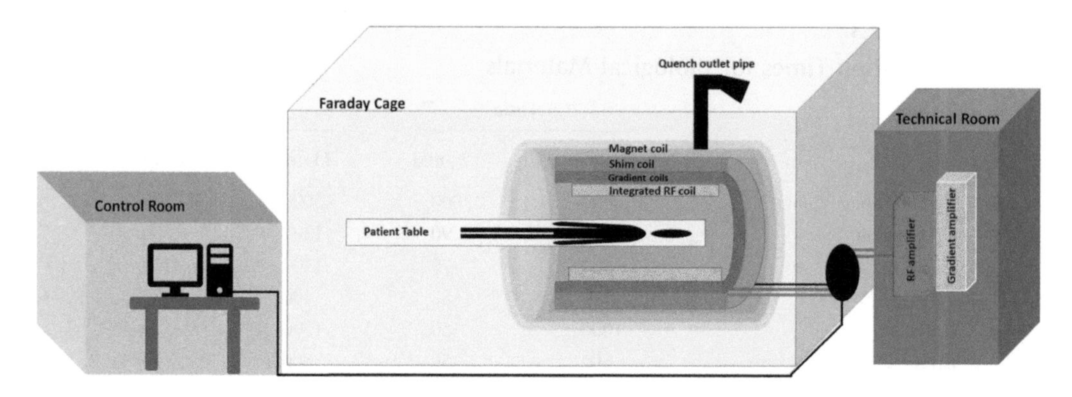

FIGURE 2.4 Basic components of a clinical whole-body MRI system, which includes a superconducting static magnet coil, magnetic gradient coils, radio-frequency coil and computer.

TABLE 2.4
MR Imaging Examples

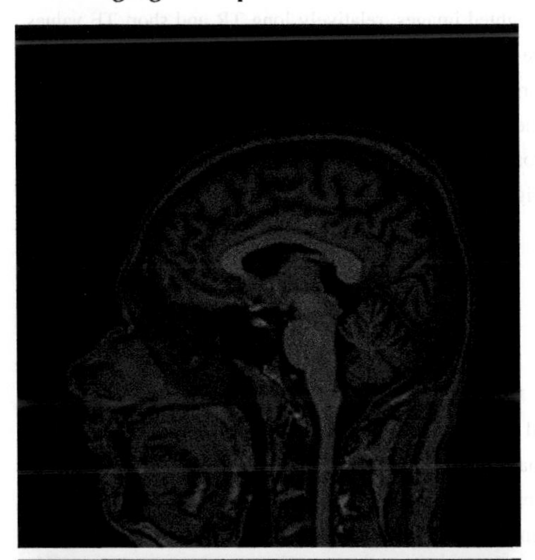

T1-weighted sagittal brain scan: In the T1 weighted scan, the TR and TE parameter settings in a gradient echo sequence are relatively short compared to the T1 and T2 of the tissues. Hence, the contrast and brightness of the image are mostly dependent on the T1 properties of the tissue. Generally, in T1-weighted images, tissues with short T1 values (e.g. subcutaneous fat) appear bright and tissues with long T1 values (e.g. CSF) appear dark. The T1-weighted image of the brain shows white matter as bright and CSF as dark. T1-weighted images are particularly useful for displaying anatomy but are not very sensitive to pathology.

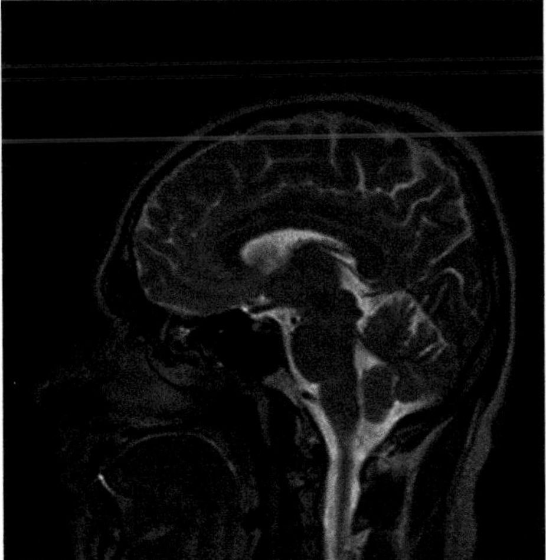

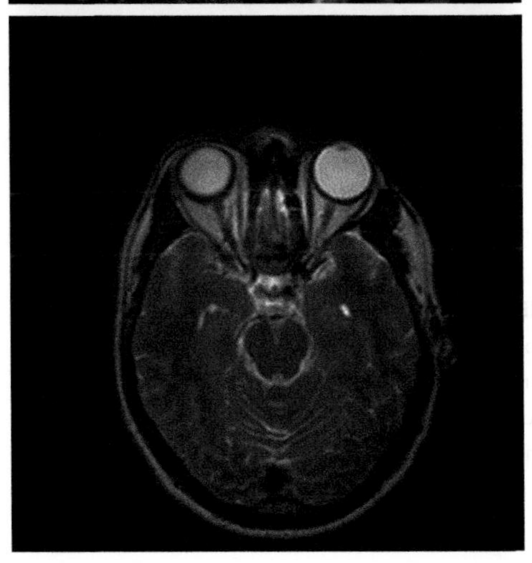

T2-weighted sagittal and axial brain scans: For T2-weighted scans, the TR and TE parameter settings in a spin echo sequence are relatively long, compared to the T1 and T2 values of the tissues. Hence, the contrast and brightness of the image are mainly dependent on T2. Tissues with short T2 appear dark. T2-weighted images of the brain show CSF as bright. Generally, tissue containing mainly fat (e.g. subcutaneous fat) can be distinguished from tissue containing mainly water (e.g. CSF) by comparing T1-weighted and T2-weighted images. Regions that are bright on the T2-weighted images but dark on the T1-weighted images is fluid-based tissue. T2-weighted images are quite sensitive to pathology. For example, oedema, which is an abnormal accumulation of water in the tissue, has a long T2 and therefore appears brightly in a T2-weighted scan.

(*Continued*)

TABLE 2.4 (Continued)

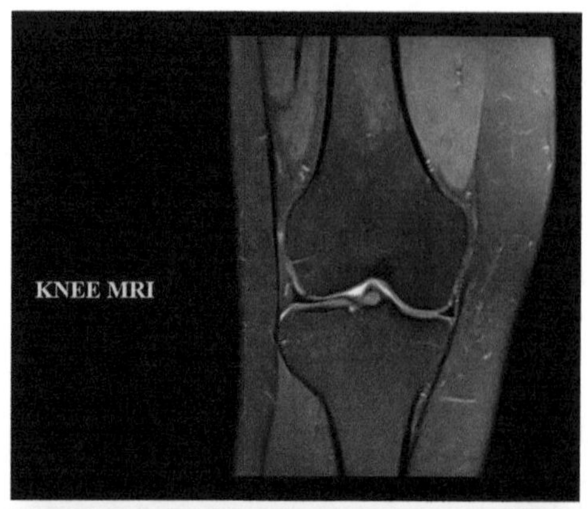

Proton density weighted scan of the knee: In proton-density weighted images, relatively long TR and short TE values are used in the sequence, to minimise the impact of variations in tissue T1 and T2 values on image contrast. The tissues with the greatest concentration of hydrogen atoms, thus with the greatest density of protons, will appear brightest in the image.

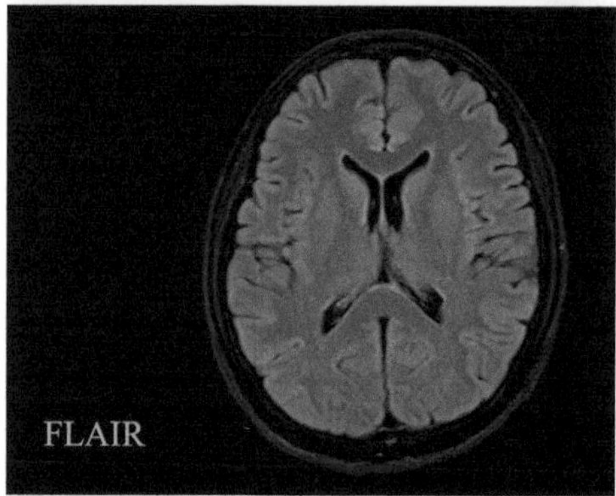

Fluid Attenuated Inversion Recovery (FLAIR): FLAIR images of the brain are often acquired to show pathology more clearly by suppressing the appearance of CSF in the ventricles.

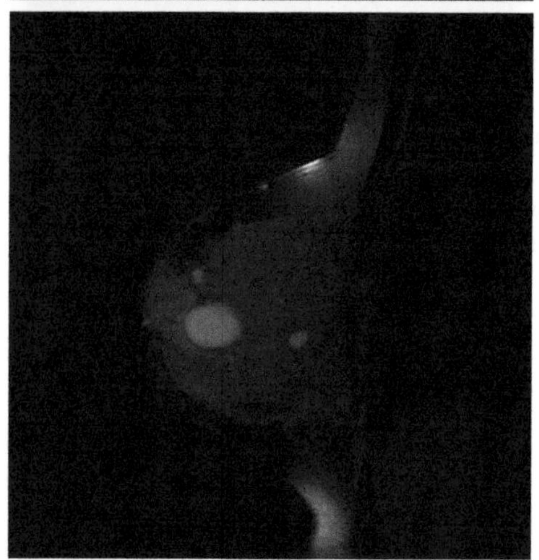

Short Tau Inversion Recovery (STIR): STIR images show the brightest regions where there is most water by suppressing signal coming from fat. MRI breast screening is carried often carried out using a STIR sequence because the images are particularly sensitive to water-filled cysts.

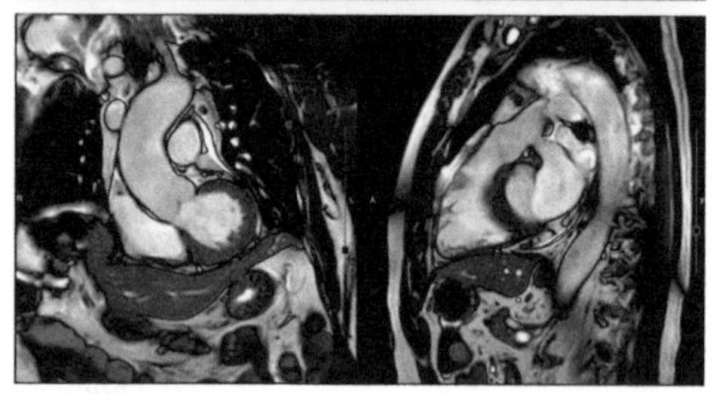

Cardiac MRI "cine" scan: Images of the beating heart can be produced in "cine" scans that typically repeat acquisitions of the heart at a single slice location throughout the cardiac cycle. Approximately 10–30 cardiac phases are usually sampled to allow diagnosis of cardiac function.

(*Continued*)

TABLE 2.4 (Continued)

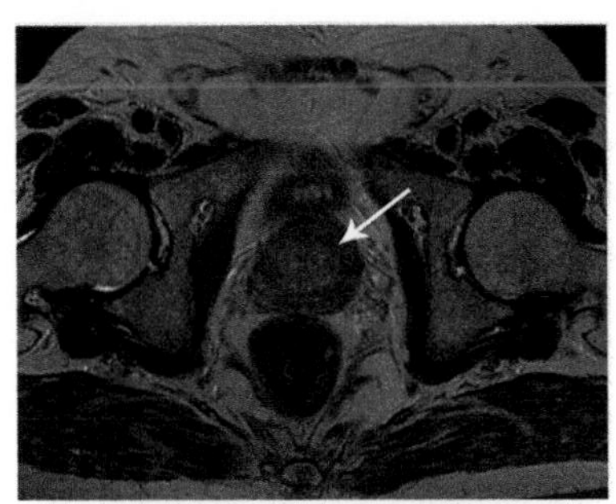
T2-weighted imaging (T2WI)

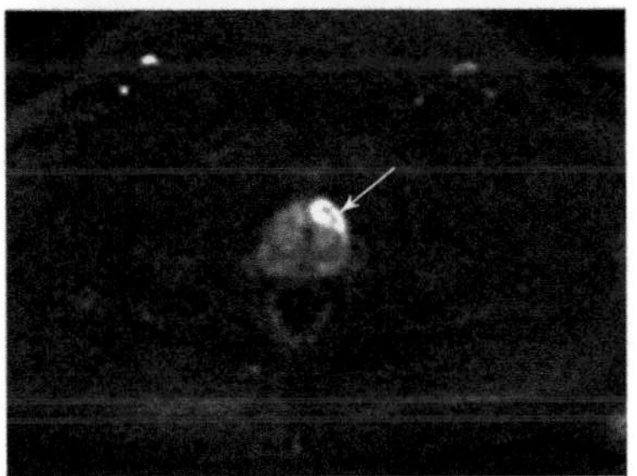
Diffusion-weighted imaging (DWI)

Diffusion-weighted imaging (DWI): Apparent diffusion coefficient (ADC) maps can be produced from DWI scans (Baliyan et al. 2016). Applications include imaging of prostate cancer.

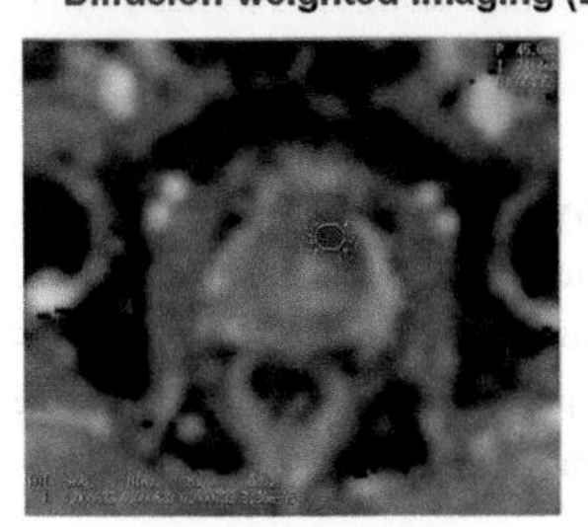
Apparent diffusion coefficient (ADC) map

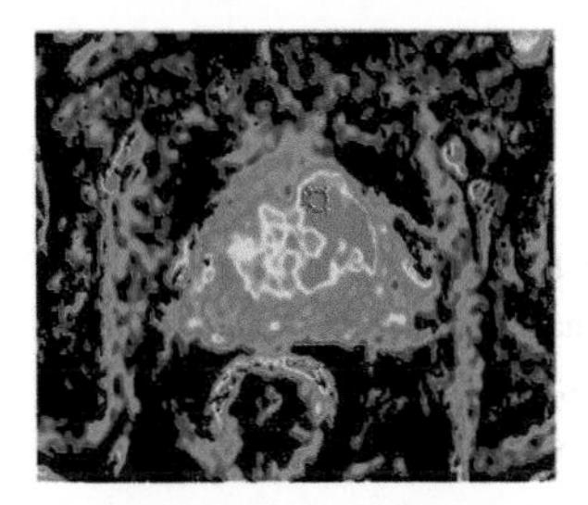
Dynamic contrast-enhanced imaging (DCE-MRI)

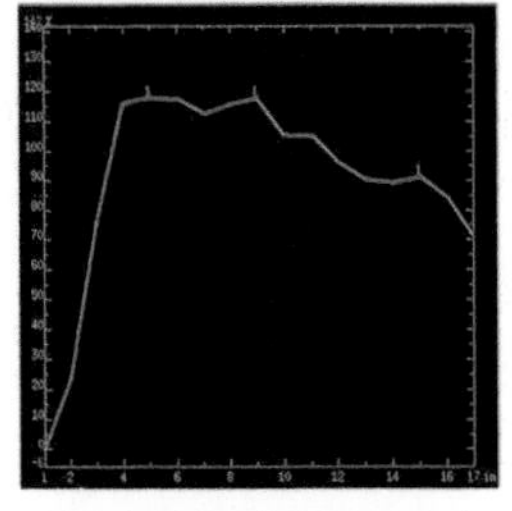
Contrast enhancement curve

Dynamic contrast enhancement (DCE): In DCE MRI, scans are optimised for acquiring T1-weighted fast gradient echo images before, during and after intravenous injection of a gadolinium chelate contrast agent. In the prostate, the gadolinium contrast agent passes more rapidly through cancerous lesions compared to normal tissue because of increased permeability of tumour vessels; hence, cancerous lesions initially appear hyper-intense in DCE images (Berman et al. 2016).

It is important not to confuse DCE with dynamic susceptibility contrast (DSC) MRI. In DSC methods, a gadolinium contrast agent is used but the MRI scans are optimised for acquiring a series of T2* weighted images, to calculate perfusion parameters.

(Continued)

TABLE 2.4 (Continued)

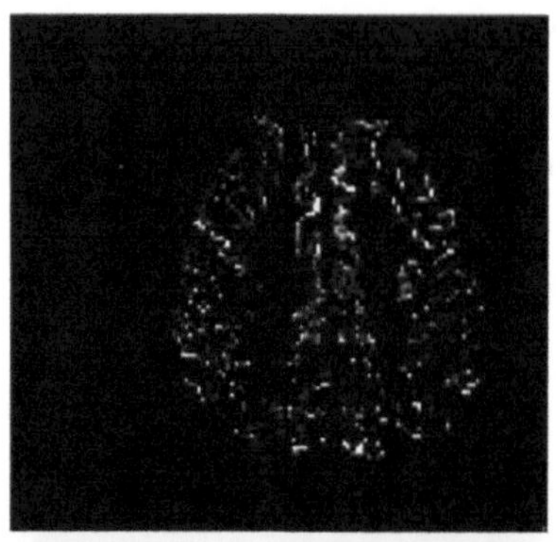

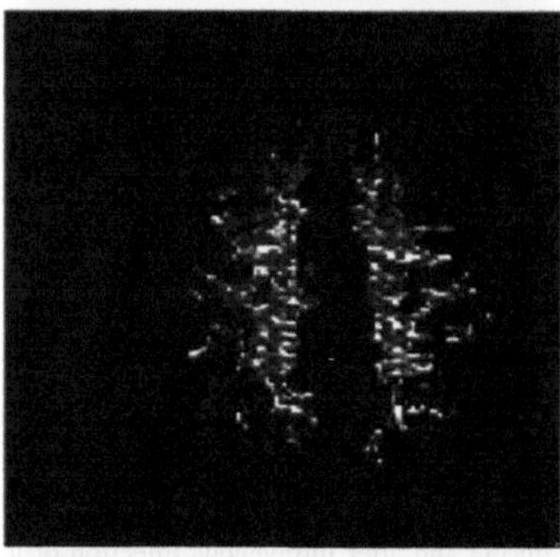

An arterial spin labelling (ASL): An ASL sequence allows perfusion measurements to be carried out by using water as an endogenous contrast agent (Ferré et al. 2013). Grey matter and white matter cerebral blood flow (CBF) can be measured using ASL methods.

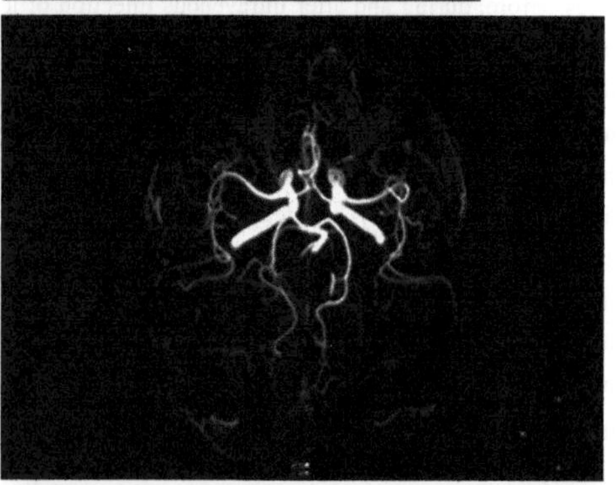

Time of flight angiography: Time of flight (TOF) MRI techniques can be used to produce a maximum intensity projection (MIP) map showing areas of blood flow. A time-of-flight angiography sequence could be used to image the arteries supplying the brain (Chowdhury et al. 2003).

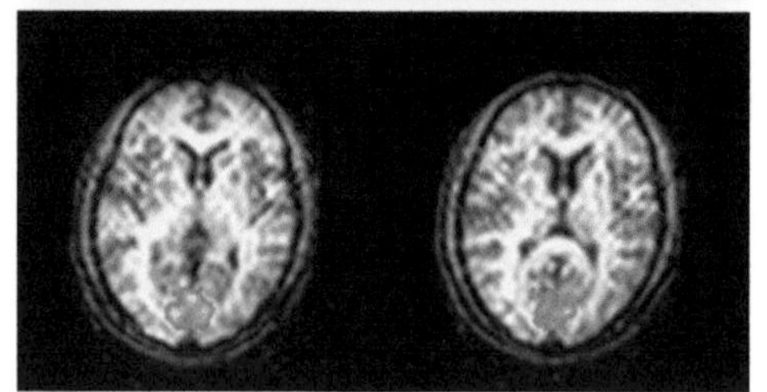

Functional MRI (fMRI): Blood oxygen level dependent (BOLD) contrast MR images are sensitive to brain activity. In these example images, the occipital region of the brain is showing activation in response to a visual stimulus.

(*Continued*)

TABLE 2.4 (Continued)

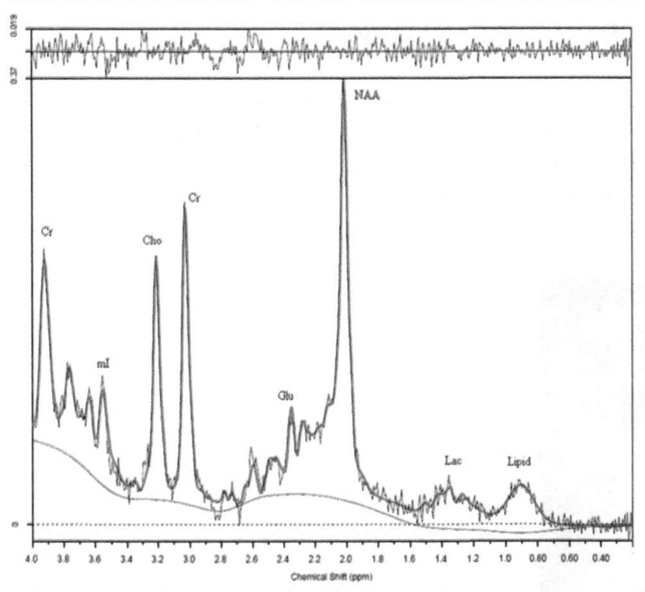

Point-RESolved Spectroscopy (PRESS): A PRESS sequence can be used to measure brain metabolites such as *N*-acetyl aspartate (NAA), Creatine (Cr), Choline (Cho), Myo Inositol (Ins), glutamate (Glu) and Glutamine (Gln).

diffusion tensor imaging (DTI) sequence optimised for white matter tractography (Assaf and Pasternak 2008). An expert MR physicist may be found programming and testing novel MR sequences that are not in routine clinical use, or that might be optimised for use "off label" in research. For example, a conventional pulse sequence called Point RESolved Spectroscopy (PRESS), as described in Table 2.4, might be developed to measure Glutathione (Rai et al. 2018).

TABLE 2.5
Different Trade Names for Common Sequences

	Siemens	GE	Philips
Spin Echo Sequences	**SE**	**SE**	**SE**
Turbo Spin Echo or Fast Spin Echo	TSE	FastSE	TSE
Single-Shot TSE	HASTE	Single-Shot FSE	Single-Shot TSE
3D TSE with variable Flip Angle	SPACE	CUBE	VISTA
Gradient Echo Sequences	**GRE**	**GRE**	**Fast Field Echo (FFE)**
Spoiled Gradient Echo	FLASH	SPGR	T1-FFE
Coherent Gradient Echo	FISP	GRASS	FFE
Steady-State Free Precession	PSIF	SSFP	T2-FFE
True FISP Balanced steady-state gradient echo	TrueFISP	FIESTA, COSMIC	Balanced FFE
2D Ultrafast Gradient Echo with preparation pulse	TurboFLASH	Fast GRE, Fast SPGR	TFE
3D Ultrafast Gradient Echo with preparation pulse	MPRAGE	3D FGRE, 3D Fast SPGR, BRAVO	3D TFE
Arterial Spin Labelling	ASL	ASL	ASL
Inversion Recovery	**IR, Turbo IR (TIR)**	**IR, MPIR, FastIR**	**IR-TSE**
Short-Tau IR	TIRM, STIR	STIR	STIR
Long-Tau IR	TIRM, Dark Fluid	FLAIR	FLAIR

CASE STUDY: A TYPICAL MRI SCAN

A Clinical Scientist is asked to set up an MRI scan protocol for a clinical research trial involving a 30-minute MR scan investigating adult subjects with Autosomal Dominant Polycystic Kidney Disease. A healthy volunteer is scanned to test out the necessary sequences including timing and image quality (Figure 2.5). The following sequences are included in the protocol and optimised.

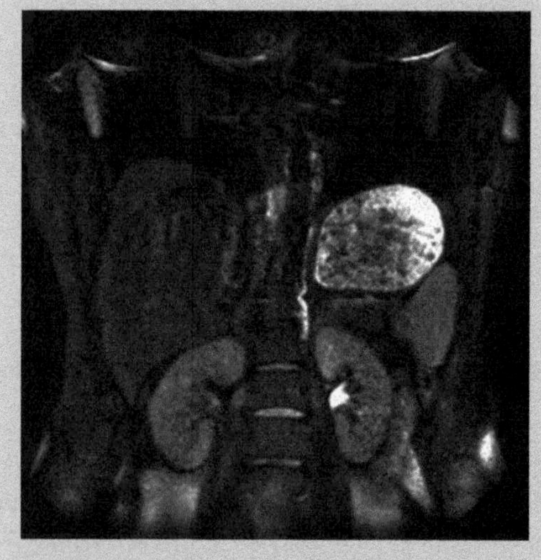

FIGURE 2.5 A coronal MR image showing the kidneys.

1. Localiser scan – Acquire images in three orthogonal planes
2. T1 3D Spoiled Gradient Echo – Scan should be acquired without fat saturation
3. T2 Single Shot Turbo Spin Echo – Acquire in the coronal plane with fat saturation and breath holding.
4. T2 Single Shot Turbo Spin Echo – Acquire in the coronal plane without fat saturation and breath holding.

The first scan in an MRI examination is usually a quick localiser scan, involving rapid acquisition of low-resolution images for each of the three orthogonal directions. These are not informative for clinical diagnosis, but are important for visualising the patient's position, to enable the user to plan the more time-consuming acquisition of other images. Images for this application need to be of adequate quality to be able to draw regions of interest around each kidney delineating renal cysts, which appear brighter in T2-weighted images and darker in T1-weighted images. Regions of interest around each kidney are drawn on the T1-weighted images, using the T2-weighted scans to aid identification of the kidney boundaries.

In order to optimise, the SNR scans are acquired using a "flexible body" RF coil rather than using just the "built in body" RF coil. The main important sequence parameters to consider for optimising the image quality are as follows:

i. The image slice positioning of the main scans needs to be correctly planned using the localiser scans and an appropriate field of view (FOV) is required to ensure that the entire kidney is captured in the images, to allow accurate kidney and cyst volume measurements.

ii. The image slice thickness needs to be carefully considered, balancing the image SNR, spatial resolution and total scan time.
iii. The image voxel size, thus the nominal spatial resolution, is dependent on the FOV and the image matrix size as well as the slice thickness. The better the spatial resolution, the more accurately tissue borders can be delineated.
iv. The sequence parameters including mainly the flip angle, TR and TE have to be correctly set to produce the necessary contrast in the image and differentiate between different types of tissues.
v. The number of signal averages to use should be based on an assessment of the necessary SNR and total scan time, considering that the breath-hold time is limited. If the number of averages is increased, then the image SNR is increased, which means the images will potentially be of better quality. However, increasing the number of averages also means scanning for longer, which might exceed the 30-minute scan duration. Ideally, the participant would be able to hold their breath to minimise movement artefacts; however, not all participants can tolerate breath holding for long.

2.5 MR SAFETY

MRI involves several hazards, including strong magnetic fields, the use of liquid helium, contrast agent side-effects and potential hazards to hearing. Trainee Clinical Scientists need to develop expertise in all aspects of MRI safety, which includes reading and familiarising themselves with relevant guidelines, local rules and standard operating procedures (SOP). A key role of the Clinical Scientist is to review MR local rules and SOPs, providing necessary recommendations for updating these safety documents.

The Medicines and Healthcare Products Regulatory Agency (MHRA) document, "Safety Guidelines for Magnetic Resonance Imaging Equipment in Clinical Use" is particularly important reading for trainees planning to develop an understanding of MRI safety. Important safety information is also provided by (i) the American College of Radiology (ACR), (ii) IPEM MR-SIG Working Party Reports, and (iii) the International Commission on Non-Ionising Radiation (ICNIRP) guidelines. The website "MRIsafety.com" is also a good resource, particularly for finding MRI safety information about various types of medical devices (Shellock 2020). Important guidelines are listed in Table 2.6.

The risks of having an MRI scan are quite different compared to other imaging modalities that use ionising radiation. MRI scans are generally perceived to be safer than scans involving X-rays, but this is only true if MRI is carried out with strict safety control measures in place. It is important for Clinical Scientists to be aware of new scientific publications related to MRI safety and adjust local rules and procedures in accordance to the latest scientific and regulatory information.

TABLE 2.6
MR Safety and Best Practice Guidelines

"Safety Guidelines for Magnetic Resonance Imaging Equipment in Clinical Use" (MHRA 2021)	Essential reading related to clinical MRI safety
"American College of Radiology (ACR) Guidance Document on MR Safe Practices: Updates and Critical Information 2019" (ACR Committee on MR Safety 2020)	Provides additional information related to screening patients for MRI safety, with more details for understanding spatial field gradients and labelling of implanted devices

During an MR scan, individuals being scanned, and those in the immediate vicinity of the scanner, will be exposed to (i) a static magnetic field, (ii) relatively low-frequency time-varying magnetic fields, and (iii) RF electromagnetic fields. The safety implications of these are summarised in the subsequent paragraphs.

2.5.1 Static Magnetic Field

Almost all clinical MRI scanners contain a superconducting magnet that produces a static magnetic field, either at 1.5 Tesla or at 3 Tesla. The superconducting magnet is permanently activated even when the scanner is not being used for an examination. The static magnetic field is always present in the MR Environment and can interfere with electrically, mechanically or magnetically active devices. Due to the risk of life-threatening malfunction of implanted devices, all patients, participants and visitors must be properly screened before being allowed entry into the MR Environment. Credit cards, or other cards that contain magnetically stored information, can be wiped at field strengths above 1 Tesla, so should be securely stored outside of the MR Environment. There are also extremely steep fringe magnetic field gradients present, so a high level of vigilance must be maintained to prevent anything being taken into the magnet room that might result in a projectile incident. Mechanical attraction of metallic, especially ferromagnetic objects (e.g. scissors, pins, paperclips, fire extinguishers, oxygen cylinders, etc.), can become projectiles when accelerating towards the magnet at high speed, causing serious injury and damage to the scanner. Embedded metal fragments, aneurysm clips, or other implanted medical devices can also be displaced or rotated due to an attractive force and/or torque, causing injury to the patient and damage to the implanted device.

2.5.2 Time-Varying Magnetic Fields

Moving around close to the static field effectively results in exposure to time-varying magnetic fields. Some individuals may experience acute effects such as vertigo, dizziness or nausea when moving close to the scanner. Slow movement rather than fast movement of the head is advised to minimise the risk of acute biological effects, particularly when positioning a test object inside of the magnet bore. During scanning, rapidly changing magnetic field gradients are used to encode spatial position in the image, which also can result in exposure to time-varying magnetic fields if an individual is inside the MR Environment. Time-varying magnetic fields can induce electric currents that could be sufficiently large in tissues to interfere with the normal function of nerve cells, generating peripheral nerve stimulation (PNS) and direct muscle stimulation. While potentially uncomfortable during the scan, the effects have not been shown to be long lasting.

2.5.3 RF Fields

MRI involves the use of RF electromagnetic energy that is applied to the body in order to generate MRI signals in the form of RF waves. RF signals are applied and detected using RF coils placed near the body part being scanned. One potential hazard associated with using RF electromagnetic fields in this way is the heating of the tissue. The specific absorption rate (SAR) is a term used to estimate the amount of RF dose received by the patient and is expressed as Watts of power per kilogram of the patient's body weight.

SAR limits have been set for specific areas of the body (e.g. the head) and must not be exceeded. Scanners include safety features to limit the SAR so that RF deposition is within standard limits set by organisations such as the International Electrotechnical Commission (IEC). For the SAR to be correctly calculated by the scanner software, it is critical to enter the patient's correct weight and to follow the manufacturer's safety instructions. Failure to do so may result in incorrect calculation of SAR limits, which may result in patient heating or burns. The SAR applied to the body is also dependent on the temperature and humidity within the bore.

Electrical currents can be induced in metallic leads and electrodes (such as ECG leads used to monitor heart rate) potentially causing burns in adjacent tissues. Electrical conducting elements, such as wires, must never be looped across the participant's body or contact the participant's skin. In addition, the patient's skin should never be allowed to touch either the sides of the RF coil, or the bore of the scanner. If necessary, padding (supplied by the manufacturer) should be used to ensure adequate insulation between the participant's body and RF coil. Heating can occur from induced currents when a conductive loop is formed within the patient's body, for example, by skin contact between the patient's thighs, feet and arms.

Certain types of make-up contain metallic compounds, often found in eye make-up. These potentially cause heating during scanning leading to discomfort or injury. Metallic make-up may also significantly affect MR image quality. Patients are asked to remove all make-up to avoid heating effects during scanning. If patients have tattoos or permanent make-up, and the presence of metallic inks cannot be excluded, then the patient can be scanned after a risk assessment but must be advised to press the alarm-call button if any heating is sensed in the region of the tattoo.

An experienced MR physicist may get involved with scanning research volunteers and may learn how to safely operate the scanner to scan human participants as well as test objects. Before scanning an individual, an MR safety questionnaire needs to be completed. An example MRI safety screening questionnaire is shown in Figure 2.6.

2.5.4 The MR Controlled Access Area

One of the first control measures that a trainee Clinical Scientist needs to be aware of is the MRI unit area layout. The MRI unit layout includes three main areas: (i) the MR Controlled Access Area, (ii) MR Environment and (iii) MR Projectile Zone. Figure 2.7 shows an example drawing of an MRI unit layout after taking measurements using a Gauss meter to check that the static magnetic field outside of the MR Environment does not exceed the 0.5 mT (or 5 Gauss) limit, produced from help by a trainee Clinical Scientist. Understanding the correct definitions and usage of the terminology to describe MRI safety is an important component of Clinical Scientist training.

It is vital to appropriately control access of staff, patients and equipment to the "MR Controlled Access Area". All entrances to the MR Controlled Access Area should be restricted, possibly with entry via self-locking doors that can only be unlocked by suitably trained, "MR Authorised Personnel". The trainee scientist should expect to see warning signs, providing notice that very high levels of magnetic field are produced by the MRI scanner. The scanner is kept switched on 24 hours a day and 7 days a week, regardless of whether patients are being scanned. The high magnetic field strengths associated with the MRI scanner are the main hazard, including risk of projectile effects from devices containing ferromagnetic materials. Figure 2.8 shows the example of a warning sign that the trainee might encounter at the entrance of the MR Controlled Access Area.

Based on definitions from the ASTM international standard F2503-13, implanted devices and ancillary equipment can be categorised into one of three groups: (i) MR Safe, (ii) MR Conditional and (iii) MR Unsafe. It is very important that the Clinical Scientist develops expertise to accurately identify the correct safety status of the device and often they would carry out audits in MRI facilities to check that equipment within the "MR Controlled Access Areas" that might enter into the "MR Environment" has been correctly categorised and labelled. Moreover, for implanted devices and equipment that are categorised as "MR Conditional", it is necessary to understand and correctly record the actual conditions of safety.

MRI SAFETY QUESTIONNAIRE

Name... Date of Birth............................

Address...

Height..Weight................................

Do you or have you **ever** had a pacemaker, heart valve or heart stent? Details..
Have you **ever** had an operation on your head, eyes or ears? Details..
Have you **ever** had any operations elsewhere in the body? Details..
Do you have a hydrocephalus shunt? Is it programmable?......................
Do you have any implants? (e.g. joint replacement, clips, stents, coils, electrical, GI video capsules, magnetic, cochlear) Details..
Have you ever had any metal fragments, splinters, shrapnel, or gunshot wounds anywhere in the body especially your eye?
Do you have metal dentures in or are you wearing a hearing aid?
Are you wearing any drug patches on your skin, any tattoos or any body piercing in?
Do you suffer from fits?
Do you suffer from diabetes?
Do you suffer from asthma or have any allergies?
Do you have any kidney problems?
Are you, or is there any possibility that you may be pregnant, or are you breast feeding?

Please remove all loose metal objects including your watch, money, credit cards, keys, all jewellery, hair clips, wigs or hair pieces, heavy eye make-up, coloured contact lenses and metal on clothing.
The MRI procedure has been explained to me. I understand that the Doctor supervising my study may wish me to have an injection of an intravenous contrast.
I consent to the MRI examination and contrast injection if required.
I confirm that I have answered all of the above questions correctly.

Patient / Parent / Guardian / Volunteer

Signature...Date ....

I confirm that all metallic items, e.g. ECG electrodes, external pacing wires, have been removed from the patient, and that there are no visible sutures or clips that may be unsafe for MRI scanning.

Radiographer / Assistant Practitioner ....Date ....

FIGURE 2.6 Example of an MRI safety screening questionnaire.

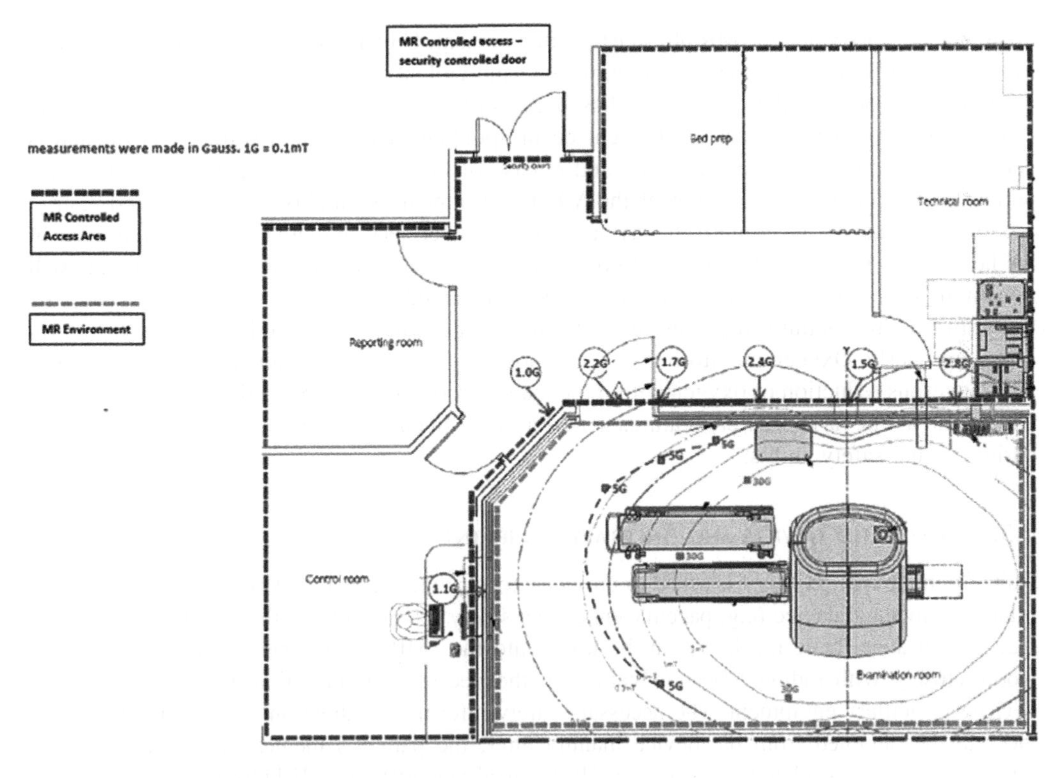

FIGURE 2.7 Example layout of an MRI unit.

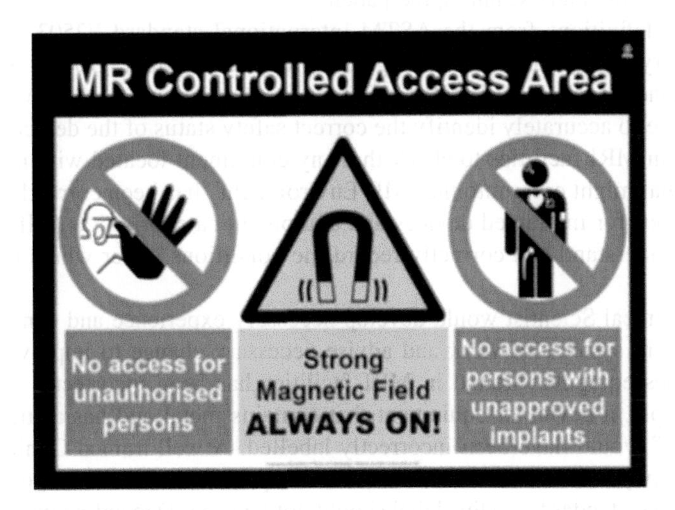

FIGURE 2.8 Example of a warning sign at the entrance of the MR Controlled Access Area. Produced by the Institute of Physics and Engineering in Medicine (available at www.ipem.ac.uk).

2.5.5 Use of Liquid Helium

The liquefied helium used by the MRI superconducting magnet poses potential hazards. The equipment on top of the magnet and pipes leading from it can be extremely cold. Touching these surfaces can result in severe cold burns. This risk is reduced during normal scanning because covers are in place to make these areas less accessible, but these covers are removed during maintenance work when the MR physicist is often present. Collapse of the magnetic field (known as "quenching") can occur, causing an immediate and violent boil-off of very cold helium gas. A quench might occur either because it has been activated by the scanner operator during an emergency or on a rare occasion it might occur spontaneously without warning. A quench is generally accompanied by a loud bang and release of large quantities of helium gas. It is important to evacuate everybody out of the MR Environment as quickly as possible and restrict access to the magnet room until the arrival of the scanner service engineers. Although most of the helium gas is vented through a purpose-built pipe, there is a risk that cold helium gas will remain in the scanner room. This can cause cold-related injuries and potentially cause asphyxiation as the helium displaces oxygen within the room if the ventilation system fails to operate properly. Oxygen monitoring systems should be installed and maintained to alert staff to any dangerous reduction of oxygen within the magnet room. Clinical Scientists will be trained to understand about various safety features necessary for MRI and will check that these are in place during annual safety audits.

2.5.6 Implanted Devices and Ancillary Equipment

Quite often the MR physicist is asked to provide expert advice related to whether a patient with an implanted medical device (e.g. pacemaker) can be safely scanned or whether ancillary equipment (such as an anaesthetic machine) can be taken into the "MR Environment". The MR physicist works closely with radiographers to gather all the necessary information about the implanted device or ancillary equipment. The necessary safety information about an implanted medical device can be obtained from the device manufacturer, the patient's medical history records and relevant publications related to the device. The Clinical Scientist with MRI physics safety expertise can help to analyse the information about the implanted device and provide advice including possible conditions for safely scanning the patient.

Based on the definitions from the ASTM international standard F2503-13, implanted devices and ancillary equipment can be categorised into one of three groups, which are (i) MR Safe, (ii) MR Conditional and (iii) MR Unsafe. It is very important that the Clinical Scientist develops expertise to accurately identify the correct safety status of the device. Scientists often carry out audits in MRI facilities to check that any equipment located within "MR Controlled Access Areas" that might enter into the "MR Environment" has been correctly categorised and labelled. Moreover, for implanted devices and equipment categorised as "MR Conditional", it is necessary to understand and correctly record the conditions under which that equipment is safe to operate.

The trainee Clinical Scientist would develop necessary experience and confidence to be able to identify errors in systems of work and advise necessary change to improve safety and efficiency. Sometimes equipment found in MRI facilities has been incorrectly categorised or labelled. For example, a device or equipment that contains metal and has a sticker label stating "MR Compatible" would have been incorrectly labelled. A well-trained Clinical Scientist specialising in MRI physics would know that "MR Compatible" is not a category defined by the ASTM international standards, so this label should not be used. Depending on the materials used in the device or equipment, it would be possibly correct to label the equipment as "MR Conditional", with a recorded statement that provides information about the actual conditions for safe use.

2.5.7 Hearing Protection

Permanent hearing loss has been reported following an MR examination where hearing protection is not used. The MRI scanner has currents flowing through gradient coils within the high static magnetic field; hence, strong forces are exerted on the gradient coils making them move. The movement of the coils generates a high level of unavoidable noise within the scanner. As gradient strength and gradient switching rate increase on clinical scanners, the risk of hearing damage increases. Rapid gradient switching in MR systems causes a great deal of acoustic noise (usually between 80 and 120 dB) during scanning. Noise from MR systems usually exceeds recommended safety levels. To comply with the Control of Noise at Work Regulations 2005 personnel exposed to acoustic noise above 80 dB should be offered hearing protection and an exposure limit of 85 dB should not be exceeded. It is essential for all individuals to wear hearing protection if they are ever in the magnet room during scanning.

2.5.8 Safety of MR Contrast Agents

In some MRI examinations, contrast agents are used (Table 2.4). The risk of significant morbidity from intravenous paramagnetic MR contrast agents (gadolinium chelates) is considerably less than for iodinated contrast media often used for CT, but nevertheless medical supervision is mandatory following MR contrast injection. Administration of intravenous MR contrast media must be authorised and supervised by a medical practitioner/Radiologist familiar with the contraindications and potential side-effects, such as anaphylaxis. Caution should be exercised in subjects with severe renal impairment since gadolinium is excreted via glomerular filtration in the kidneys. The risk of nephrogenic systemic fibrosis (NSF) must be taken into consideration before using any gadolinium-based contrast agents (GBCAs).

2.5.9 MR Safety during Pregnancy

Data concerning the safety of the static magnetic field, time-varying magnetic fields and electromagnetic RF fields for the developing foetus are still being gathered. To date, there has not been any study showing evidence of direct harm from the magnetic fields associated with clinical MRI to mother or baby during pregnancy. A study by Reeves et al. (2010) provides evidence that exposure of a foetus to 1.5 T imaging during the second and third trimesters is not associated with an increased risk of substantial neonatal hearing impairment. However, pregnant women should only be scanned when there is a clear clinical benefit to either the mother or baby. GBCAs should be avoided during pregnancy. Pregnant staff may work within the MR Environment after carrying out a risk assessment, but they should avoid being in the magnet room during scanning.

2.6 MRI QC

MRI QC is important for minimising the risk of significant deterioration of the MRI system while in clinical use. This could potentially compromise clinical diagnostic quality and ultimately safety. Clinical Scientists perform a number of performance tests to advise whether a detected deterioration requires (i) immediate corrective maintenance, (ii) corrective maintenance during a future planned service and (iii) clinical awareness and monitoring. It is not always feasible or cost effective to carry out corrective maintenance whenever deterioration is detected by QC testing, so experienced judgement and a risk assessment may be required. Sometimes it can take extensive systematic testing by both Clinical Scientists and Engineers to determine exactly what component in the system has caused a deterioration in image quality.

Scanner down-time while carrying out corrective maintenance is costly and leads to longer patient waiting times. The clinical scientist's input based on QC tests can sometimes help to determine the root cause of the underlying fault in the system and get things running again sooner. Excellent guidance for MR QA can be found in IPEM report 112, “Quality Assurance and Artefacts in Magnetic Resonance Imaging”. The American College of Radiology (ACR) also

FIGURE 2.9 An ACR MRI Quality Control phantom.

provides useful information and QC guidelines. Figure 2.9 shows an ACR phantom that is used for QC at many clinical and research MRI sites.

During a typical QC testing session, the MR physicist should first be aware of the safety issues, making sure that they have been screened properly and ensure that they do not carry any MR unsafe items into the MR environment. The physicist would need to correctly position the test object within the centre of scanner bore by checking the orientation using a positioning laser system. It is important not to stare into the laser beam as this could potentially cause mild eye injury. After the test object has been correctly positioned within the centre of the scanner bore, it is necessary to set up the QC sequences that need to be run to acquire data. Usually, MR sequences are set up and saved in the system by an experienced physicist during acceptance testing and these same sequences can then be repeated in subsequent QC sessions. The QA programme might include short daily and weekly QC tests that can be run by the Radiographers, plus annual tests that are much longer but more thorough and run by a Clinical Scientist.

Annual MRI QC tests typically take 2–4 hours depending on how many tests need to be carried out and the number of RF coils to be tested. Performance tests for MRI include checking the following parameters.

SNR: SNR measurements are carried out often using a “flood field” phantom, which is a cylindrical or spherical container filled with an aqueous or oil fluid. The phantom provides a homogenous proton density distribution within the phantom. It is useful to compare measured SNR values with an appropriate benchmark value established during the scanner’s acceptance. SNR measurements, if carried out consistently and frequently, can sensitively detect both acute and gradual MR system deterioration. For example, gradually developing electrical spike noise resulting from a loose connection within the scanner hardware might result in a significant decrease in the image SNR, which could be detected at a relatively early stage, with corrective maintenance carried out before the fault worsens. There are several different methods for measuring SNR. As part of a QC programme, the simplest “background” measurement method typically takes a few minutes and can be carried out on a daily or weekly basis. Although SNR measurements can alert staff that there may be a potential problem with the scanner, a reduction in the SNR compared to the benchmark value is not very specific as to the cause. Figure 2.10 shows an image that has been acquired using an ACR phantom to provide a quick measurement of SNR.

Uniformity: Signal intensity uniformity measurements are usually also carried out using images acquired with a “flood field” phantom. Significant variations in the image signal intensity

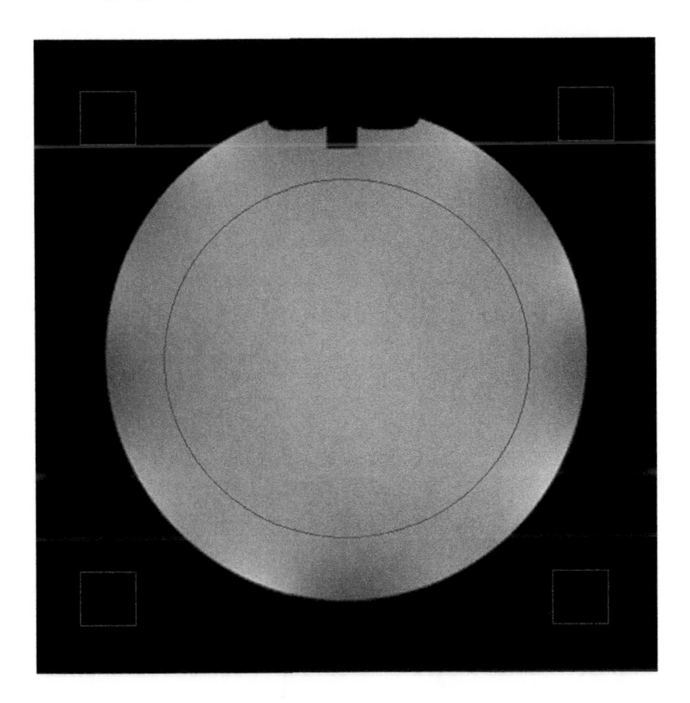

FIGURE 2.10 ACR phantom image for SNR calculations. Example regions of interest that could be used for the calculations are indicated by the large circle for measuring the mean signal in the phantom image and the squares for measuring the standard deviation of the background noise.

acquired from a homogenous test object indicate a possible fault in the RF coil but the results need to be interpreted alongside other performance tests carried out.

Geometric linearity and distortion: This measurement is carried out by measuring distances in the MR image that represent markers within the test object, known as benchmark distances (Figure 2.11). By calculating variations across the measured distances compared to the benchmark distances, it is possible to quantitatively and objectively assess how accurately the test object shape is being represented in the image. Failure of this test can indicate poor static field shimming or a fault in the gradient system.

Slice position: The slice position test assesses the positional accuracy for obtaining the desired image slice based on the localiser scan. Large errors are indicative of poor gradient performance or shimming but can also be due to human error in positioning the phantom, or the slices, during planning, so it is best to repeat these measurements before calling in the Engineers.

Slice thickness: The width of the image slice profile is compared to the prescribed width in the scan settings. Significant errors can be caused by faults in the gradient system or RF coils.

Spatial resolution: This tests the scanner's ability to resolve small objects in a high contrast-to-noise ratio image acquired with a prescribed field of view and matrix size. Failure of this test can be caused by too much vibration, a poor gradient system, or static field inhomogeneity. Figure 2.12 shows the ACR phantom image that can be used to assess spatial resolution.

Ghosting: A ghosting artefact is a faint image of an object that is displaced relative to the main position of the object in the image. A ghosting artefact performance test quantifies the level of ghosting in the image. Failure of this test is non-specific to the cause, but the test is useful for visually indicating a deteriorating system.

Contrast-to-noise ratio: In scanners that fail the contrast-to-noise ratio test, it will be more difficult to discern objects with low contrast compared to the background and or resolve objects in the image that are located close together. This could impact clinical diagnosis. Figure 2.13 shows

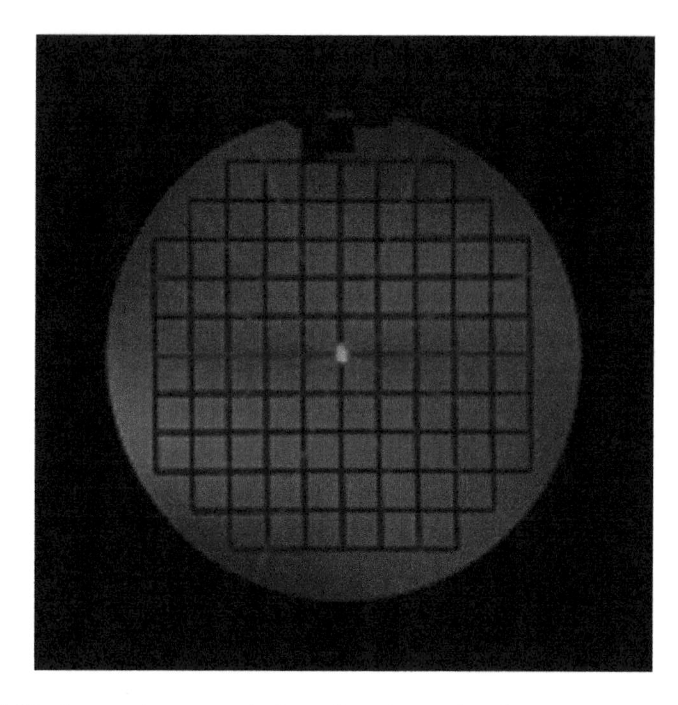

FIGURE 2.11 ACR phantom image acquired to allow image linearity and distortion test measurements. Example regions of interest that could be used for the linearity and distortion calculations are indicated by multiple straight lines going through the middle of the phantom to measure the internal diameter of the cylindrical phantom.

images acquired using the ACR phantom to assess the contrast-to-noise ratio based on how discernible the small circular disks appear relative to the background.

After phantom images have been acquired on the scanner, it is often necessary to carry out quantitative image analysis offline to determine objectively whether there has been any significant deterioration in the system. This can involve several more hours of work, including writing a summary report. Often the analysis is made less labour intensive by using software to automate some of the analysis procedures required. MR physicists may work on writing computer

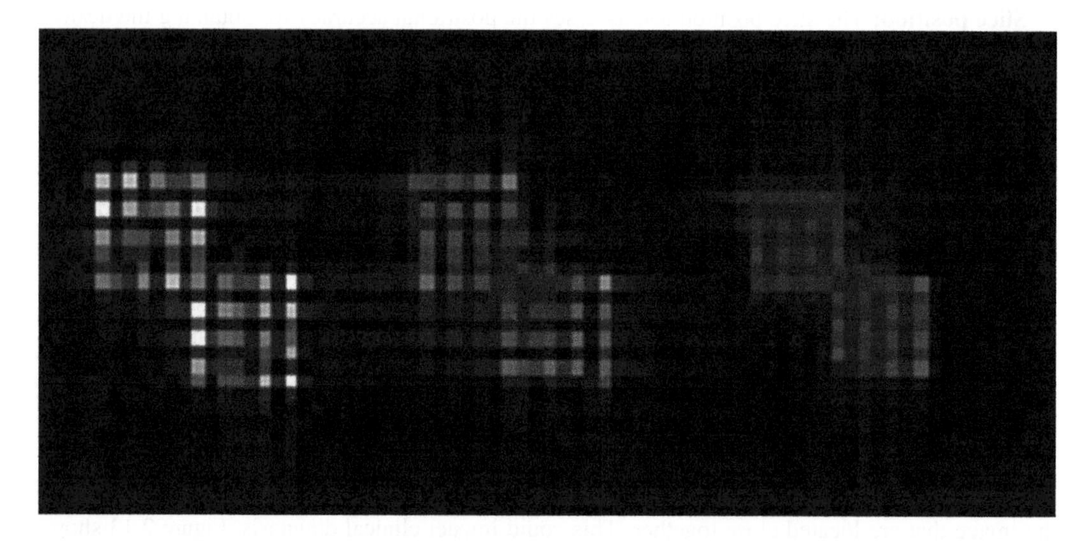

FIGURE 2.12 Image acquired using the ACR phantom to assess spatial resolution.

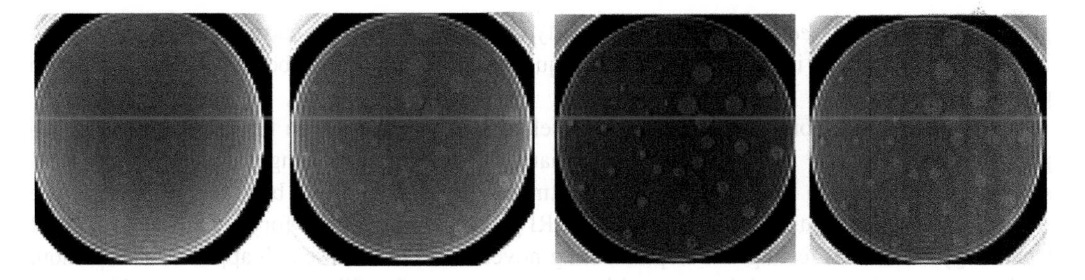

FIGURE 2.13 Images acquired using the ACR phantom to evaluate contrast-to-noise ratio.

algorithms and the development of software to automate the QC image analysis process. It is often worth investing time in software development rather than spending hours manually sorting and analysing data.

The Clinical Scientist's role includes developing MR QA processes and QC protocols, data acquisition, data processing, data analysis and reporting. Reviewing QC test pass criteria is also important as testing needs to remain relevant and fit for purpose as MRI technology develops.

2.7 THE FUTURE OF MRI

MRI has already proven to be an extremely powerful diagnostic tool in the past 40 years. In the next 10–20 years, it is very likely that there will be a continuation of significant developments that will have a substantial increase in the range of clinical and research applications using MRI. Physics support requirements from specialists in MRI are likely to exceed the current capacity of staffing within many hospitals; hence, it is extremely important to continue recruitment, retention, training and career development in this field.

Future areas of development are likely to include MRS allowing metabolite profiling and molecular imaging within different anatomical organs. MRS is likely to find a huge number of applications in routine diagnosis and research. The temporal and spatial resolution of MRS is relatively low compared to conventional MRI. However, with improved MRI hardware and computing systems, MRS imaging is likely to be developed for an increasing number of applications. Whole-body clinical MRI systems are probably also going to be introduced for detecting signals from nuclei of interest in addition to ^{1}H, such as ^{13}C, ^{23}Na and ^{31}P. For example, ^{31}P cardiac MRS is an area of interest that allows non-invasive assessments of cardiac energy metabolism, to evaluate the diseased myocardium. Cardiac MRS is an area that will most likely be developed for routine clinical applications in future, with help from MR physicists to overcome some of the current technical challenges.

MR methods using hyperpolarisation, which can potentially increase the SNR by a factor of 10,000, are an area of great interest. With necessary MRI physics support, MRI hyperpolarisation methods could possibly be developed for routine clinical applications in the future, for example, allowing signal enhancement from nuclei such as ^{13}C to achieve real-time metabolic imaging (Apps et al. 2018).

MRI methods will continue to become ever more sophisticated with advances in hardware, increasingly powerful computers, and artificial intelligence (AI). MRI can be used to produce many different types of image contrast, but analysis and interpretation of the enormous amounts of data that can be generated can be challenging and time consuming. Data processing and analysis with AI is an area that will see rapid developments, requiring assistance from MR physicists to trial and validate new methods.

It is very likely that powerful MRI systems operating at 7 Tesla will start to enter into routine clinical applications. Clinical Scientist input will be pivotal to ensure safety. Moreover, hybrid MRI systems integrating other imaging modalities and guided treatment delivery are also likely to be increasingly developed. For example, MR-guided high-intensity focussed ultrasound is a particularly promising *technology* (Health Quality Ontario 2018) that is currently in its infancy but is likely to have a huge impact in the future, requiring a combination of both MRI and ultrasound physics expertise.

If the current trajectory of technological development continues, then the number of MRI applications will continue increasing, and with that there will be increasing numbers of new MRI systems installed within hospital imaging departments. There is likely to be a huge demand in the future for Clinical Scientists with expertise in MRI physics, to help drive forward advances in MRI technology, as well as testing and implementing novel clinical and research applications. Without doubt, there are huge potential opportunities and challenges for Clinical Scientists working with MRI, with prospects of a very exciting career that can significantly impact the future of healthcare.

REFERENCES

ACR Committee on MR Safety (2020). "Americal College of Radiology (ACR) Guidance Document on MR Safe Practices: Updates and Critical Information 2019". *J Magn Reson Imaging* **51**(2): 331–338.

Apps A, Lau J, Peterzan M, et al. (2018). "Hyperpolarised Magnetic Resonance for In Vivo Real-Time Metabolic Imaging". *Heart* **104**: 1484–1491.

Assaf Y and Pasternak O (2008). "Diffusion Tensor Imaging (DTI)-Based White Matter Mapping in Brain Research: A Review". *J Mol Neurosci* **34**: 51–61.

Baliyan V, Das CJ, Sharma R and Gupta AK (2016). "Diffusion Weighted Imaging: Technique and Applications". *World J Radiol* **8**(9): 785–798.

Berman RM, Brown AM, Chang SD, et al. (2016). "DCE MRI of Prostate Cancer". *Abdomin Radiol* **41**(5): 844–853.

Bernstein MA, King KF and Zhou XJ (2004) *Handbook of MRI Pulse Sequences*. Elsevier Academic Press, London, UK.

Bizino MB, Hammer S and Lamb HJ (2014). "Metabolic Imaging of the Human Heart: Clinical Application of Magnetic Resonance Spectroscopy". *Heart* **100**: 881–890.

Bojorquez JZ, Bricq S, Acquitter C, Brunotte F, Walker PM and Lalande A (2017). "What Are Normal Relaxation Times of Tissues at 3 T?". *Magn Reson Imaging* **35**: 69–80.

Brown RW, Cheng Y-CN, Haacke M, Thompson MR and Venkatesan R (2014). *Magnetic Resonance Imaging: Physical Principles and Sequence Design*, 2nd edition. John Wiley and Sons, UK.

Chowdhury A, Harding SG, Carpenter TA, Price CJ, Guadagno JV and Baron JC (2003) "First Results of a Multi-Slice 2D Time-of-Flight (TOF) Magnetic Resonance Angiography (MRA) Sequence on a 3T Whole Body Scanner for Stroke Diagnosis". BC ISMRM proceedings.

de Graaf RA (2019). In Vivo *NMR Spectroscopy: Principles and Techniques*, 3rd edition. John Wiley and Sons, UK.

Elster AD (2020). "Questions and Answers in MRI". Online Available at: http://www.MRIquestions.com/ [accessed 11 September 2020].

Ethofer T, Mader I, Seeger U, et al. (2003). "Comparison of Longitudinal Metabolite Relaxation Times in Different Regions of the Human Brain at 1.5 and 3 Tesla". *Magn Reson Med* **50**(6): 1296–1301.

Ferré J-C, Bannier E, Raoult H, Mineur G, Carsin-Nicol B and Gauvrit J-Y (2013) "Arterial Spin Labeling (ASL) Perfusion: Techniques and Clinical Use". *Diagn Interv Imaging* **94**(12): 1211–1223.

Haaga JR and Boll D (2016). *CT and MRI of the Whole Body*, 6th edition. Elsevier, Germany.

Health Quality Ontario (2018). "Magnetic Resonance-Guided Focused Ultrasound Neurosurgery for Essential Tremor: A Health Technology Assessment". *Ont Health Technol Assess Ser* **18**(4): 1–141.

Hornak JP (2020) "The Basics of MRI". Online Available at: http://www.cis.rit.edu/htbooks/mri/ [accessed 11 September 2020].

Ilangovan R and Stewart O, eds. (2020) "MRImaster.com". Online Available at: https://mrimaster.com/ [accessed 11 September 2020].

Jalali R, Chowdhury A, Wilson M, Miall R and Galea J (2018). "Neural Changes Associated with Cerebellar tDCS Studied Using MR Spectroscopy". *Exp Brain Res* **236**(4): 997–1006.

Kelley LR and Petersen CM (2018) *Sectional Anatomy for Imaging Professionals*, 4th edition. Elsevier, Germany.

McRobbie D and Semple S, eds. (2017). *IPEM Report 112. Quality Control and Artefacts in Magnetic Resonance Imaging*. Institute of Physics and Engineering in Medicine, York. UK.

McRobbie DW, Moore EA, Graves MJ and Prince MR (2017). *MRI from Picture to Proton*, 3rd edition. Cambridge University Press, UK.

MHRA (2021). *Safety Guidelines for Magnetic Resonance Imaging Equipment in Clinical Use*. Medicines and Healthcare Products Regulatory Agency, London.

Paulson F and Waschke J (2011). *Sobotta Atlas of Human Anatomy, Package*, 15th edition. Elsevier, Germany.

Perman WH, Hilal SK, Simon HE and Maudsley AA (1984). "Contrast Manipulation in NMR Imaging". *Magn Reson Imaging* **2**(1): 23–32.

Rai S, Chowdhury A, Reniers RLEP, Wood SJ, Lucas SJE and Aldred S (2018). "A Pilot Study to Assess the Effect of Acute Exercise on Brain Glutathione". *Free Radic Res* **52**(1): 57–69.

Reeves MJ, Brandreth M, Whitby EH, et al. (2010). "Neonatal Cochlear Function: Measurement after Exposure to Acoustic Noise during In Utero MR Imaging". *Radiology* **257**: 802–809.

Rinck PA (2018). *Magnetic Resonance in Medicine. A Critical Introduction. The Basic Textbook of the European Magnetic Resonance Forum*, 12th edition. BoD, Germany.

Shellock FG (2020). "MRIsafety.com". Online Available at: http://www.mrisafety.com/ [accessed 11 September 2020].

Westbrook C (2016). *MRI at a Glance*, 3rd edition. John Wiley and Sons, Ltd. UK.

3 Ultrasound Physics

Emma Chung and Justyna Janus

CONTENTS

Ultrasound imaging is different from other forms of medical imaging. If you imagine "shining a torch" into the body; the sound "bouncing back" from soft tissue, fluids, and surfaces, enables these structures to be visualised. Diagnostic ultrasound machines were first developed in the early 1950s, leading to the publication of the first obstetrics ultrasound paper in the Lancet, showing an image of the fetal head (Donald, MacVicar and Brown 1958). In response to evidence suggesting that X-ray exposure during pregnancy could be harmful, ultrasound has now almost completely replaced X-rays as a much safer form of imaging for scanning in pregnancy. Today, ultrasound is the second most performed imaging investigation after planar X-rays. As ultrasound becomes more portable and affordable, the medical uses of ultrasound are likely to continue to grow.

Over the past decade, computational advances have enabled rapid acquisition and processing of ultrasound data, leading to an explosion of novel imaging methods. It is important to keep up to date with latest advances through attending manufacturers' exhibits and demonstrations. Hospital Clinical Scientists specialising in ultrasound are usually responsible for ultrasound safety and quality assurance (QA). They are often also involved in research and innovation, using their laboratory or programming skills to support the development and evaluation of new ultrasound techniques.

You may also work with patients through accompanying emerging ultrasound techniques into clinics and theatres and supporting clinical trials. Scientists specialising in ultrasound can offer practical and scientific support to help new users get the best out of their equipment, joining clinical teams for several months to help them establish new techniques.

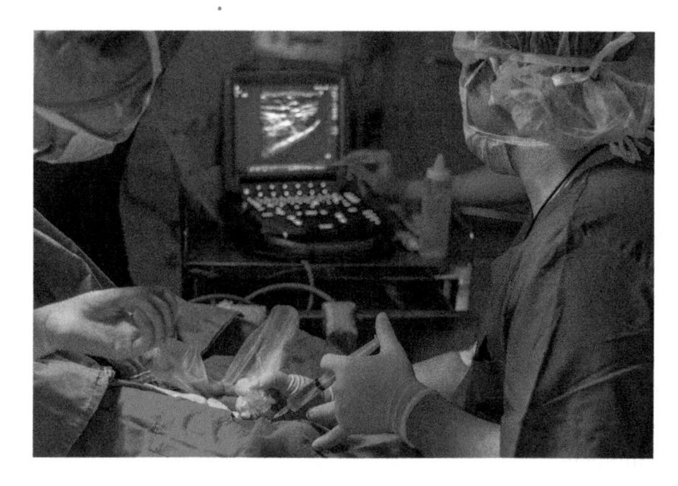

FIGURE 3.1 Ultrasound-guided injection of a nerve block.

Portable ultrasound machines and "pocket" handheld devices are increasingly attractive to non-imaging specialists (such as emergency physicians and GPs) who want to use ultrasound for a specific imaging purpose. For example, guided insertion of a chest drain or biopsy needle, or targeted anaesthesia (Figure 3.1). Ultrasound equipment can range from specialist medical equipment, such as bladder scanners and foetal heart-rate monitors, to versatile ultrasound scanners, offering full diagnostic imaging. At higher energies, ultrasound can be used therapeutically for physiotherapy, brain stimulation and tumour ablation. For example, High-Intensity Focused Ultrasound (HIFU) can be used to treat localised prostate cancer as an alternative to surgery. Ultrasound physics training and safety and QA processes therefore need to cover a wide range of clinical applications.

Hospitals typically own several hundred diagnostic ultrasound scanners. This poses logistical challenges for managing ultrasound QA. Compared to annual QC checks of magnetic resonance imaging (MRI) scanners, which typically involve assessing a handful of machines, QC of ultrasound scanners continues throughout the year. Timetabling of ultrasound QC therefore requires good organisation and communication with users to track down equipment and avoid unnecessary disruption of clinical services. Between QC tests, you need to encourage Sonographers and Radiologists to check their equipment regularly to identify common probe faults. If a fault is found, repairs need to be swiftly organised through Clinical Engineering, or as part of local service agreements. Given the number of scanners involved, accurate documetation of ultrasound QC results is vital for comparing results and monitoring scanner performance over time. As a Clinical Scientist specialising in ultrasound you might be involved with:

- Designing and managing ultrasound QA programmes
- Acceptance testing of new ultrasound equipment
- Reviewing local protocols to ensure that ultrasound QC follows national and international guidelines
- Teaching and training of NHS staff and university students
- Development of laboratory phantoms and QC test objects for unusual or emerging ultrasound applications
- Ultrasound-related research and service development
- Ultrasound output power measurements (e.g. for use of non-CE marked ultrasound equipment)

- Software development for ultrasound image processing and signal analysis
- Simulations (e.g. using an ultrasound propagation simulation package, such as k-wave (Treeby and Cox 2010)).

3.1 HOW DOES ULTRASOUND WORK?

Medical ultrasound technology was developed following the Second World War based on Sound Navigation Ranging (SONAR) methods developed for detecting submarines. In SONAR, a "ping" of sound is emitted, and an echo intensity map is then generated by analysing the direction, intensity and timing of the resulting echoes. This uses the same principle of echolocation that bats and dolphins use naturally to navigate their surroundings. As the speed of sound in water is known, measuring the time taken for echoes to return to the transducer, makes it possible to plot echo intensities as a function of depth (Figure 3.2).

In medical applications, the sonar "ping" is replaced by a pulse of high-frequency sound: ultrasound. Ultrasound has a shorter wavelength than audible sound (<1 mm), so is suitable for visualising anatomy.

The speed (*c*) of ultrasound propagation in tissue varies with each tissue's density and stiffness. As most soft tissues in the body possess similar speed of sound values (Table 3.1) this makes it possible to convert echo time to depth in the image with reasonable accuracy. Virtually all ultrasound scanners assume an average speed of sound in soft tissue of $c = 1540$ m/s in order to construct the ultrasound image. If the speed of sound deviates from this, such as in fat or bone, depths will appear truncated ($c > 1540$ m/s) or elongated ($c < 1540$ m/s) in the image.

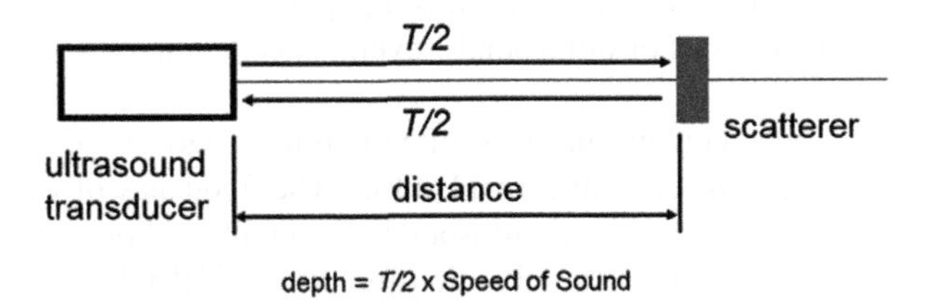

FIGURE 3.2 The principle of echolocation; sound is scattered back to the transducer with a characteristic echo time (T) that can be used to deduce the depth of the target from the probe.

TABLE 3.1
Speed of Sound (*c*) and Density (*ρ*) of Common Biological Materials

Material	*c* (m/s)		Density, *ρ* (kg/m³)
Air	343		1.2
Fat	1450	The average speed of sound in soft tissue is 1540 m/s	952
Liver	1550		1060
Kidney	1560		1038
Muscle	1580		1080
Blood	1575		1057
Bone	3500		1912

Medical ultrasound scanners are designed to generate brief pulses of ultrasound with frequencies ranging between 2 and 20 MHz. Assuming a speed of sound of 1540 m/s, and using $c = f\lambda$, this corresponds to ultrasound wavelengths (λ) ranging between 1 and 0.1 mm; perfect for imaging structures in the body.

Conventionally, each pulse-echo cycle generates a single line in the image and a two-dimensional (2D) echo map is then built up "line by line". Pixels along each line display the echo intensity received from a particular sample volume. Sample volume location and size depends on the echo time and pulse duration, as well as the physical dimensions of the ultrasound beam (beam width and slice thickness). As a pulse of ultrasound moves through the body it is scattered and reflected and the intensity of the ultrasound beam becomes progressively weaker with depth. Typically only a small fraction of the beam's original intensity is detected back at the probe as an echo. To obtain an image of uniform brightness, the scanner amplifies weak ultrasound echoes obtained from deeper within the body as a function of echo time (Figure 3.3).

By applying these principles, it becomes possible to build up an image revealing ultrasound echoes generated within and between materials. As long as the speed of sound approximation is valid, the resulting images bear a close resemblance to anatomy and can be updated in real time.

An example of a conventional 2D ultrasound image is shown in Figure 3.3. The shape of the surface of the probe can be seen at the top of the image. The brightness of each pixel in the image is related to the depth compensated echo intensity; white indicates the strongest echoes (echogenic, or "echo generating" structures), whereas black indicates that echoes are absent (anechoic regions). Anechoic regions can occur either due to the absence of ultrasound (acoustic shadowing beyond a brightly reflecting object) or due to the absence of scatterers (as occurs in fluids). Interpretation of subtle variations in ultrasound image appearance is aided by the operators ability to orient the probe to confidently distinguish tissue structures and pathological features from artefacts. For more information about the physics of ultrasound, several excellent texts and online resources are available (Table 3.2).

Ultrasound imaging probes contain an array of vibrating (piezoelectric) crystals that can be precisely controlled to emit and receive ultrasound pulses. The frequency of vibration of the crystal governs the central frequency of the emitted ultrasound. The number of pulses emitted by the probe per second occurs at a rate called the pulse repetition frequency (PRF). In conventional ultrasound imaging, information from adjacent beamlines are combined to form a 2D "echo map". However, manufacturers have become increasingly creative in finding new ways of manipulating the

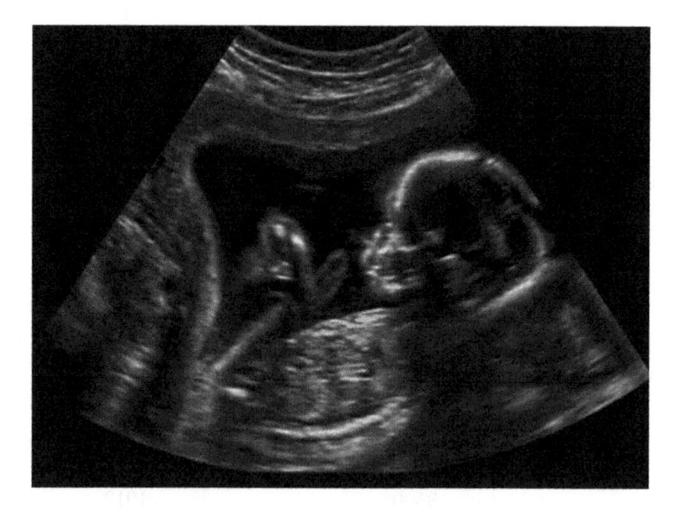

FIGURE 3.3 A typical brightness-mode (B-mode) pregnancy scan.

TABLE 3.2
Recommended Ultrasound Physics Texts. Many of These Texts Are Freely Available for Pdf Download (see, for example, https://am-medicine.com/)

Allisy-Roberts and Williams (2007) "Farr's Medical Imaging (2nd Ed)"	A condensed brief introduction to ultrasound physics aimed at trainee Radiologists.
Hoskins et al. (2019) "Diagnostic Ultrasound: Physics and Equipment (3rd Ed)"	A comprehensive introduction to the physics and technology of medical ultrasound suitable for trainee Clinical Scientists.
ter Haar (2012) "The Safe Use of Ultrasound in Medical Diagnosis (3rd Ed)"	This important safety text is freely available as part of the British Institute of Radiology's Open Access initiative. It covers key concepts for the safe use of ultrasound.
Barrie-Smith and Webb (2011) "Introduction to Medical Imaging: Physics, Engineering and Clinical Applications"	This textbook includes chapters introducing ultrasound physics and imaging at a level suited to physics undergraduate and MSc students.
Dendy and Heaton (2012) "Physics for Diagnostic Radiology (3rd Ed)"	The ultrasound chapter is detailed and well explained, providing a strong foundation for all Clinical Scientists working in ultrasound. Currently freely available as a pdf.

emission and processing of ultrasound pulses to construct ultrasound images using combinations of plane waves and other pulse sequences capable of ultra-fast and 3D image acquisition.

The maximum penetration depth achieved by a given transducer is governed by the choice of central ultrasound frequency of the pulse and attenuation rate of the tissue being imaged. The axial resolution (ability to separate two point-like objects located in the direction of propagation of the pulse) is governed by the pulse duration. Lateral resolution and slice thickness are governed by the physical dimensions of the beam (Figure 3.4). An acoustic lens is often added to the face of the transducer to help provide a roughly constant slice thickness. The beam width in the lateral

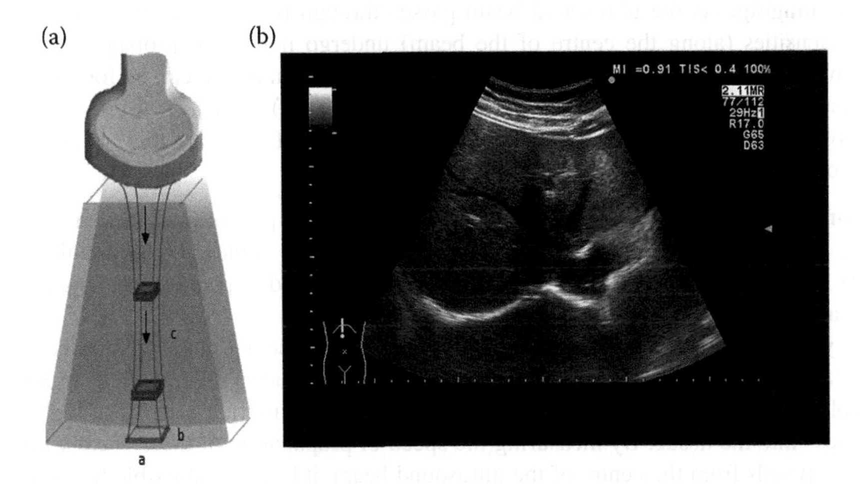

FIGURE 3.4 Pulses travel along each beamline from a convex array: label (a) indicates the beam width, which governs the lateral resolution; label (b) indicates the slice thickness; and label (c) indicates the focal point of the beam (with best lateral resolution). Arrows show the direction of propagation of the ultrasound pulse. The axial extent of each shaded sample volume is governed by pulse duration, which defines the axial resolution of the beam. (b) Example image of the liver.

direction can be focused and manipulated by selecting groups of piezoelectric elements and controlling their transmission times to improve resolution around a specific region of interest. Overall, the dimensons of the beam are shaped according to Huygens's principle and ultrasound propagation can be accurately modelled using simulations.

As ultrasound moves through the body, tissue structures with dimensions smaller than the ultrasound wavelength scatter ultrasound isotropically (equally in all directions). These sub-resolution scatterers generate echoes that combine (or interfere) to produce the speckle pattern characteristic of the appearance of tissue in the ultrasound image. Note that the speckle pattern seen in an ultrasound image is not the result of "random noise", as described in Figure 1.8 of Chapter 1. The speckle pattern is the ultrasound interference pattern generated by sub-resolution structures within the tissue and is therefore an intrinsic property of the ultrasound image.

The ultrasound beam is also reflected from any transitions it encounters between materials, especially where there is a pronounced difference in the physical properties of materials either side of a boundary (such as between tissue and fluid, or tissue and bone). Since the angle of reflection from boundaries is equal to the angle of incidence, the appearance of large smooth (specular) reflectors is also highly angle dependent; some surfaces only become visible in the image at angles close to 90°. This can be tested by the operator angling the probe so that the surface reflects ultrasound directly back to the transducer. The position of objects in an ultrasound image can also be affected by refraction. Refraction of sound occurs where there is a pronounced change in the speed of sound between materials and is also angle dependent. An ultrasound examination is therefore a dynamic process where the operator deliberately utilises the angle dependent nature of echoes from boundaries and adjusts probe position and angle to help identify tissue structures based on their speckle, reflection and refraction characteristics.

Ultrasound imaging is routinely used for foetal screening and diagnosis and monitoring of a wide range of medical conditions. Ultrasound is also increasingly used to guide real-time interventions, such as needle biopsies and brachytherapy. Examples of common types of ultrasound image display are provided in Table 3.3.

Additional modes of operation available to improve the appearance of ultrasound images include:

Harmonic imaging: As the ultrasound beam passes through tissue, areas exposed to higher ultrasound intensities (along the centre of the beam) undergo non-linear propagation to generate echoes at harmonics (multiples) of the transmitted frequency. An image can be formed using only the harmonic frequencies, especially the second harmonic ($2f$), which is useful for improving lateral resolution. As harmonics only occur close to the centre of the beam this effectively reduces the width of the beam, improving lateral resolution.

Spatial compound imaging: High frame rates and faster signal processing times make it possible to capture multiple ultrasound images from slightly different angles. By combining (or compounding) multiple images the speckle becomes smoothed and appearance of angle-dependent artefacts can be reduced, producing a better quality image.

Recent years have also seen increasing clinical use of ultrasound elastography techniques for the diagnosis of tissue stiffness changes (e.g., associated with cancer). For example, in Shear Wave Elastography (SWE) the scanner generates shear waves by transmitting a series of high amplitude "push pulses" into the tissue. By measuring the speed of propagation of these shear waves, which propagate outwards from the centre of the ultrasound beam, it becomes possible to quantify tissue stiffness as an estimate of Young's Modulus. In ultrasound systems capable of elastography, a transparent colour map is superimposed over a B-mode image to indicate regions of stiffness (or low strain) in the image.

TABLE 3.3
Ultrasound Imaging Examples

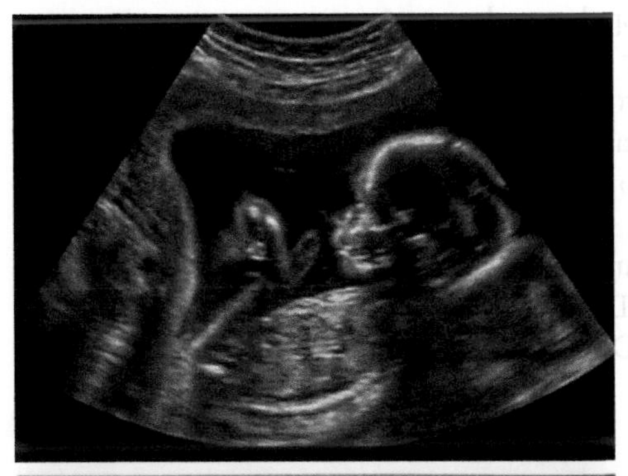

2D (B-mode): A standard 2D "brightness mode" ultrasound image.

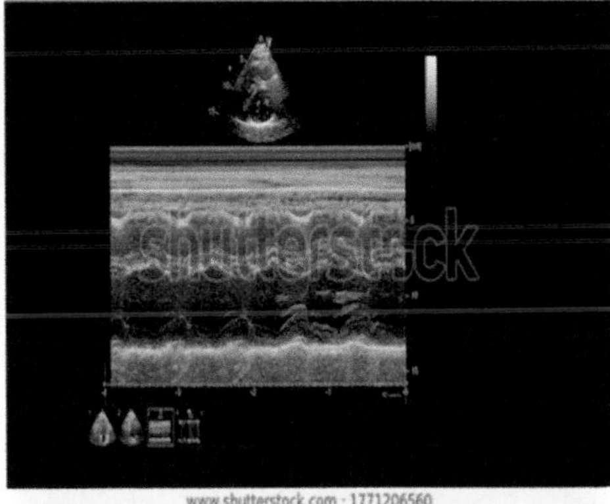

M (M-mode): This provides an additional plot that follows a single-beamline over time. The beamline is selected by the user. This can be useful for visualising heart valves or tissue motion.

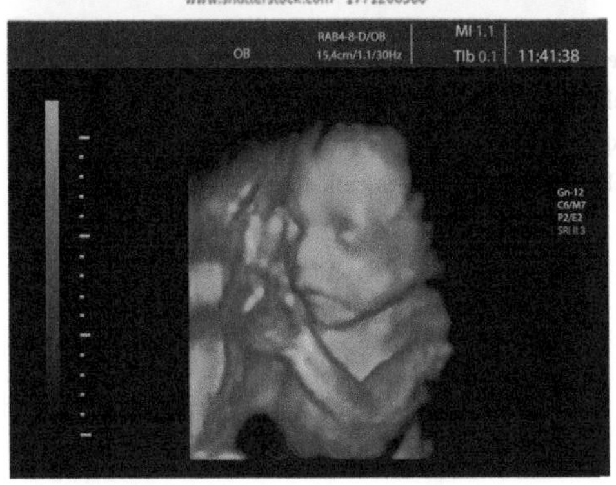

3D real-time imaging: Some ultrasound equipment also offers 3D imaging, or 3D real-time imaging (also known as 4D ultrasound). This involves obtaining 3D data that can be surface rendered to show volumes and surfaces of interest. May be useful in the assessment of foetal pathology and congenital malformations but is generally not widely used clinically.

3.1.1 Doppler Ultrasound

A major advantage of ultrasound over other types of imaging is that images can be displayed and updated in real-time. The high temporal resolution of ultrasound techniques enables imaging to be combined with visualisation and real-time monitoring of blood flow. A phenomenon known as the "Doppler effect" is used to estimate the velocities of moving red blood cells by measuring the shift in frequency (or phase shift) between transmitted and received ultrasound signals:

$$f_D = f_r - f_t = \frac{2 f_t v \cos \theta_D}{c}$$

where, f_D is the Doppler frequency, f_r and f_t are the received and transmitted ultrasound frequencies, v is the velocity of the blood, c is the speed of sound, and θ_D is the angle between the beam and direction of blood flow, known as the Doppler angle.

When ultrasound imaging of vessels and Doppler measurements are combined, this provides a powerful tool for displaying blood flow velocity profiles from specific vessels or regions of interest. Doppler ultrasound is commonly used to image the heart (Echocardiography) and vasculature. (Table 3.4). By manually adjusting a cursor to coincide with the direction of flow, scanners combining pulse-wave (PW) Doppler with B-mode imaging can accurately estimate blood flow velocities in the heart and major vessels after adjustment for Doppler angle. B-mode and Colour Doppler are often combined with spectral (PW) Doppler to estimate flow in major vessels. Manufacturers have also started offering vector Doppler imaging where the magnitude and

TABLE 3.4
Examples of Doppler Modes

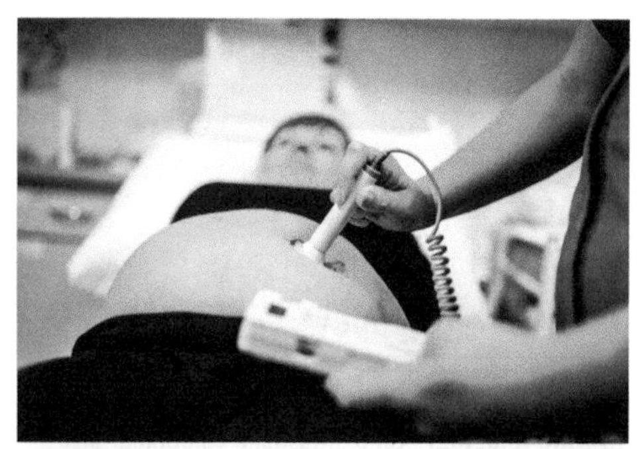

Continuous Wave (CW) Doppler: The simplest Doppler devices emit and receive ultrasound continuously to provide an audible Doppler signal for monitoring foetal heart rate. CW Doppler mode can also be used to measure high blood flow velocities in echocardiography.

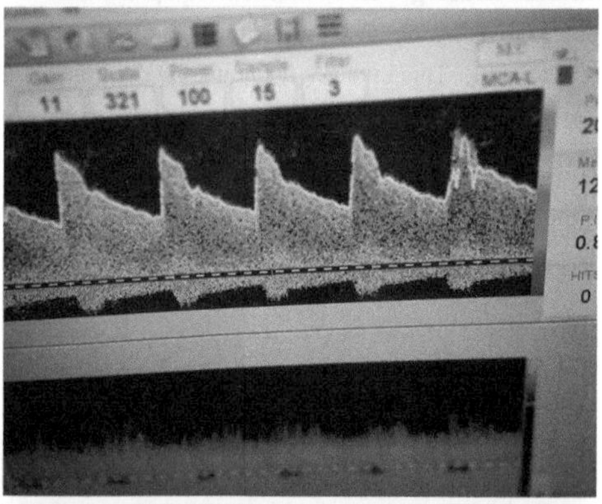

Pulse wave (PW) Doppler: This shows the Doppler shift (or velocity) as a function of time with pixel intensity indicating the number of scatterers moving at each velocity. This display is called spectral Doppler.

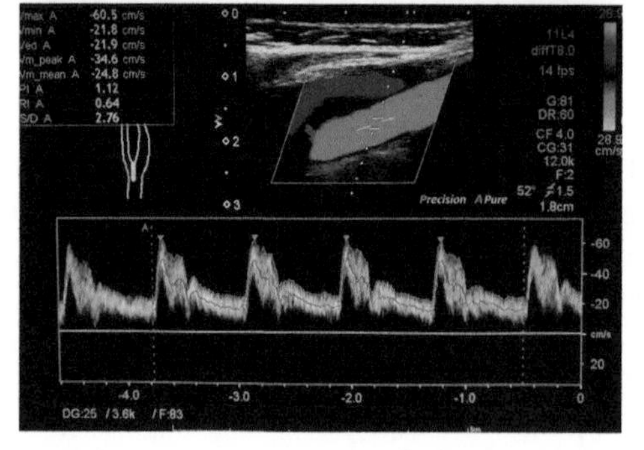

Colour Doppler: Doppler information is presented as a colour map and superimposed over a region of interest in the B-mode image. Velocities are colour-coded (red or blue) to indicate whether flow is moving towards or away from the probe. Colour Doppler is often combined with spectral Doppler information (lower panel).

direction of flow is indicated by coloured arrows. More recently, with the advent of "ultrafast" Doppler and "microvascular imaging", perfusion and microvascular flow measurements have also become possible.

Ultrasound contrast agents: Vascular ultrasound contrast agents take the form of tiny "microbubbles" that are introduced to the blood stream to enhance the ultrasound signal from blood flow. Microbubbles can be generated through agitation of saline and are commercially available as vials of encapsulated microbubbles. Each vial contains millions of 1–10 μm diameter bubbles containing a biocompatible inert gas, encapsulated in a lipid shell. These improve the visibility of vessels and can potentially be targeted to adhere to thrombus or tumours. Contrast-enhanced ultrasound (CEUS) imaging is often used as an adjunct to conventional B-mode imaging to distinguish between benign and malignant tumours through identifying hypervascularity associated with tumour growth.

3.2 LINKS TO OTHER PROFESSIONS

Most ultrasound practitioners are qualified Radiographers, Radiologists, Vascular Technologists or Midwife Sonographers who will have received lectures in ultrasound physics as part of their training. Although there is no legal requirement to hold a recognised qualification to practice ultrasound in the UK, specialist training is highly recommended.

Clinical Scientists specialising in ultrasound physics should be familiar with the underlying physics, instrumentation, safety and QC procedures associated with medical ultrasound imaging. It is important for Clinical Scientists to be able to make this material accessible and understandable to non-experts, and to encourage Sonographers and Radiologists to use their knowledge of physics to improve their imaging practice. Clinical Scientists need good communication skills and are able to move between medical and technical terminology and settings. Ultrasound physicists are therefore uniquely placed to be able to make a difference to patient care by introducing new ultrasound technologies to clinical areas.

3.3 ULTRASOUND EQUIPMENT

Diagnostic ultrasound procedures typically require an experienced operator to obtain accurate measurements and report imaging findings for medical diagnosis. Ultrasound operators need to understand the impact of equipment settings on image quality. Operators should notify Medical Physics or Clinical Engineering staff if they think there might be a technical issue with their equipment so that problems can be quickly rectified.

3.3.1 Ultrasound Transducers

For a given clinical examination, it is important to make the correct choice of ultrasound transducer. Each hand-held transducer contains an arrangement of piezoelectric elements that collectively vibrates to transmit controlled pulses of ultrasound into the body. Each type of probe has been optimised for a specific imaging application, with the size and arrangement of elements, and range of ultrasound frequencies that can be emitted and received by the transducer, carefully selected to provide an appropriate field of view and suitable image quality.

Ultrasound scanners are usually equipped with a choice of probes. Each probe has been designed and optimised to visualise a particular field of view. For example, the C1–5 transducer pictured in Figure 3.5 is a 2–5 MHz (wide bandwidth) curved linear-array probe, which would have a penetration depth and field of view suitable for abdominal and obstetrics applications. Most probes also include a raised mark, or "dimple", on the side of the probe to help guide orientation.

FIGURE 3.5 Ultrasound probes contain an arrangement of transducer elements optimised for particular clinical applications.

3.3.2 The Scan Engine – Knobology

Scientists need to be able to describe how the patient, user and scanner controls combine to govern the appearance of an ultrasound image. Often, staff will have very little interest or enthusiasm for physics or engineering so the more relevant and accessible you can make this information the better. Gaining an understanding of how to manipulate ultrasound system controls (known as "knobology" – a knowledge of knobs) is the first task of anyone working in ultrasound. A typical ultrasound control panel is shown in Figure 3.6.

The vast array of "knobs and buttons" for controlling advanced imaging options and pre-sets may initially seem daunting, but one of the advantages of ultrasound, compared to other types of medical imaging, is that it can be very "hands-on". Trainees have ample opportunity to practice and to try out different scanners and familiarise themselves with the controls.

Numerous buttons allow the user to switch between, or combine, different ultrasound acquisition modes to suit their clinical examination. Clinical Scientists should be familiar with the controls of different makes and models of ultrasound machine so that they can confidently "drive the scanner" as well as being able to explain and demonstrate scanner functions to new users.

Ultrasound equipment that is used within a single clinic, or for specific applications, will have common configurations pre-set by the users. These presets provide a handy short-cut for performing common types of scans.

The display can be adjusted using brightness and contrast display settings, as well as time-gain control (TGC) settings used to improve uniformity of the B-mode image. Other gain controls, persistence settings, and modes such as harmonic and compound imaging, are available for specialist applications. It is worth being aware that changes to some settings, such as depth and scale (for Doppler information) will automatically adjust other settings, such as the pulse repetition frequency, to achieve faster sampling rates. Often, a range of colour maps, a zoom function and axis scaling options are available to optimise the display. Callipers and other tools for measuring distances, or delineating a region of interest are also available.

3.4 SAFETY OF DIAGNOSTIC ULTRASOUND

Ultrasound has an exceptional safety profile compared to other imaging modalities and is safe enough for use during pregnancy. However, interactions of sound waves with tissue, through

TABLE 3.5
Examples of Linear, Curvilinear, Phased Array and Endocavity Ultrasound Probes

Type of a Transducer	Clinical Applications
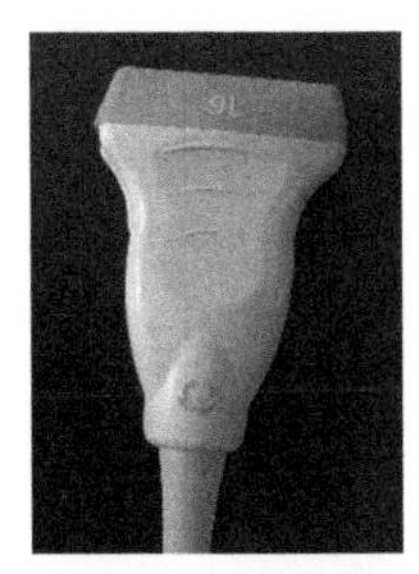	**Linear "Vascular probe"** Vascular, "small parts" and musculoskeletal applications
	Curvilinear "abdominal probe" General obstetrics and abdominal applications
	Phased array "cardiac probe" Echocardiography applications
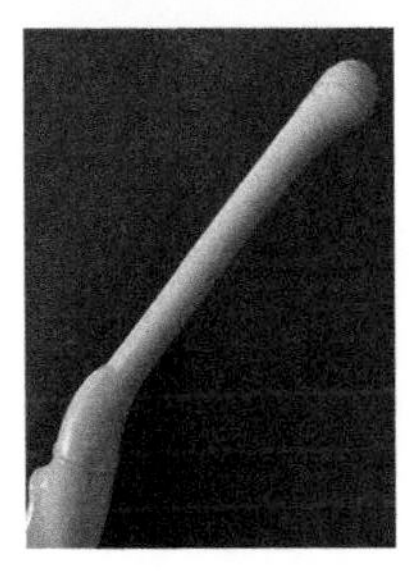	**Endocavity probe** Vaginal and rectal applications

thermal and non-thermal mechanisms, do have the potential to lead to biological effects if not used prudently. In therapeutic applications, such as ablation of tissue using HIFU, ultrasound power output is several orders of magnitude higher than diagnostic levels. At present, HIFU is still under evaluation and is only available at specialist centres. This section is limited to the safety of diagnostic rather than therapeutic ultrasound applications.

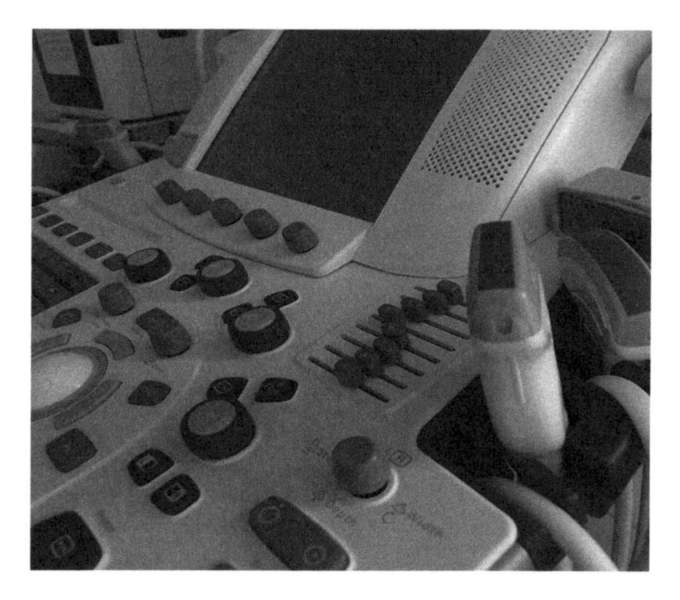

FIGURE 3.6 Ultrasound scanners offer a wide arrange of ultrasound modalities, imaging pre-sets, and image acquisition, processing, and display options.

In situations where tissue heating or cavitation effects have potential to be non-negligible, ultrasound safety indices, thermal index (TI) and mechanical index (MI), are displayed, e.g. Figure 3.7. It is the responsibility of the user to ensure that MI and TI values displayed on the screen are kept within recommended limits. In situations where the user is unable to obtain high enough image quality for accurate diagnosis by any other means, it is recognised that recommended MI and TI limits may need to be exceeded for short periods.

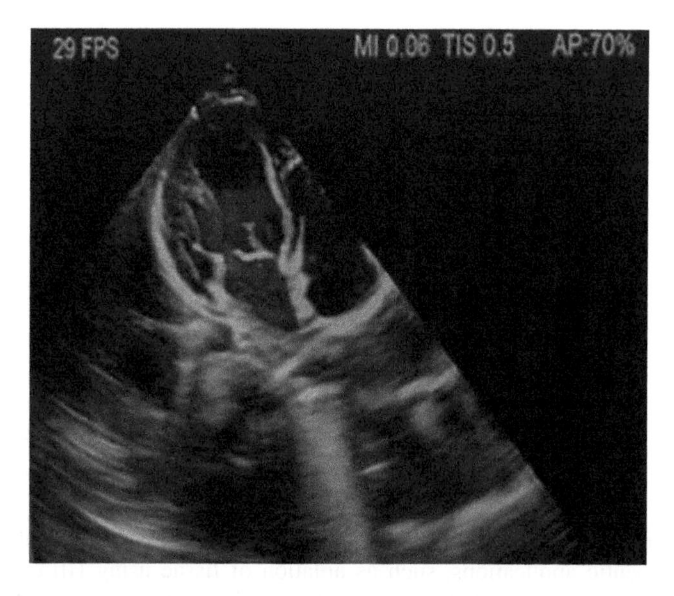

FIGURE 3.7 Ultrasound display showing mechanical index (MI) and thermal index for soft tissue (TIS) values.

3.4.1 Thermal Index

As ultrasound vibrations travel through tissue, much of the mechanical energy of the wave is absorbed and converted to heat. The TI for soft tissue (TIS) describes the potential temperature rise expected to occur in soft tissue under the current operating conditions of the scanner. This is given by:

$$\text{TIS} = W_0 / W_{\text{deg}}$$

where W_0 is the current acoustic output power from the transducer and W_{deg} is the estimated power necessary to raise the target tissue temperature by 1°C (Szabo 2014). The TIS can therefore be thought of as an estimate of the increase in temperature that may occur in soft tissue in the region of the ultrasound beam.

Heating is usually greatest at the surface of the probe (due to probe self-heating, especially if left radiating into air) and near the beam's focus (due to ultrasound absorption). The TI accounts only for heating due to ultrasound absorption. To prevent heating operators are advised to "freeze" image acquisition whenever the transducer is not in use. A special case occurs in applications such as transcranial ultrasound where bone is present near to the transducer. Bone absorbs heat more readily than soft tissue so when scanning through the skull a "thermal index for cranium" (TIC) is monitored. If bone may be present close to the beam's focus, such as in obstetrics scanning, the TI for bone (TIB) should be monitored. The temperature rise in tissue also depends on beam width, ultrasound frequency, and the duration of exposure. Excessive heat generated by ultrasound is hazardous to sensitive organs, embryo and foetus and has been shown in animal models to lead to various forms of tissue damage including membrane dysfunction, apoptosis, abnormal cell migration and altered gene expression (Izadifar et al. 2017; ter Haar 2012).

3.4.2 Mechanical Index

In response to variations in pressure generated by the ultrasound beam, bubbles have the potential to resonate, oscillate or implode (cavitate). Cavitation effects have been demonstrated in animal models to lead to rupture of alveoli in the lungs after short duration exposures in rodents, swans and monkeys (Dalecki et al. 1997, 1999; Frizzell et al. 2003; Tarantal and Canfield 1994). It has also been reported that exposure to ultrasound can cause petechial haemorrhages in the mouse intestine (Miller and Thomas 1994).

The mechanical index MI describes the probability of ultrasound-induced cavitation of bubbles. As bubbles are not naturally present in the majority of organs, cavitation is only possible when scanning the neonatal lung or bowel, or when imaging in the presence of microbubble contrast agents.

$$\text{MI} = \frac{P_{-\text{derated}}}{\sqrt{f_0}}$$

The MI is derived by measuring the peak rarefaction pressure (P) in water using an ultrasound beam plotting system; this value of P obtained in water is then adjusted (or "derated") to estimate the peak rarefaction pressure that would be likely in soft tissue (Szabo 2014). For example, for soft tissue with an attenuation of 0.3 dB/cm/MHz, the derated pressure would be:

$$P_{-0.3} = P \times 10^{\frac{0.3 \times \text{dist} \times f_0}{20}}$$

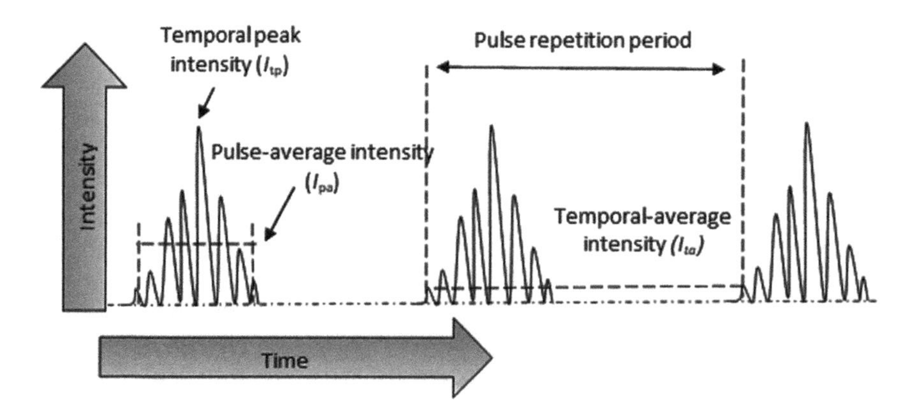

FIGURE 3.8 Definitions relating to measures of the temporal intensity of an ultrasound beam.

A beam plot can be used to estimate further safety measures of ultrasound intensity, such as the spatial average and peak intensity measured over a cross-section of the ultrasound beam. Temporal averages, (estimated over time) are summarised in Figure 3.8. These include the pulse-average and peak intensity of the ultrasound pulse, as well as the temporal average intensity of the beam, which accounts for the 'off-time' between pulses.

The British Medical Ultrasound Society (BMUS 2010) recommends limiting the duration of ultrasound exposure based on the MI and TI. For an example of time recommendations for general diagnostic ultrasound scanning, see Table 3.6. Note that additional tables are available through BMUS for fetal and neonatal scanning.

3.4.3 Safety Tests of Non-CE Marked Ultrasound Equipment

Most commercially available ultrasound systems do not require independent assessment of ultrasound output as they have been confirmed to be within safe diagnostic limits through FDA and "CE" mark processes. Nonetheless, an understanding of ultrasound power and the properties of the ultrasound beam form a key competency for Clinical Scientists training in ultrasound physics.

If prototype ultrasound equipment is being used for research studies, or CE-marked systems are proposed for use outside of their intended purpose, it may be necessary to perform independent output measurements and QC checks before the system can be used on patients. This involves measuring the acoustic output of the system using a power balance and mapping the pressure field generated by the transducer using a calibrated ultrasound beam plotting system. If this cannot be achieved "in house", commercial and NHS services are available for the evaluation of ultrasound output against international calibration standards.

3.5 ULTRASOUND QUALITY CONTROL

The goal of ultrasound quality control (QC) checks is to ensure that equipment is fit for purpose and operating at consistent and acceptable levels of performance. A number of professional organisations and screening bodies have produced recommendations for assessing imaging performance, such as the Institute of Physics and Engineering in Medicine (IPEM 2010), the Royal College of Radiologists and Society and College of Radiographers (RCR 2014), the European Federation of Societies for Ultrasound in Medicine and Biology (EFSUMB 2020), the British Medical Ultrasound Society (BMUS 2010), and the NHS Breast Screening Programme (Public Health England 2011). Although UK Hospitals are not legally required to implement ultrasound QA recommendations, healthcare institutions do have an obligation to ensure all medical equipment is safe and fit for purpose and most centres will have developed QA procedures incorporating one or more of the above guidelines.

TABLE 3.6
BMUS Guidelines Limiting the Duration of Ultrasound Exposure Based on TI and MI Values for General Ultrasound Scanning in Adults (BMUS 2010)

Application	Values to Inspect	TI Value >1.0	MI Value >0.7
Adult transcranial (imaging and stand-alone)	TIC and MI	Restrict time to: 0.7 < TIC ≤ 1.0: 60 min 1.0 ≤ TIC ≤ 1.5: 30 min 1.5 < TIC ≤ 2.0: 15 min 2.0 < TIC ≤ 2.5: 4 min 2.5 < TIC ≤ 3.0: 1 min TIC > 3.0 not recommended	Risk of cavitation with contrast agent
Eye	TIS and MI recommended	Scanning of the eye is not recommended	Risk of cavitation with contrast agent
General abdominal	Usually TIB and MI	Restrict time to: 1.0 ≤ TIB ≤ 1.5: 120 min 1.5 < TIB ≤ 2: 60 min	Risk of cavitation with contrast agent
Peripheral vascular	TIC and MI if bone closer than 1 cm	2.0 < TIB ≤ 2.5: 15 min 2.5 < TIB ≤ 3: 4 min 3.0 < TIB ≤ 4: 1 min	
Other applications	TIS and MI only if bone does not come into the image	4.0< TIB ≤ 5: 15 sec 5.0 ≤ TIB ≤ 6: 5 sec TIB > 6 not recommended	
Peripheral pulse monitoring	TIS and MI are not usually available for dedicated peripheral pulse monitors.	The output from CW Doppler devices intended for monitoring peripheral pulses is sufficiently low that their use is not contra-indicated on safety grounds.	

Tests are generally divided into "acceptance testing" on receipt of new equipment, annual or twice yearly "QC checks" performed by a Medical Physics or Clinical Engineering technician, and "user-checks" performed periodically by clinical users. Clinical Scientists are responsible for determining the appropriateness of QC checks, providing advice, and performing more advanced QC as needed. They should be able to train users and supervise technical staff in conducting routine general ultrasound QC. These acceptance, user-led and QC tests are described in more detail in the following sections.

3.5.1 Acceptance (Baseline Tests)

On acceptance of a new system, it is good practice to obtain a set of baseline measurements; this is called acceptance testing. Baseline measures are useful for identifying degradation of the system over time. When accepting a new ultrasound system into clinical service, it may also be important to establish what the machine will be used for to verify that the QC protocol assigned for that system and frequency of QC is appropriate.

Current QC recommendations and test objects are targeted almost exclusively at assessing image quality associated with general diagnostic imaging procedures. For some specialist probes or applications, such as Doppler measurements, general ultrasound QC checks may not be adequate; in this case, a bespoke QC schedule using specialist phantoms may be needed. A separate evaluation of the performance of the device in clinical practice through a research study or service/product evaluation may also be warranted. QC guidelines and suitability of phantoms and test objects should be regularly reviewed.

3.5.2 Annual QC Tests

After assessing whether there is a need for general, or specialised, ultrasound QC, and acceptance testing, periodic QC checks should be performed at least annually to check for any deterioration in image quality or performance over the lifetime of the scanner. The recommended frequency of more detailed QC checks depends on the clinical application and likelihood of wear. For most systems, checks are conducted annually, but for equipment in constant use, and in clinical applications where image quality is paramount, such as breast screening, QC tests may need to be performed more frequently.

QC tests for image quality are typically performed using a well-characterised commercially available Tissue Equivalent Test Object (TETO). This contains a series of targets embedded in an ultrasound tissue mimic. The targets and surrounding "tissue" are carefully selected to mimic the acoustic properties of soft tissue and other biological structures. The speed of sound of the tissue-mimicking material is usually 1540 m/s with impedance and attenuation similar to those of soft tissue. Different regions and objects within the phantom are used to quantify the axial, lateral and contrast resolution, maximum penetration depth, smallest cyst-like object that can be detected, tissue elasticity and calliper accuracy. The design and imaging appearance of a typical general ultrasound TETO for assessment of ultrasound image quality is shown in Figure 3.9.

QC test objects have a lifetime of approximately 10 years if there is no damage done to the airtight container and the TETO is stored appropriately at room temperature. The integrity of the TETO should be checked annually by weighing the TETO and checking the surface membrane for damage.

3.5.3 User-Led Tests

The aim of user-led QC is to promptly identify any potential problems with equipment while in everyday use. Regular "user-led" QC tests should be encouraged to ensure probe faults are picked

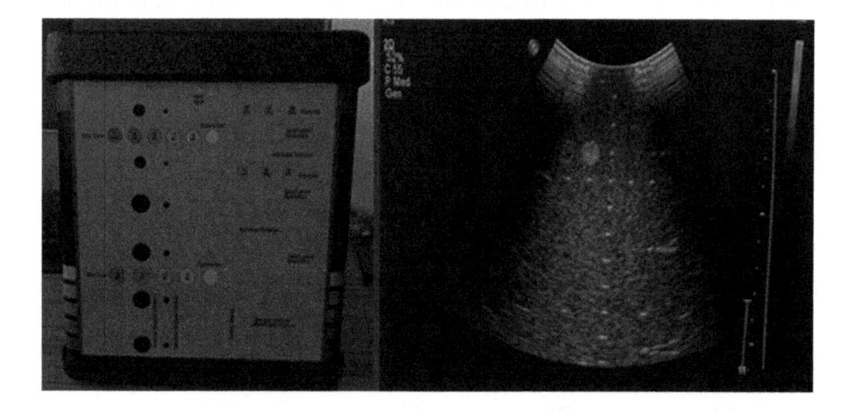

FIGURE 3.9 A commercially available multi-purpose general ultrasound Tissue Equivalent Test Object (TETO) from CIRS, constructed from Zerdine to simulate the acoustic properties of human soft tissue. The test object includes a series of wire targets embedded in the "tissue". The phantom includes two different attenuation levels: 0.5 dB. cm^{-1}. MHz^{-1} and 0.7 dB. cm^{-1}. MHz^{-1}. The scanning surface is marked with a centre line to show where the attenuation change occurs. It is recommended to use the 0.5 dB/cm/MHz side for ultrasound frequencies under 5 MHz.

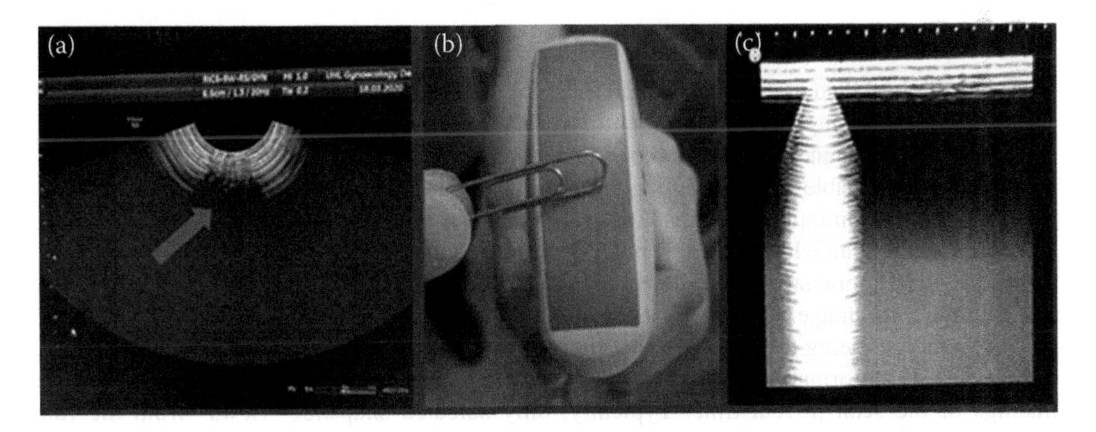

FIGURE 3.10 User-led QC tests can identify any obvious signs of probe or cable damage between QC checks. (a) A faulty "in-air" reverberation pattern, and (b) running a paperclip along the face of the probe and monitoring the resulting image (c) can be used to identify transducer element drop-out.

up between annual QC checks. These include regular (weekly) inspection of probes, the scan engine and cables for any obvious signs of wear and tear. Simple "in-air" reverberation tests, and a paperclip test, where the user runs a paperclip along the face of the probe, are recommended at least monthly to detect damaged transducer elements (Figure 3.10).

CASE STUDY: ULTRASOUND QC FINDINGS

A Sonographer reported bright patches in the ultrasound image, as shown in Figure 3.11, and asked for a Clinical Scientist to investigate. No obvious damage was visible to the machine or the probe and the "in-air" reverberation pattern and paperclip test did not reveal element drop-out. The probe was disconnected and connected to another port to check if there was a problem with the port. The bright patches persisted, which confirmed that the problem was with the probe rather than the scan engine. Further investigation showed that the lower part of the cable (near the connector) was damaged. The Sonographer in charge was informed of the fault and the transducer was sent for repair.

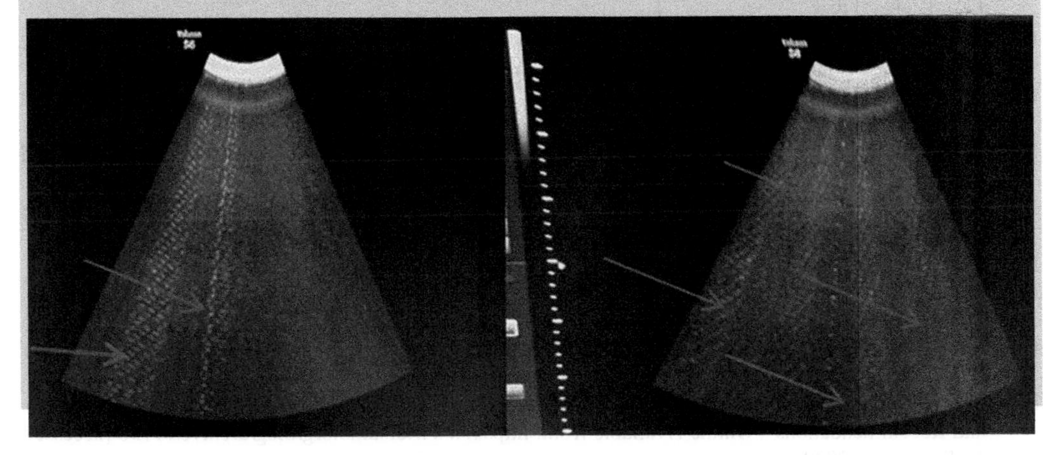

FIGURE 3.11 Unusual bright bands due to a faulty connector.

3.6 THE FUTURE OF ULTRASOUND PHYSICS

In the hands of a skilled user, ultrasound offers the highest resolution of all imaging modalities and is capable of diagnosing and excluding soft-tissue pathology with high sensitivity and specificity. The ability of ultrasound to measure blood flow and update images in real-time has made ultrasound uniquely valuable for echocardiography and vascular applications.

A key current limitation of ultrasound is the need for a highly skilled operator to perform and optimise the scan while simultaneously interpreting and reporting images. Operators need to have a good appreciation of how each patient's anatomy will interact with ultrasound to generate the final image. Ultrasound examinations are completely different to other forms of medical imaging, being dynamic and interactive. Operators rely on their skills and judgement to optimise the information that can be provided by each examination and often report the findings of the scan in real time. Typically, only selected snapshot "views" from the full ultrasound examination are captured for later radiological review. Users therefore need to be careful to work within their scope of practice, and to interpret and communicate their findings clearly and effectively. Without adequate training and experience, relevant pathology can easily be missed; with misinterpretation of images and poor reporting practice contributing to both "missed" and misdiagnosis. Recent increases in computational power available for image processing are likely to make the gathering of ultrasound data less user dependent in the future, enabling high-resolution 3D volumes and slices of ultrasound data to be acquired for retrospective review in a similar way as CT and MR scans. More reproducible modes of data gathering may also facilitate the introduction of AI image analysis methods for automated classification and reporting of images.

Exciting preclinical research in ultrasound includes the development of novel ultrasound contrast agents, functional ultrasound (fUS) imaging of brain activity, ultrafast Doppler and super-high-resolution imaging. These new techniques are likely to enable accurate imaging of the brain through the skull and targeted drug delivery. Overall, both the quality of images and ways in which ultrasound data are gathered and visualised are likely to be transformed over the next decade. The role of ultrasound physicists in introducing these technologies to hospitals over the coming years offers an exciting prospect.

REFERENCES

Allisy-Roberts PJ and Williams J (2007). *Farr's Physics for Medical Imaging*, 2nd edition. Elsevier, Philidelphia, USA.

Barrie-Smith N and Webb A (2011). *Introduction to Medical Imaging: Physics, Engineering and Clinical Applications*. Cambridge University Press, Cambridge, UK.

British Medical Ultrasound Society (BMUS) (2010). "Guidelines for the Safe Use of Diagnostic Ultrasound Equipment". *Ultrasound* **18**: 52–59.

Dalecki D, Child SZ, Raeman CH and Cox C (1999). "Hemorrhage in Murine Fetuses Exposed to Pulsed Ultrasound". *Ultrasound Med Biol* **25**(7): 1139–1144.

Dalecki D, Child SZ, Raeman CH, Cox C and Carstensen EL (1997). "Ultrasonically Induced Lung Hemorrhage in Young Swine". *Ultrasound Med Biol* **23**(5): 777–781.

Dendy PP and Heaton B (2012). *Physics for Diagnostic Radiology*, 3rd edition. CRC Press, London.

Donald I, MacVicar J and Brown TG (1958). "Investigation of Abdominal Masses by Pulsed Ultrasound". *Lancet* **271** (7032): 1188–1195.

Dudley N, Russell S, Ward B and Hoskins P (2014). "BMUS Guidelines for the Regular Quality Assurance Testing of Ultrasound Scanners by Sonographers". *Ultrasound* **22**(1): 8–14.

EFSUMB (2020). European Federation of Societies for Ultrasound in Medicine and Biology "Guidelines and Recommendations". Online Available from: https://www.efsumb.org/blog/archives/1156 [accessed 16 September 2020].

Frizzell LA, Zachary JF and O'Brien WD (2003). "Effect of Pulse Polarity and Energy on Ultrasound-Induced Lung Hemorrhage in Adult Rats". *J Acoust Soc Am* **113**(5): 2912–2918.

Hoskins PR, Martin K and Thrush A (2019). *Diagnostic Ultrasound: Physics and Equipment.* CRC Press, Florida, USA.

IPEM (2010). *Quality Assurance of Ultrasound Imaging Systems.* IPEM Report No. 102. Institute of Physics and Engineering in Medicine (IPEM), York, UK.

Izadifar Z, Babyn P and Chapman D (2017). "Mechanical and Biological Effects of Ultrasound: A Review of Present Knowledge". *Ultrasound Med Biol* **43**(6): 1085–1104.

Miller DL and Thomas RM (1994). "Heating as a Mechanism for Ultrasonically-Induced Petechial Hemorrhages in Mouse Intestine". *Ultrasound Med Biol* **20**(5): 493–503.

Public Health England (2011). "Guidance Notes for the Acquisition and Testing of Ultrasound Scanners for Use in the NHS Breast Screening Programme". Online Available from: https://www.gov.uk/government/publications/breast-screening-ultrasound-scanners-standards [accessed 16 September 2020].

Szabo, T (2014). *Diagnostic Ultrasound Imaging: Inside Out.* Elsevier, Philidelphia, USA.

RCR (2014). *The Royal College of Radiologists "Standards for the Provision of an Ultrasound Service"*. The Royal College of Radiologists, London, UK. Available online at: www.rcr.ac.uk/publication/standards-provision-ultrasound-service, accessed 11/3/21.

Tarantal AF and Canfield DR (1994). "Ultrasound-Induced Lung Hemorrhage in the Monkey". *Ultrasound Med Biol* **20**(1): 65–72.

ter Haar G. (2012). *The Safe Use of Ultrasound in Medical Diagnosis.* The British Institute of Radiology, London.

Treeby BE and Cox BT (2010). "k-Wave: MATLAB Toolbox for the Simulation and Reconstruction of Photoacoustic Wave-Fields". *J Biomed Opt* **15**(2): 21314.

US Food and Drug Administration (2019). "Information for Manufacturers Seeking Marketing Clearance of Diagnostic Ultrasound Systems and Transducers" US Department of Health and Human Services Food and Drug Administration, MD, USA. Available online: www.fda.gov/media/71100/download, accessed 11/3/21.

Part II

Imaging and Therapy Using Ionising Radiation

The term "ionising radiation" refers to high-energy particles or electromagnetic radiation capable of causing atoms or molecules to become charged, or "ionised". Although ionising radiation can be extremely useful for both medical imaging and therapy, ionisation of tissue within the body can also be harmful. Figure 2.1.1 summarises different forms of ionising radiation.

X-ray and gamma photons differ only in the way the photons are produced and controlled; gamma photons are produced by radioactive decay of unstable nuclei and cannot be "turned off" or tuned in the same way as an X-ray tube source. Although gamma radiation is often depicted in diagrams of the electromagnetic spectrum as being associated with higher energies than X-rays, in fact, X-ray and gamma photon energy ranges overlap.

For imaging purposes, an X-ray tube can be used to generate detailed images of the body in the form of 2D (planar) projection of X-rays onto a detector plate. X-rays can also be used to generate 3D information using axial computed tomography (CT). The role of the Clinical Scientist working with X-ray imaging equipment is described in Chapter 4.

Nuclear Medicine involves using radioactive sources to generate gamma photons that can then be imaged using a gamma camera, as described in Chapter 5.

Chapter 6 describes how the interactions of high-energy photons with tissue can be used for cancer treatment using radiotherapy and brachytherapy. In external beam radiotherapy, a linear accelerator is used to generate high-energy (MeV) X-rays using accelerating potentials of MV, rather than the kV potentials used to generate the keV photons suited to imaging. These high-energy X-rays are used to shrink the tumour while sparing surrounding healthy tissue and organs. Medical Physicists are involved with planning and delivering treatment to ensure that radiotherapy is delivered as safely and effectively as possible.

The process of ionisation is useful for imaging and therapy but it also carries unwanted biological side effects. Radiation safety requirements for all aspects of imaging using ionising radiation are discussed in Part 3, which focuses on radiation safety (Chapter 7).

Historically, Clinical Scientists specialised in either Nuclear Medicine or the uses of X-ray imaging in diagnostic radiology and radiation safety. However, with increasing use of hybrid

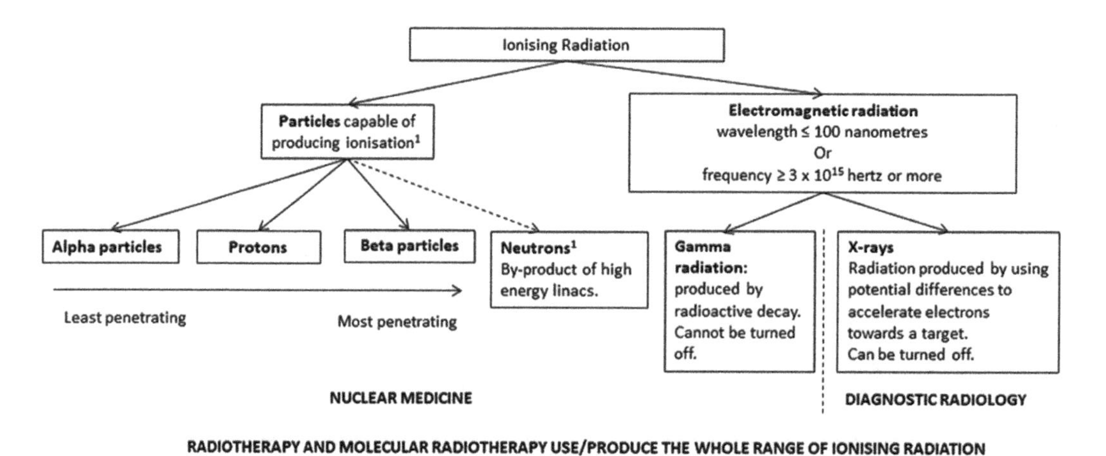

FIGURE 2.1.1 A taxonomy of ionising radiation.

imaging techniques, such as PET-CT, the boundary between disciplines has become increasingly blurred. This has led to the adoption of "Imaging with Ionising radiation" and "Radiotherapy Physics" as the two main ionising radiation specialties currently offered by the NHS Clinical Scientist training programme.

The next few sections revise some useful definitions and quantities that apply to all specialisms working with ionising radiation.

2.1.1 IONISING RADIATION – QUANTITIES, UNITS AND WEIGHTING FACTORS

The "fluence" (ϕ) describes the number of photons (or other particles), N, entering an imaginary sphere with cross-sectional area perpendicular to the beam direction (dA) (ICRU 2011).

$$\varphi = \frac{dN}{dA}$$

The fluence has units of photons (or particles) per cm^2.

In medical uses of radiation, the fluence is an important quantity, but it is also useful to understand the energy that is imparted by the radiation in terms of the concept of radiation "dose". Dose can be measured, either directly or indirectly, or estimated using a mathematical model. As "dose" can be estimated in various ways, the main methods for estimating dose are briefly described later.

Historically, radiation dose was measured using an ionisation chamber. Ionisation chambers are still sometimes used in radiotherapy, although for most applications, solid-state detectors have taken their place. Attix (1986) clearly describes the operation of the ionisation chamber and other radiation detectors together with quantities and units used in radiation dosimetry. This classic text remains relevant today for understanding the basic physics and instrumentation used in ionising radiation.

The ionisation chamber contains electrodes which collect the ion pairs a particle, or photon, generates as it passes through the ionisation chamber. Measurements are usually obtained in air, or water, with the electric charge associated with ionisation (or "exposure") measured in Coulombs per kilogram (C/kg). Field instruments are usually calibrated against a secondary standard, which in turn are calibrated against a primary standard held by a national laboratory to provide an absolute measurement of exposure.

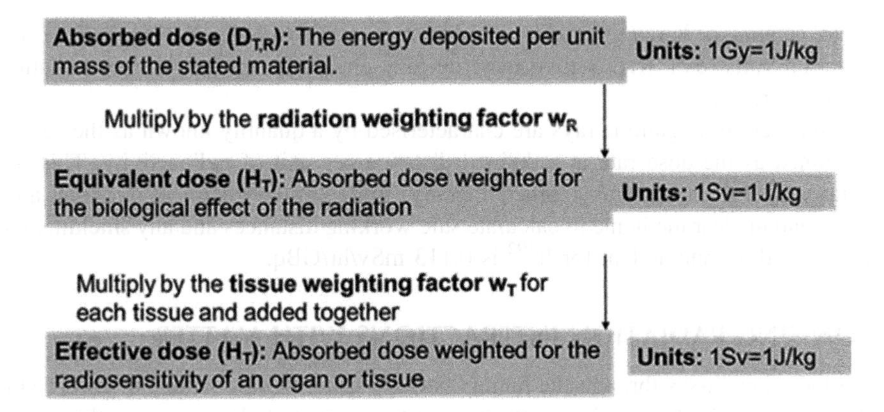

FIGURE 2.1.2 Quantities and units of ionising radiation.

Using Bragg–Gray cavity theory (Attix 1986) exposure measured by the ionisation chamber can be converted to the energy imparted per unit mass. This quantity is known as the absorbed dose and has units as summarised in Figure 2.1.2. The absorbed dose is related to the material in which the ionisation is measured. For an ionisation chamber designed to be equivalent to air, this quantity is described as Kinetic Energy Released in a Mass of Air (KERMA; ICRU 2011). KERMA is a specialised type of indirect absorbed dose measurement.

For the types of radiation other than photons and electrons (i.e. protons or neutrons), a radiation weighting factor needs to be applied to account for differences in how densely ionisation occurs for the paths of differing particles. As neutrons and protons are more densely ionising, ionisation will be more heavily concentrated, and therefore the risk of inducing cancer will be higher. This is achieved by applying a radiation weighting factor, which varies from 1 for photons and electrons, to 2 for protons, and with continuous values of up to 20 for neutrons (depending on neutron energy). Adjustment for the type of particle using a radiation weighting factor enables equivalent dose to be estimated for a range of particles.

While imaging or treating patients, it is important to be able to estimate the absorbed dose received by an individual organ, such as the breast or the skin. To be able to compare the dose from different examinations of different body parts in diagnostic radiology and nuclear medicine and as organs have differing sensitivity to radiation, a "tissue weighting factor" is applied to estimate the "effective dose" received by the whole body. The use of the concept of effective dose makes it possible to compare the doses, for example, between a chest and pelvis X-ray. The effective dose adjusts for the relative radiosensitivity of each organ as well as the dose imparted (Figure 2.1.2). There are other dose quantities used within the individual specialisms that are described in individual chapters.

2.1.2 RADIOACTIVE MATERIAL – QUANTITIES AND UNITS

Radioactive material is quantified by its Activity, A, in Becquerels (Bq). All radioactive materials decay, emitting alpha particles, beta particles and gamma rays. Radioactive material decays with a characteristic half-life ($t_{1/2}$) describing the time it takes for the activity to decrease by half.

The activity, A, at time, t, relates to the initial activity (A_0) of the sample and its half-life, $t_{1/2}$.

$$\text{where } A = A_0 e^{(-0.693t/t_{1/2})}$$

Each particle, gamma, or X-ray emitted by a radiation source possesses an energy, E, expressed in units of electron volts (eV). Radioactive material may emit a variety of types of radiation, each with differing energies.

Radionuclides emitting gamma rays are characterised by a quantity known as the "gamma ray constant" defined as the dose rate at a defined distance per unit of radioactivity. This is a very useful quantity used in radiotherapy (brachytherapy) to determine treatment duration, and in radiation safety and nuclear medicine to calculate safe working distances and any shielding required. For example, the dose rate at 1 m for Ir^{192} is 0.113 mSv/hr/GBq.

2.1.3 IONISING RADIATION INTERACTIONS WITH MATTER

As ionising radiation passes through the human body, it can either be scattered, absorbed or pass through the body entirely. Indeed, X-ray imaging in diagnostic radiology requires there to be clear differences in transmission of X-rays through the body to generate contrast and enable an image to be formed. Similarly, to generate images in Nuclear Medicine, gamma radiation needs to pass out of the human body to a radiation detector. In radiotherapy, an ideal treatment would involve depositing dose solely within the target volume, with no energy deposited elsewhere.

The intensity of a photon beam reduces exponentially as it passes through a medium. Each medium has differing attenuation properties, described in the form of a linear attenuation coefficient, μ, in cm^{-1}. For a medium of thickness, t, the intensity of collimated X-rays after passing through the medium is given by

$$I = I_0 e^{-\mu t}$$

where I is the transmitted intensity and I_0 is the initial intensity of the beam. Note that because of the exponential nature of the function, the intensity of a beam of photons never fully decreases to zero. Particles such as electrons (or beta particles, which are electrons or positrons emitted by radioactive decay) can be entirely absorbed within the tissue, which can be useful for treatment. Useful quantities for describing the thickness of material required for shielding include the half-value layer (HVL) or tenth-value layer (TVL) describing the thickness of material required to reduce the intensity of a beam of photons by a half, or a factor of 10, respectively.

Ionising radiation can be used to treat cancer but is also associated with an increased long-term cancer risk. Precise mechanisms relating to long-term risks associated with the use of ionising radiation are not well understood. The initial physical interaction of radiation with tissue takes of the order of 10^{-13} seconds, resulting in ionisation and excitation of atoms and molecules. This interaction can result in direct damage to chemical bonds as well as the production of highly reactive free radicals. DNA has been shown to be the most critical cellular target for radiation damage, although the exact mechanism of DNA damage is unclear. It is thought that lesions involving one arm of the DNA double-helical structure can be successfully repaired (single-strand breaks), whereas breaks in the DNA structure that occur in both strands, or in close proximity, are far more likely to result in cell death or carcinogenic mutations. Mutations and cellular changes occurring as a result of radiation exposure can result in changes and biological effects that emerge over many months or years and that can even be inherited by future generations.

Photons in the form of X-ray beams and gamma rays interact with tissue at a sub-atomic level. The probability of these interactions depends on the energy of the beam, and the atomic number and physical or electron density of the material.

As photons pass through an atom (Figure 2.1.3), the majority pass straight through without changing direction or losing energy. Low-energy photons can be scattered elastically (without losing energy), as shown in Figure 2.1.3(a). Photons with energies greater than the binding energy of inner-shell electrons might result in the ejection of a photoelectron and production of

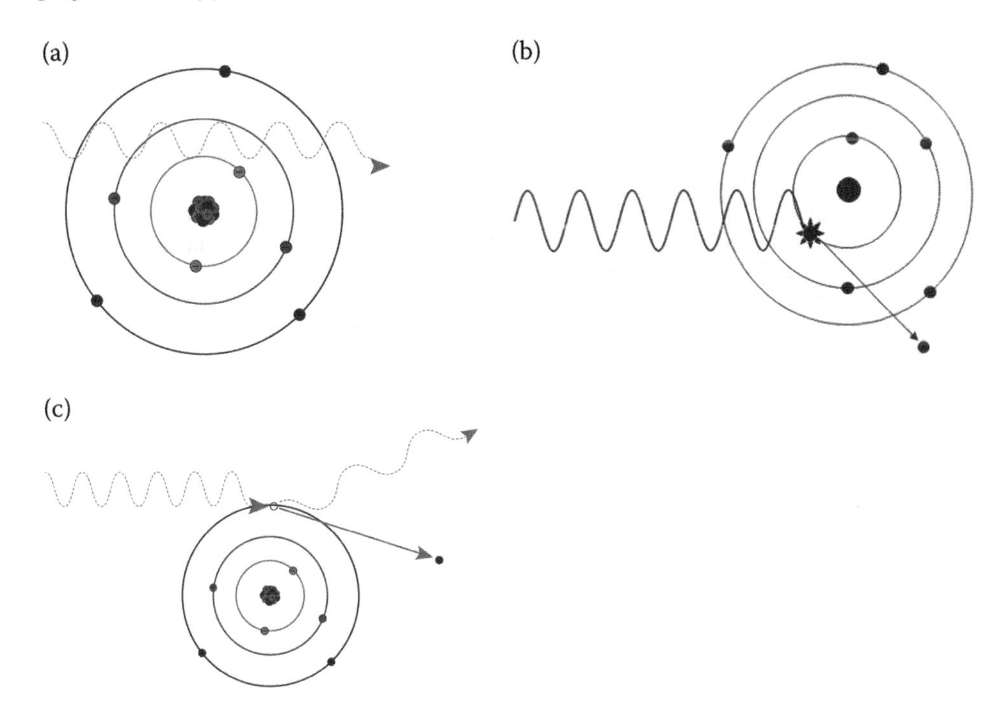

FIGURE 2.1.3 As photons pass through an atom: (a) some pass straight through, (b) some are absorbed and the energy ejected as a photoelectron, (c) some eject an outer-shell electron and are then scattered.

characteristic X-rays associated with redistribution of electrons between orbitals (Figure 2.1.3(b)). Higher energy photons can also eject outer-shell electrons. The photon loses energy to the electron and changes direction, known as Compton scattering (Figure 2.1.3(c)).

The probability (P) of the photoelectric interaction occurring increases with atomic number Z and is proportional to the material's density (ρ).

$$P = \rho Z^3 E^{-3}$$

where E is the photon energy.

The effective atomic number and density of human tissue and common building materials have been calculated (see, for example, Kurudirek (2014) and Sutton et al. (2012)). Air has a low density and relatively low atomic number, whereas lead has a high density and a high atomic number, so at low photon energies the photoelectric effect is less likely to occur in air, allowing X-rays to pass through. In contrast, lead will efficiently absorb X-rays, making lead suitable for shielding. Elastic scattering and the photoelectric effect tend to be dominant at diagnostic X-ray energies, typically up to 150keV.

Compton scattering becomes more prevalent as the X-ray energy increases. The probability of this interaction is proportional to the physical density of the material and the energy of the beam. Compton scatter also occurs at diagnostic energies but is not useful for imaging, as the scattered radiation makes images appear "foggy". The Compton effect is the dominant interaction of X-rays with tissue at the energies used in radiotherapy.

There are other interactions, such as pair-production, which take place at higher energies, but the photoelectric effect and Compton scatter are the most important interactions underlying imaging, radiotherapy and radiation safety.

REFERENCES

Attix FH (1986). *Introduction to Radiological Physics and Radiation Dosimetry*. Wiley-Interscience, New York.

ICRP (2007). "Publication 103: The 2007 Recommendations of the International Commission on Radiological Protection". *Ann ICRP* **37**(2–4).

ICRU (2011). "Report 85: Fundamental Quantities and Units for Ionizing Radiation". *J ICRU* **11**(1).

IRR17 (2017). "Ionising Radiations Regulations". SI. 1075. Online available from: http://www.legislation.gov.uk/uksi/2017/1075/contents/made [accessed 13 September 2020].

Kurudirek M (2014). "Effective Atomic Numbers and Electron Densities of Some Human Tissues and Dosimetric Materials for Mean Energies of Various Radiation Sources Relevant to Radiotherapy and Medical Applications". *Radiat Phys Chem* **102**: 139–146.

Sutton DG, Martin CJ, Williams JR and Peet DJ (2012). *Radiation Shielding for Diagnostic X-rays*, 2nd edition. British Institute of Radiology (BIR), London, England.

4 Diagnostic Imaging Using X-rays

Debbie Peet, Richard Farley and Elizabeth Davies

CONTENTS

In the NHS in England, roughly 23 million conventional (two-dimensional [2D] planar) X-ray images and 5.5 million X-ray computed tomography (CT) scans are conducted each year (NHS England 2019). As with magnetic resonance imaging (MRI) and ultrasound examinations, patients can be referred for X-ray imaging by their general practitioner (GP) or hospital doctor. Referral guidelines are decided by doctors specialising in imaging (Radiologists) and are published by the Royal College of Radiologists (RCR 2020). Most patients receiving X-ray scans are adults; however, there may also be paediatric patients.

X-ray imaging encompasses a number of "modalities" that all use X-rays generated in an X-ray tube to form an image based on the transmission of X-rays through the body. As different tissues and structures attenuate X-rays by differing amounts (depending on their density, atomic number and thickness), this provides contrast to leave an image of the body on X-ray film, or a digital detector. Further development of X-ray techniques to allow three-dimensional

(3D) imaging using X-ray CT has provided doctors with the ability to clearly visualise structures and organs in 3D, rather than being limited to a 2D projection. The equipment is configured differently in the different modalities and can look very different. However, there are common elements relating to the generation of X-rays and their detection. This process is described in more detail in Farr's Physics for Medical Imaging (Allisy Roberts, 2007) and Dendy and Heaton (Dendy, 2012).

4.1 GENERATION OF DIAGNOSTIC X-RAYS

All diagnostic X-ray imaging systems generate X-rays using an X-ray tube, as shown in Figure 4.1. The X-ray tube uses a high voltage to accelerate electrons produced by thermionic emission across a vacuum tube.

When these fast-moving electrons collide with a dense target metal (e.g. tungsten), they are rapidly slowed down, releasing their energy. Most of the kinetic energy from the electrons is converted to heat with only a small fraction being converted into X-rays. These X-rays are known as Bremsstrahlung, which is German for "braking radiation". To prevent the anode from overheating, it is rotated at very high speed to help dissipate the heat. The maximum photon energy achieved is equal to the maximum kinetic energy of the electrons, driven by the peak voltage between the anode and cathode (kV_P; Figure 4.2). The flux of electrons, and therefore the intensity of X-rays, is determined by the cathode filament current (typically several amps are needed). The target material is usually tungsten, but other materials are also used for specialist imaging applications, such as breast imaging (mammography). Characteristic X-rays are also produced at an energy dependent on the target material.

To prevent X-rays being emitted from the tube in all directions, it is enclosed in a lead housing. This attenuates most X-rays, although there will still be some leakage. The useful beam is shaped and collimated to create a well-defined parallel beam of X-rays of an appropriate size and shape for imaging. The collimators might be fixed, adjustable or a combination of both. Additional filters can also be added in front of the beam to adjust the energy distribution (spectrum) of the emitted X-rays. This is done to remove low-energy X-rays that are unlikely to penetrate far enough to contribute to the image, and to optimise the energy distribution for different types of imaging. The 'quality' of X-rays refers to the shape of the X-ray spectrum; this depends on the target material, filter properties and the maximum kinetic energy of electrons (controlled by varying the tube's voltage). The intensity, or quantity, of

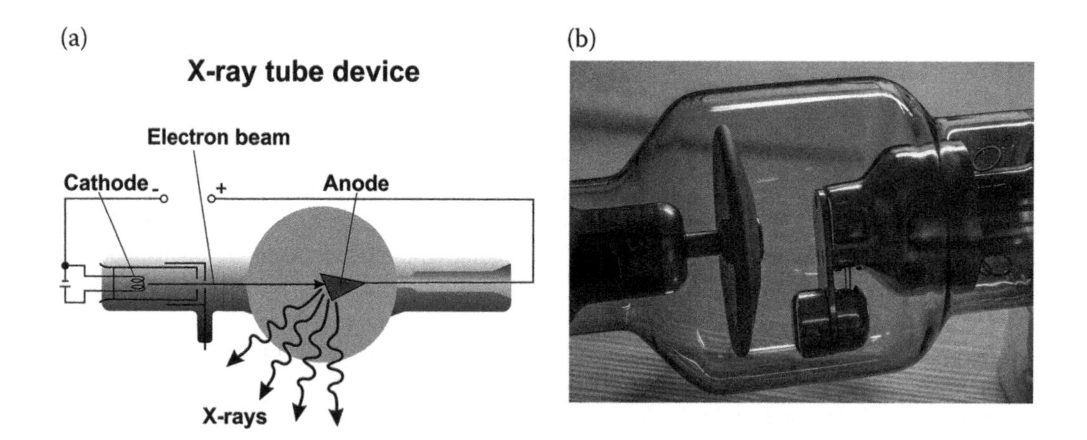

FIGURE 4.1 (a) Schematic of an X-ray tube and (b) a photograph of a tube insert showing the rotating anode to the left of the image and the cathode to the right.

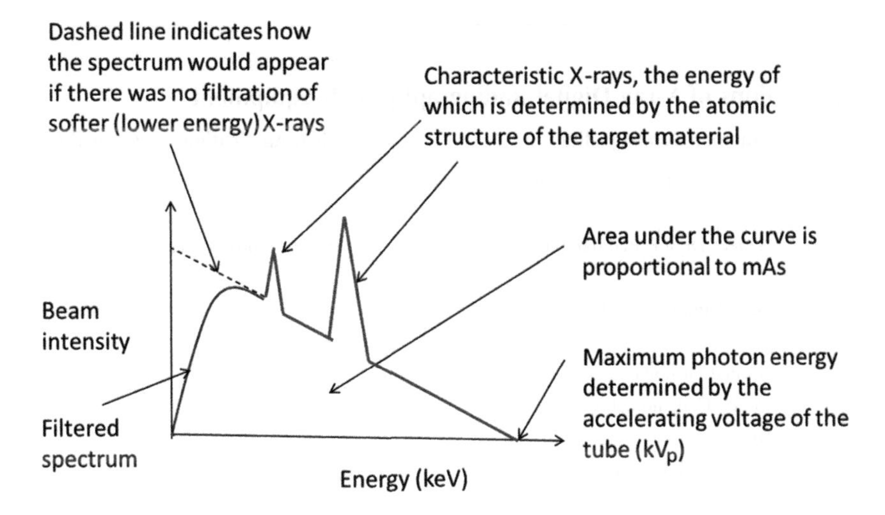

FIGURE 4.2 The spectrum of X-rays generated by a tungsten target.

photons is determined by the number of electrons generated, which is controlled by varying the X-ray tube current (in mA) and exposure time (in seconds). The combination of tube current and exposure time is often summarised as their product, that is, milliampere-seconds (mA.s).

X-rays tubes are used in different ways to obtain the information required to diagnose and treat patients with the lowest possible dose. There are many types of Diagnostic Radiology equipment (summarised in Table 4.1) and part of the challenge of the role is understanding the vast array of equipment available from different manufacturers.

4.2 X-RAY IMAGING MODALITIES

4.2.1 Conventional 2D X-ray Projections

Although X-ray film is now rarely used, X-ray imaging continues to be referred to as plain film, planar or 2D projection radiography. A conventional planar X-ray room comprises an X-ray tube connected to an X-ray generator, a digital X-ray detector [Figure 4.3(a)] and a control panel operated from behind a screen. The X-ray tube can be directed towards a detector placed in a tray below an X-ray table for examinations where the patient is supine, or the image can be obtained by resting a limb on the table. A vertical assembly (or Bucky) can also be used to hold the detector vertically so that the patient can stand against it, for example, for a chest X-ray.

The most common form of diagnostic X-ray imaging used in hospitals has evolved very little from Roentgen's early cathode-ray tube experiments, as described in Chapter 1. The X-ray tube is typically operated at accelerating voltages between 50 and 150 kVp and tube current of between 0.5 and a few hundred mA. On passing through the body, X-ray photons are preferentially absorbed by materials with high atomic number (e.g. the calcium in bone, or the metal of an artificial hip), generating a shadow on a flat detector, as shown in Figure 4.4(a). This produces a 2D projection of X-rays through the body onto a detector. If the body part to be imaged is relatively thin, such as a hand or an ankle, the limb can be placed directly on the imaging plate. Thicker body parts, such as the abdomen, generate a significant number of scattered X-rays [Figure 4.3(b)]. Scattered X-rays cause the image to appear "foggy" but can be reduced by placing an anti-scatter grid in front of the detector. The anti-scatter grid is made

TABLE 4.1
Clinical Applications of X-ray Digital Radiography (DR) Equipment

DR Modality	Common Sub-Modality	Clinical Application	Clinical Disciplines
Planar X-ray imaging	Computed Radiography (CR)	Primary diagnostic	Radiology (General) and mammography
	[1]Direct Digital Radiography (DDR)	Therapeutic procedures	Dental Radiology
			Rheumatology
	Dental Intra Oral (IO)		Oncology
	[2]Dental Orthopantomography (OPT)		MSK
	Dental Cephalometric		
	DEXA		
Fluoroscopy	[3]Fluorography – single- or multi-planar acquisitions	Primary diagnostic	Radiology (General)
			Angiography and vascular
	Cine modes		Cardiology
	Digital subtraction angiography (DSA)	Interventional procedures	Urology
			Oncology
	Lithotripsy	Therapeutic procedures	MSK
CT	Non-contrast	Primary diagnostic	Radiology (General)
	Single phase		
	Dual phase		
			Angiography and vascular
	CT-Fluoroscopy	Interventional procedures	Cardiology
	SPECT-CT (hybrid)		
	PET-CT (hybrid)	Therapeutic procedures	Oncology

Notes

1 DDR is sometimes shortened to DR. Be careful not to confuse this with DR (as in Diagnostic Radiology).

2 Panoramic scanning radiograph is also referred to as OPG (Orthopantomogram), which is derived from an early trade name.

3 Fluorography is an older term and is now more commonly referred to as "acquisition imaging" or "spot imaging".

up of layers of attenuating material to filter out any scattered X-rays. A device known as an automatic exposure control (AEC) is often embedded in the grid assembly to terminate the exposure once a certain intensity of radiation has been detected [Figure 4.3(c)]. This corrects for patient thickness and different densities of tissues within the body, such as for bone or lung (Figure 4.4a). X-ray energy, exposure factors and technique, collimation, the presence or absence of the anti-scatter grid and the settings for the AEC all need to be optimised to achieve the desired appearance and characteristics of the image for various clinical applications.

One major change in planar radiography has been the move from film to digital detectors, mirroring the switch from film cameras to digital optical photography. Further development of X-ray techniques to allow 3D imaging using X-ray CT has provided clinicians with the ability to clearly visualise structures and organs in 3D, rather than being limited to a 2D projection, therby adding more information which may be beneficial to a diagnosis and patient management.

One application of planar radiography is mammography which uses X-ray projection to image the breast (Figure 4.4b). The breast has very little inherent contrast as it is mainly formed of glandular tissue and fat. Imaging therefore needs to be optimised to enable detection of small

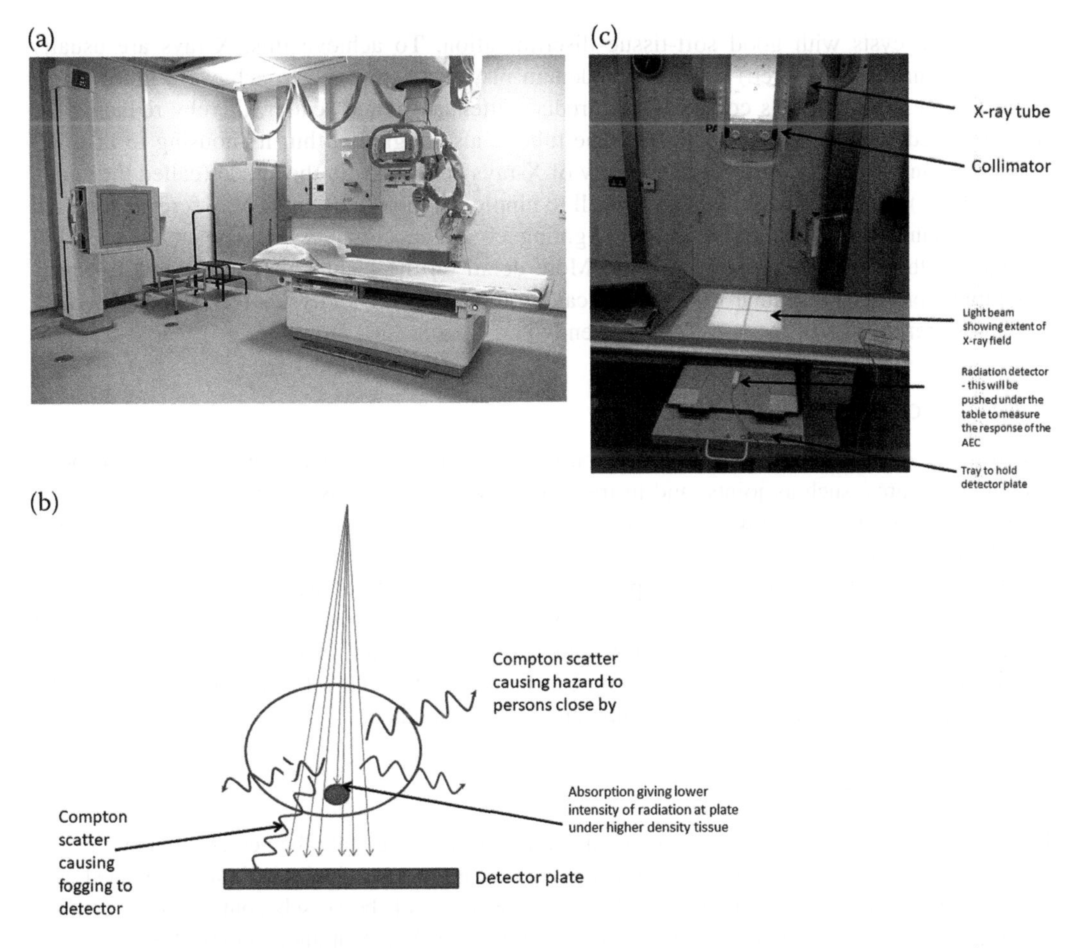

FIGURE 4.3 (a) A typical plain film X-ray room with a table and vertibucky, (b) production of an image of a hand and (c) measurements on the AEC device.

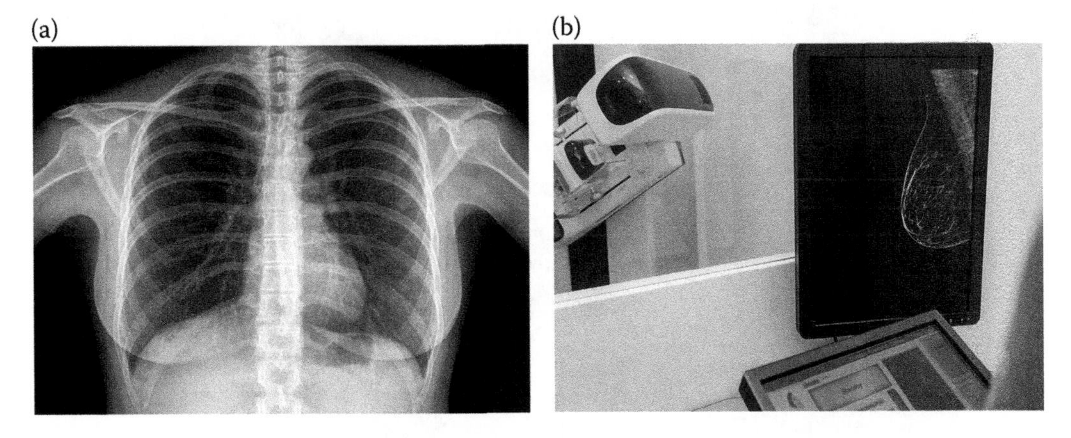

FIGURE 4.4 (a) A chest X-ray, (b) mammography. The X-ray set used for breast screening looks quite different from that of a conventional X-ray system.

tumours and cysts with good soft-tissue discrimination. To achieve this, X-rays are usually generated (and often filtered) using molybdenum instead of tungsten to achieve a much lower energy range. The breast is compressed to reduce attenuation and scatter, thereby reducing the dose required to provide a good image. The tube is also angled within its housing to take advantage of angular variations in the intensity of X-rays emitted from the anode (called the anode "heel effect") to better image from chest wall to nipple. High-resolution display screens are used to view mammographic images in dim lighting conditions to help prevent small or subtle changes in the breast from being missed. More detail on the technical requirements of mammography compared to planar radiography can be found in Farr's Physics for Medical Imaging (Allisy-Roberts and Williams 2007) and Dendy and Heaton (2012).

4.2.2 Fluoroscopy

Fluoroscopy enables X-ray imaging to be performed in real time. This can be used to capture moving structures, such as joints, and to track the passage of contrast through the circulatory or digestive system. Fluoroscopy is also used to guide interventional procedures in cardiology and "key-hole" surgery.

Fluoroscopy systems are similar to "plain film" X-ray systems but use short pulses of X-rays to form a moving image on a detector. Pulses of X-rays are used to minimise dose to the patient but makes the resulting images relatively noisy compared to "plain film" X-rays. Fluoroscopy systems are often mounted on a C-arm so that they can be oriented in almost any direction around the patient (Figure 4.5), keeping the X-ray tube and detector aligned.

4.2.3 Angiography

In angiography, a contrast agent (usually iodine-based) is introduced into the vessels to improve their visibility (Figure 4.6). Where organs are relatively stationary, such as the brain, images obtained with and without contrast can be subtracted to leave a clear image of the vessels containing the contrast agent alone [Figure 4.6(b)]. This technique is called digital subtraction angiography (DSA).

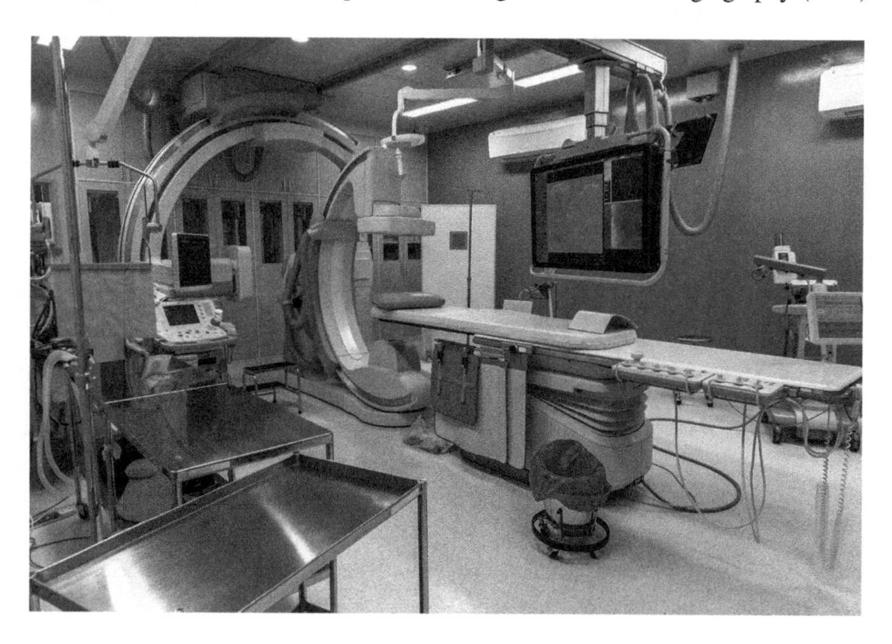

FIGURE 4.5 An interventional X-ray suite. A C-arm gantry is used to provide access and versatile movement around the patient to acquire dynamic images displayed on large monitors.

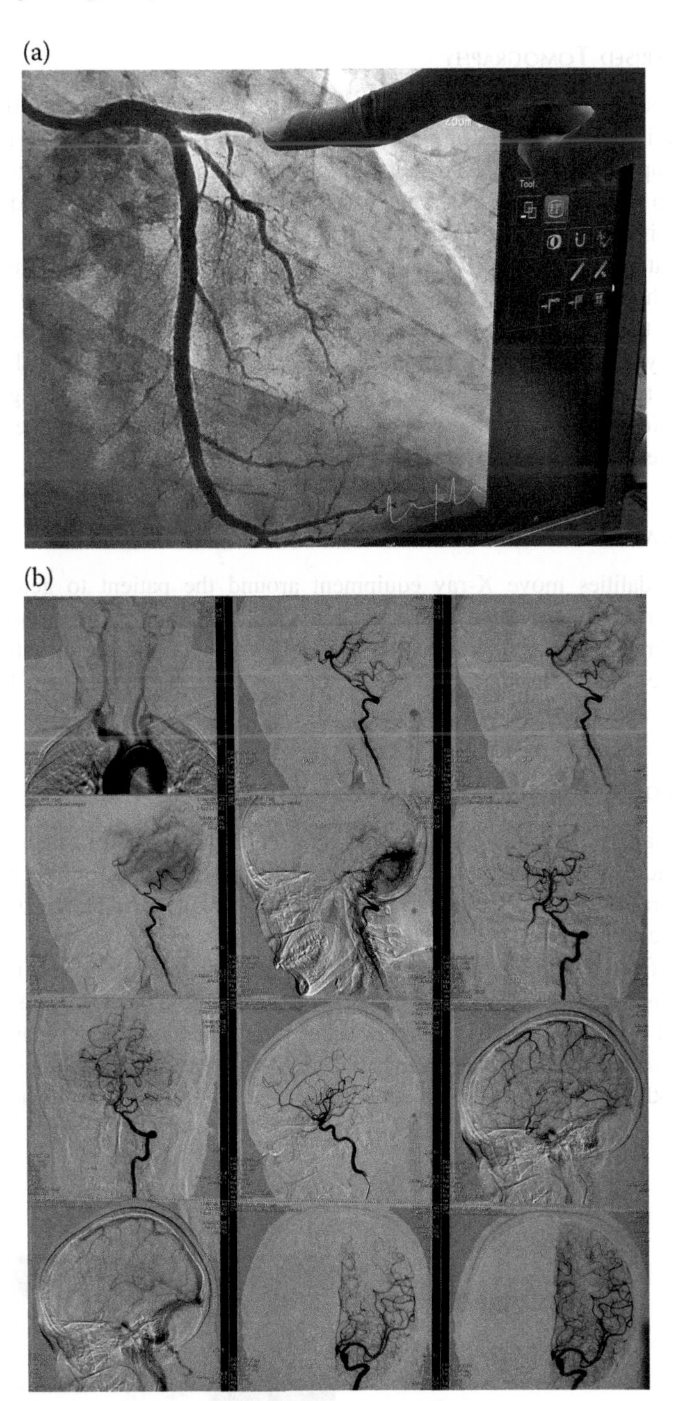

FIGURE 4.6 (a) Coronary angiography showing iodine contrast in the heart, and (b) neuroangiography showing subtracted images of vessels in the brain. The background is removed from the image to show the vasculature more clearly. This is not possible in the heart as the background moves with each heartbeat.

4.2.4 Computerised Tomography

X-ray CT scanners (Figure 4.7a) are used to map regions of the body with differing attenuation, rather than providing a simple 2D projection. Narrow beams of X-rays with relatively high kV values (typically up to 140 kV) are generated (Kalendar 2001). Solid-state detectors surrounding the patient pick-up these X-rays over a range of angles, spiralling around the body to generate a 3D dataset showing differing regions of X-ray attenuation.

The level of attenuation is usually mapped in shades of grey varying from black (which represents air) to white (which represents bone). Images can also be artificially coloured and rendered to highlight the surfaces of structures or organs of interest (Figure 4.7b). CT is a relatively high dose modality; therefore, optimisation is an important part of the Clinical Scientist's role to ensure that doses received by patients are as low as reasonably practicable, given the intended clinical purpose. Advances in image reconstruction are constantly improving the quality of images, which also provides opportunities for lower doses to be delivered.

4.2.5 Cone Beam CT (CBCT)

Several X-ray modalities move X-ray equipment around the patient to generate tomographic images. Cone beam CT is a relatively new modality used in dentistry, Diagnostic Radiology and Radiotherapy. There are few textbooks on the subject, but some good information for Diagnostic Radiology can be found through the American Association of Medical Physics (AAPM), Sedentexct (2020) and Wong et al. (2011).

4.2.6 Other Forms of Equipment and Hybrid Imaging

All the X-ray modalities described here are available as mobile units for imaging in theatre or by the patient's bedside (although mobile CT is rare due to dose implications for those in the area). Other common X-ray modalities include Dental Radiology (Waites and Drage 2013) and the use of X-rays for bone density scans.

Often it is beneficial to combine different modalities to generate hybrid scanners combining for example single positron emmission tomography with CT (SPECT-CT) or positron emission tomography with CT (PET-CT). Hybrid imaging is becoming increasingly popular as many forms of imaging provide complementary information. Imaging is increasingly used to guide treatment. For example, in Radiotherapy, an "on board" cone-beam CT can be used to check patient positioning.

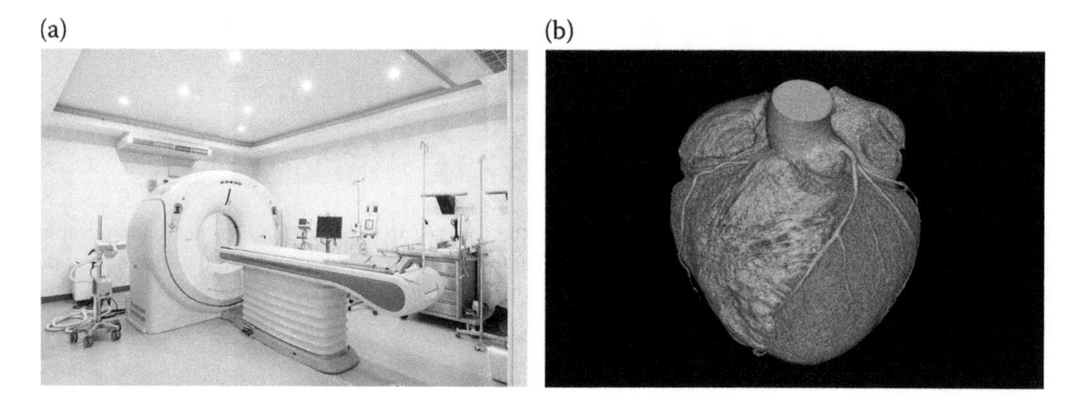

FIGURE 4.7 (a) A CT scanner, and (b) a 3D reconstruction of the heart generated from CT data.

4.3 PERFORMANCE MEASUREMENTS

Given the complexity of X-ray equipment, it is important to ensure performance efficacy in terms of radiation safety. This involves assessing the function of all components along the imaging chain, from the X-ray tube to final presented image. Clinical Scientists are involved in:

- Determining the required tests
- Advising on the requirements of local quality control (QC) carried out by Radiographers
- Carrying out more complex QC with technician colleagues
- Analysing results
- Problem solving and decision making when issues arise

Current UK legislation requires Diagnostic Radiology X-ray equipment to be assessed as outlined in IRR17 and IR(ME)R17; for more information, see Chapter 7. Clinical Scientists may work alongside a Radiation Protection Advisor (RPA) or Medical Physics Expert (MPE) to provide advice on high-dose interventional procedures, develop suitable QC tests, especially for new X-ray equipment or emerging technologies. RPAs are formally certified to give advice on compliance with IRR17. MPEs are Clinical Scientists who have obtained a specified level of experience of working with radiation equipment and in the understanding of patient exposure. It is important to be aware of current UK guidance, as well as other international references that may be of use when developing such a QA program and QC tests.

There are several IPEM references that describe performance measurements for diagnostic X-ray equipment (Table 4.2).

The UK Medical Dental Guidance Notes (IPEM 2002) provides guidance around the requirements of a critical examation to look at the safety features of the installation of X-ray equipment. Periodically, new reports are published and updated to reflect developing technologies. Clinical Scientists are often involved with writing guidance and agreeing appropriate methods of assessment. Any deviations from these references would need to be justified based on new evidence or additional information. Other helpful references covering equipment performance testing include the following:

- The American Association of Physicists in Medicine (AAPM) task group report series
- The European Federation of Organisations in Medical Physics (EFOMP) mammography and CBCT protocol
- The International Atomic Energy Agency (IAEA) Medical Physics resource

TABLE 4.2
IPEM Resources for Assessing the Performance of Diagnostic X-ray Equipment

Reference	Title
IPEM Report 91	Routine performance testing of most modalities of equipment
IPEM Report 89	Mammography systems
IPEM Report 107	Critical examination
IPEM Report 32	Part I. Tube and generators
	Part II. Fluoroscopy
	Part VI. Fluorography
	Part VII. Digital imaging systems

Different types of performance assessments are carried out on diagnostic X-ray equipment, involving varying levels of complexity and a range of measurements. These are required during the equipment's operational life, with some assessments performed more frequently than others. The core types of practical assessment are summarised in Table 4.3. These form part of the specialist part of the equipment life cycle outlined in Chapter 1. Senior scientists may also work as part of a multidisciplinary team involved in the selection of equipment, by evaluating commercially available scanners and imaging systems in terms of expected performance, technical capability, safety and potential patient dose.

All X-ray modalities will require these core practical evaluations, although the specific method and survey equipment used for these tests may vary. For example, mammography system assessment requires different specialised survey equipment and phantoms compared to CT systems.

Performance measurement of X-ray equipment is usually an integral part of local Quality Management Systems, aimed at delivering consistently high levels of performance. Ensuring

TABLE 4.3
Different Types of X-ray Equipment Performance Assessment

Assessment	Why is This Necessary?	How is This Assessed?
Critical Examination	To ensure equipment is safe for operation. This is a legal requirement of IRR17.	Evaluation of radiation safety systems, controls and warning devices against set standards. This is performed for new installations and following the replacement of components, such as X-ray tubes. Critical examination excludes mobile X-ray equipment.
Acceptance	To verify the system functions as expected and meets the specifications as set out in the purchase contract.	Check that all components and systems are present, such as anti-scatter grids and compression paddles. Verify the efficacy of performance against standards and specifications. This is only performed on new installations.
Commissioning	To check the system is suitable for clinical use and determine baseline values for future comparison with routine testing.	Measurement of output and image quality performance, to produce baseline values for a range of operational conditions. This is performed for new installations and following clinical adjustments or optimisation work.
Planned Routine Testing	To assess performance reliability and efficacy against standards and commissioning values. This is a legal requirement of IR(ME)R17.	Periodic measurements at relevant intervals of functional operation, output, and image quality performance. Performed during the lifetime of the system.
Post Equipment maintenance	To check the system is safe and acceptable for clinical use following planned equipment service, or other works or adjustments.	To detect changes following maintenance procedures, a sample of measurements, or full assessment, could be performed to check against standards and baseline values.
Ad Hoc Testing	To investigate or confirm performance of equipment following erroneous or suspicious clinical use.	Relevant output or image quality measurements to investigate a query, such as using water phantom to investigate artefacts seen on clinical CT images.

quality and performance of diagnostic X-ray equipment is part of quality assurance (QA), with the practical assessment and measurements being known as QC. Note that these terms are not interchangeable; QA programs include QC tests, but are more wide ranging than the tests themselves, incorporating all processes and procedures in place to ensure safety and good practice.

For example, to ensure QA robust procedures must be in place for handover of equipment after maintenance work. This should include clear communication and record keeping indicating aspects of the system that have been changed. Based on this handover, the Scientist can decide whether local QC checks, specific medical physics QC checks or a full annual QC check is required prior to returning the equipment to clinical use. It should be noted that each organisation will approach the QA process and perform QC differently, with no single universal method adopted.

It is important that the results of performance checks are analysed quickly and communicated to clinical users as early as possible, especially if issues have been found. Serious issues may mean that equipment needs to be removed from service. Patients scanned with faulty equipment may need to have their scans repeated. To speed up the analysis and reporting of performance tests, Scientists are often involved with implementing systems to automate the analysis of performance measurements. This can be achieved using a variety of different software packages. An example of a Microsoft Excel spreadsheet used to speed up the analysis of routine CT QC tests is provided in Figure 4.8.

There may be difficulty in accessing radiology equipment to perform QC measurements, particularly if the equipment is located within emergency departments, or areas of the hospital where equipment is in constant use. There can be pressures and challenges convincing clinical colleagues, or even senior management, of the necessity of having access and time. Clinical

CTDI air Measurements

Input baselines, align CT chamber with lasers and ensure it is central in first axial image. Axial imaging only, ensure each measurement is a single rotation and stop table movement.

Variation with	B-T Filter	kV	mA	Time	Collimation (mm)	Dose Reading (mGy)	$CTDI_{air,100}$ (mGy)	$_nCTDI_{air,100}$ (mGy/mAs)	Baseline (mGy/mAs)	% Diff.	Ratio	Specification
Collimation (mm)	Large	120	100	1	4 x 0.50 mm	1.959	97.950	0.980	0.940	4.2	1.50	
					4 x 1.00 mm	2.607	65.175	0.652		0.0	1.00	
					4 x 2.00 mm	3.848	48.100	0.481		0.0	0.74	
					4 x 4.00 mm	6.973	43.581	0.436	0.440	-1.0	0.67	
					4 x 8.00 mm	13.130	41.031	0.410	0.417	-1.6	0.63	
Reference for Dose Calcs			100	1								
kV	Large	80	100	1	16.0	3.498	21.863	0.219	0.218	0.3		
		100				5.152	32.200	0.322	0.339	-5.0		
		120				6.962	43.513	0.435	0.440	-1.1		
		135				8.469	52.931	0.529	0.577	-8.3		
mA	Large	120	50	1	16.0	3.466	21.663	0.433	0.426	1.7		
			100			6.993	43.706	0.437		0.0		
			200			13.99	87.438	0.437	0.440	-0.6		
			400			30.15	188.438	0.471	0.476	-1.0		
Rotation time (s)	Large	120	100	0.25	16.0	2.097	13.106	0.524	0.520	0.8		
				0.50		3.728	23.300	0.466	0.479	-2.7		
				1.00		7.022	43.888	0.439	0.440	-0.3		
				2.00		13.62	85.125	0.426	0.429	-0.8		
Off-axis	L	120	100	1.0	16.0	4.521	28.256	0.283				
	L - Centre					6.993	43.706	0.437				
	SS					3.405	21.281	0.213				
10cm down	SS-centre					5.409	33.806	0.338				

CTDI (uGy/mAs.kV^2)
0.034
0.032
0.030
0.029

Ratio	Baseline	% diff
0.65	0.58	11.47
0.63	0.62	1.53

Repeatability	Results
#1	0.435
#2	0.437
#3	0.439
Mean	0.437
Coefficient of Variation	0.4

Variation with kV	Results	Baseline
mean (µGy/mAs.kV²)	0.031	0.032
% variation	7.2	4.7
% change	-4.4	% change

Variation w/Rotation Time	Results	Baseline
mean (mGy/mAs)	0.464	0.467
% variation	9.4	8.9
% change	-0.7	% change

Variation with mA	Results	Baseline
mean (mGy/mAs)	0.445	0.447
% variation	4.0	5.8
% change	-0.6	% change

Max. variation with Collimation vs Baseline	4.2
Max. variation with Off-axis vs Baseline	11.5

FIGURE 4.8 Excerpt from a QC testing spreadsheet analysing CT data. Cells are auto-filled with alerts to show when results fall outside of expected values. Blue cells show the variable settings for the system; green and red cells show baseline and measured values, respectively. The yellow cells show automatically calculated key results. How these spreadsheets work will depend on the preference of the Medical Physics Expert at the site.

Scientists, and technicians, must be able to clearly communicate the legal requirements and the benefits for both patient and staff safety.

Scientists carrying out performance measurements should fully understand the limitations of their test equipment. Tests and tolerances are mainly based on IPEM Report 91 (IPEM 2005b), which outlines remedial and suspension levels. If a remedial level is exceeded, some action will be required but the equipment can continue to be used in the interim. If a suspension level is breached, the system, or affected sub-component, should not be used clinically until the issue has been resolved. Scientists need to be aware of the clinical settings, and any system for reporting and addressing faults, to advise staff of available options.

CASE STUDY: OUTPUT MEASUREMENT FOR A PLANAR X-RAY SYSTEM

Measurements are made of the radiation output in air by varying kV_p, mA, and exposure time for a range of exposure parameters. Results show a linear relationship with dose, which is within expected tolerances (Figure 4.9). The equipment is recorded as having passed the test.

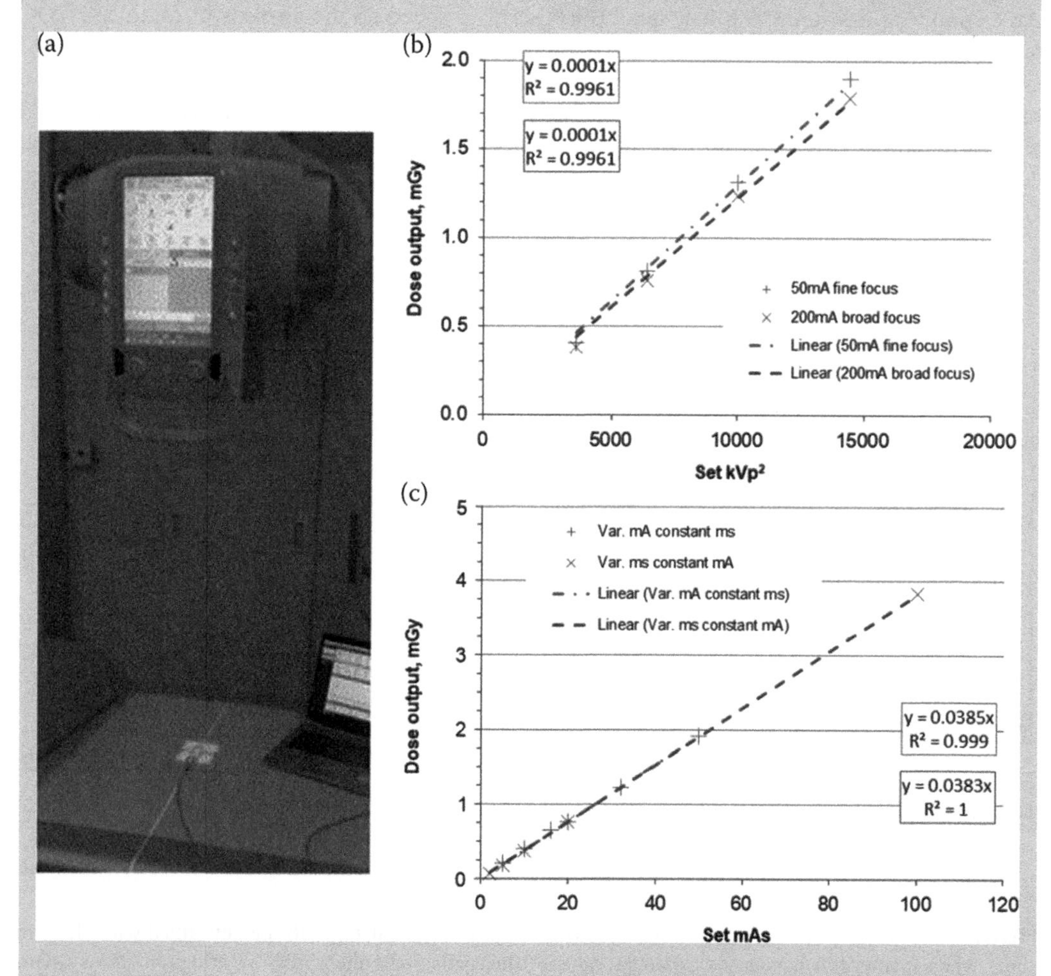

FIGURE 4.9 Setup for output measurements, confirming a linear relationship between kVp, mA.s, and dose (mGy) for low mA (small focus) and high mA (broad focus) settings.

When devising a test protocol, it is important to consider how equipment is used clinically. An example exposure chart used by radiographers for standard clinical examination protocols is shown in Figure 4.10. These charts should be developed with input from a number of professions including scientists. This provides an insight into the range of kV_p, mA, mA.s and AEC settings that the QC survey parameters need to cover. Additional readings might also be obtained to ensure the performance of the equipment has not changed. Measurement setup and equipment used for QC tests should be clearly recorded so that the measurements can be easily replicated by others if needed. One of the hardest things to absorb is the variety of equipment and how it is used. However, there are standard symbols for the different components that can be helpful when starting out (HPA 1976).

As well as tests conducted by Medical Physics, more regular user-led testing is often also required. Typically, this is performed monthly according to training and instructions issued by the Medical Physics department. Usually, Medical Physics will also provide a spreadsheet to analyse the results and provide clear feedback to the user. There are often queries based on these tests raised by staff. Sometimes the tests pick up true changes in equipment function or issues with the test protocol. It is important that Clinical Scientists provide support to frontline staff to ensure that external engineers are not called to site unnecessarily. An example of the kind of support that may be required is provided by the following case study.

Adult exposure settings

	kVp	mAs	FFD, cm	Grid	AEC
Tib/Fib					
AP & LAT	60	2.5	110	None	None
Ankle					
AP	60	2	110	None	None
LAT	60	2.5	110	None	None
Cervical Spine					
AP	75	4	110	None	C
LAT	78	5	180	None	C
Lumbar Spine					
AP	90	16	110	110	C
LAT	90	40	110	110	C
Pelvis & Hip					
AP	85	16	110	110	L & R
AP Hip	70	16	110	110	C
HBL (table)	85	80	110	Sim Grid	None
HBL (WS)	80	80	180	180	C
Chest					
AP	90	1.6	180	Sim Grid	None
PA	120	1.6	180	180	L & R
Abdomen					
AP	90	12.8	110	110	L & R
Renal	90	12.8	110	110	None
Bladder	90	16	110	110	C
Shoulder					
AP	65	5	110	None	C
Axial	65	8	110	None	None
Humerus					
AP	60	4	110	None	None
LAT	70	8	110	None	None
Wrist					
AP	60	2	110	None	None
LAT	60	2.5	110	None	None

PAEDIATRIC EXPOSURE CHART

Exam	kV	mA	ms	FFD, cm	grid	AEC
UPPER LIMB						
6-12 MONTH	60	100	10	110	-	-
LOWER LIMB						
6-12 MONTH	60	100	10	110	-	-
SKULL – AP						
6 MONTHS	70	100	10	110	-	-
12 MONTHS	80	200	10	110	✓	Y
2 YEARS	80	300	10	110	✓	Y
4 YEARS	80	300	40	110	✓	Y
7 YEARS	80	300	60	110	✓	Y
CHEST – SUPINE						
12 MONTHS	60	200	10	120	-	-
2 YEARS	65	200	10	120	-	-
CHEST – ERECT						
2 YEARS	65	200	10	180	-	-
4 YEARS	65	200	10	180	-	-
7 YEARS	70	200	10	180	-	-
HIPS						
6 MONTHS	70	50	10	110	-	-
12 MONTHS	75	50	10	110	-	-
2 YEARS	70	100	10	110	-	-
4 YEARS	70	200	10	110	✓	Y
7 YEARS	70	200	10	110	✓	Y
ABDOMEN						
6 MONTHS	70	50	10	110	-	-
12 MONTHS	75	50	10	110	-	-
2 YEARS	70	100	10	110	-	-
4 YEARS	70	200	50	110	✓	Y
7 YEARS	70	200	50	110	✓	Y

FIGURE 4.10 Adult and paediatric planar X-ray imaging exposure settings.

CASE STUDY: PLAIN FILM INCORRECT OUTPUT MEASUREMENTS

An X-ray Radiographer calls the department and says that local QC shows that the output has halved. The scientist asked the following questions:

- Has an engineer visited recently?
- Has there been any deterioration in clinical image quality?

The answer to both questions was "no". The Radiographer was then carefully asked whether they had done this test before. They had done the test before without issues and had repeated the measurement before calling. The following questions were then asked:

- Has there been any deterioration seen in previous QC results?
- Is the same detector used as usual? Solid-state detectors will exclude backscatter, whereas ionisation chambers, which are not often used for local QC, include backscatter.
- Is the protocol the same as previous tests?

The Radiographer was then asked to remove the equipment and invite another individual to perform the QC test independently. The result was found to return to its original expected value. Discussions showed that a filter was in the beam for the first Radiographer, which had reduced the output. The equipment was in fact performing as expected.

This issue was resolved over the telephone, but often a visit is required to check the performance of equipment and identify the source of any discrepancies.

4.3.1 Qualitative vs Quantitative QC

Historically, many assessments of image quality performed as part of QA processes were subjective. They used images of test objects, such as a Huttner (Figure 4.11a) to measure limiting spatial resolution or a contrast detail test object (Figure 4.11b) to measure visible objects of different sizes against a noisy background to assess contrast visibility. An observer counts the number of objects that are visible, and this number is related back to the spatial, or threshold contrast, resolution of the system. While these tests are quick and simple (Figure 4.11c), they are also subjective, and the visibility of objects can vary between individuals.

With the move to digital imaging, quantitative image analysis is now possible, and has become a standard tool for consistent and objective assessment of medical images. Developing such tools can be time-consuming and complex, but a range of freely and commercially available software is available for image analysis, such as ImageJ, a Java-based image processing package, and IQWorks. For ImageJ, there are two useful modules "DRIQ" and "SPICE-CT" that are often used for quantitative image analysis. A detailed discussion of image processing software is beyond the scope of this text, but it is recommended that any individual with an interest in this area downloads these platforms and explores how they can be used.

Simple data analysis can be performed using Microsoft Excel, for example, determining the signal transfer property (STP) relationship of an imaging receptor This relates a detector dose indicator (DDI) metric value indicated by a radiographic system to the radiation dose incident on the receptor. Typical DDI values could include Exposure Index (EI) or detector pixel value, although other manufacturer-specific versions exist. An example of STP relationship is shown in Figure 4.12. When performing quantitative analysis, DDI value ought to be linearised, as described in IPEM Report 32 Part VII for assessments of detector performance.

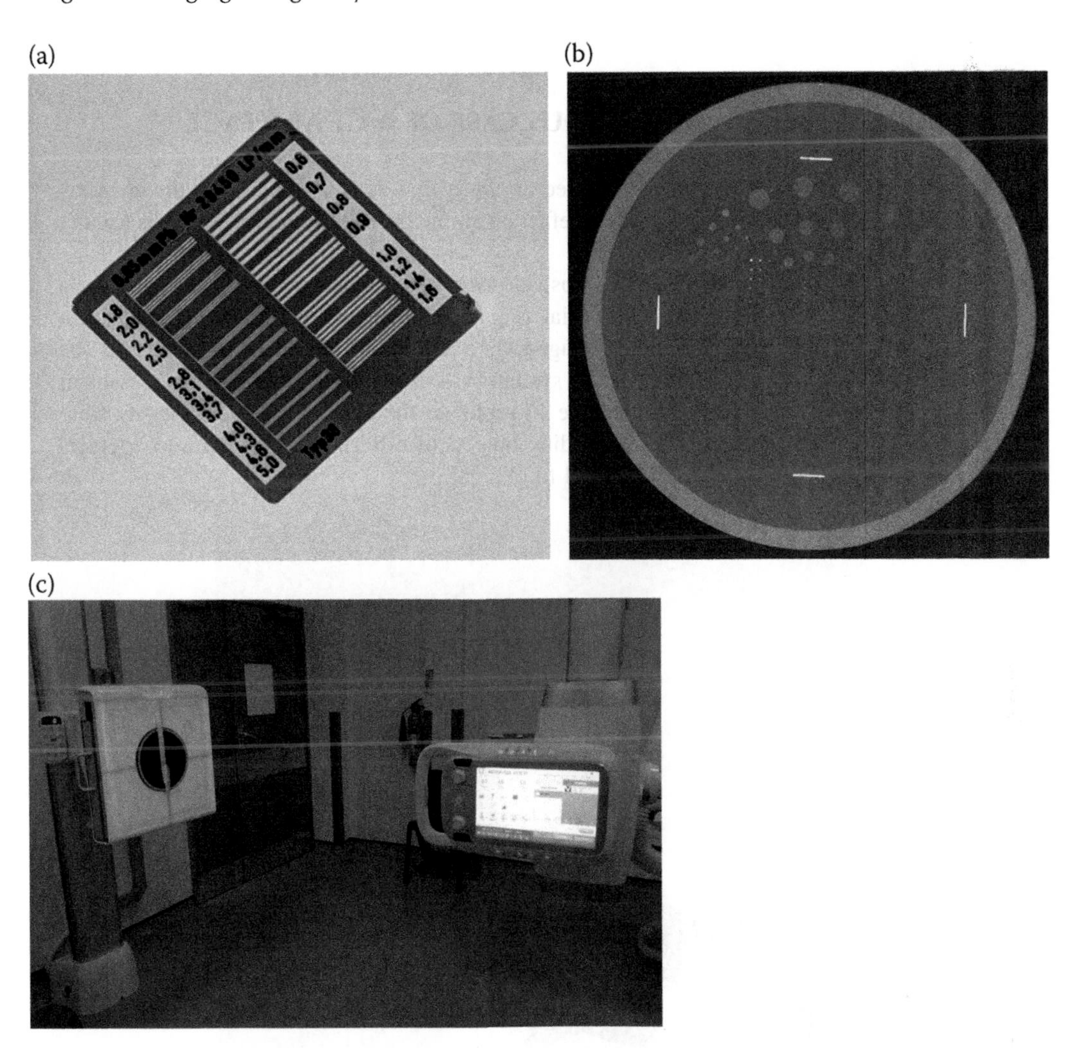

FIGURE 4.11 (a) Huttner, (b) T020 X-ray test object images, (c) the setup for the measurement of TO20.

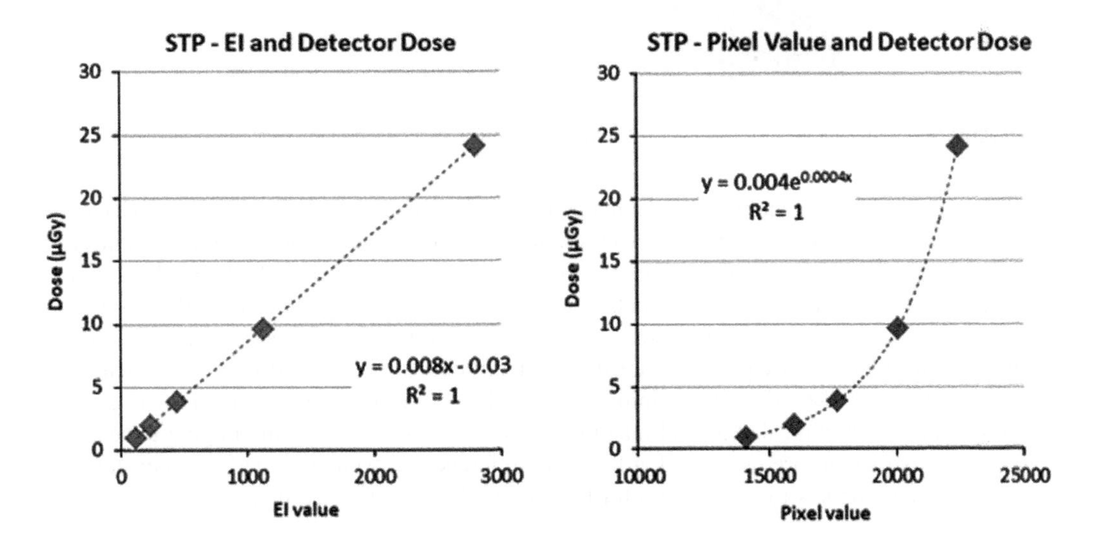

FIGURE 4.12 Relationship between dose, EI value and pixel value.

CASE STUDY: MYSTERIOUS CASE OF A CT ARTEFACT

Sometimes clinical images include unwanted artefacts that bear no relation to the anatomy being imaged. If a Radiographer sees an artefact on an image, the Clinical Scientist is usually the first person they will call.

In this case, a Radiographer reported "pseudo-ring-like" image artefacts occurring for a CT system. These were not complete radial rings, which might indicate an issue with a detector. The rings were not on every imaged slice and were always located towards the bottom of the gantry. A calibration of the system was carried out. Subsequent assessment using a phantom (Figure 4.13) was unable to replicate the artefact. Further images were taken of phantoms of differing sizes, including objects of differing density, but no artefacts were observed so the scanner was returned to clinical use.

FIGURE 4.13 Catphan® used to assess CT image quality.

The artefacts returned, and this time remained following an air calibration. On the advice of the Scientist, the unit was removed from clinical service and the manufacturer's Engineer was called to investigate further. The cause of the artefact was found to be fine debris on a circuit board connected to a detector in the gantry that was intermittently causing the artefact. This was corrected and the unit was returned to routine clinical service after testing. Intermittent issues with equipment are the most difficult to rectify and often the scientist cannot determine the origin from initial testing. Therefore, it is important to have good relationships with clinical staff and manufacturers to ensure errors are thoroughly traced.

Instances like this are useful to log for future reference, as well as providing an example of interesting artefacts to serve as useful learning tools and reference images.

QC testing continues to evolve with the advent of new technology, clinical applications and evidence-based data. It is likely that over the coming years, with the increased reliability of systems, the emphasis will move away from traditional QC towards automated image analysis with an increased focus on optimisation. Scientists will have an integral role in determining the future direction of this field.

4.4 PATIENT DOSE MEASUREMENT AND CALCULATIONS

All patients undergoing clinical examinations using X-rays will be exposed to some level of ionising radiation. Dose calculations are used for a variety of purposes that include:

- Providing risk information to support the informed consent process for participants in research trials.
- The assessment of dose in the event of an individual patient query: for example if an examination took place in error and to advise on whether this is an incident reportable to a legislative body.
- To determine the dose to the foetus when inadvertently exposed during pregnancy.
- Patient dose evaluation can be used in the standardisation or comparison of different clinical techniques using alternative modalities.

Individual patient dose cannot be directly measured and is currently estimated based on known measurable quantities, anthropomorphic models (physical or mathematical), and ionising radiation interaction and biological effect models. There are significant uncertainties associated with the calculation of patient dose. A key concept is that patient dose estimates are not intended to be analysed on an individual basis.

Representative estimates are often used at a population level to explore the stochastic effects of medical imaging with a view to reducing the incidence of cancer in the overall population.

For Clinical Scientists, common questions relating to patient dose could include "What was the patient dose and is this exposure significant?" or "What is the patient dose for this new proposed technique?" Calculation of patient dose is an integral part of the role of Clinical Scientists working with ionising radiation. Some of this work is becoming routine, other tasks are novel with potential for publication. Much of the work requires good communication with clinical colleagues and other professions such as Radiographers. There are several methods available for estimating patient dose, each with advantages and disadvantages.

4.4.1 THE BASICS OF PATIENT DOSE

If we accept that radiation can be damaging to the body, we need to develop methods to quantify the likelihood of damage taking place. This provides clinicians and patients with information about the risks associated with various imaging tests, to help decide whether exposure to this risk is clinically justified. IR(ME)R states the situations in which a Medical Physicist should be available to provide these calculations.

It is important that any Clinical Scientists specialising in ionising imaging are well versed in the specifics of dose quantification and aware of the pitfalls of how radiation doses are reported. The main measure of dose that you will come across in scientific papers is the "effective dose", which indicates the level of stochastic risk associated with an exposure, and "absorbed dose to an organ". In diagnostic radiology this organ is commonly, although not exclusively, the skin. Methods applied to the measurement and calculation of each of these dose quantities is now discussed in turn.

4.4.2 Deterministic Effects

Deterministic effects, as described in Chapter 7, involve the death of large numbers of cells. Deterministic effects only occur above a certain threshold, and then increase in severity with dose. In order to know whether a deterministic effect is likely the Radiographer and Radiologist need to have a dose indicator that they can relate back to the level of deterministic risk.

A measurement of an X-ray field obtained in air is a measure of the charge induced in an ionisation chamber, or solid-state detector calibrated to display a dose quantity known as "air KERMA". KERMA stands for kinetic energy released per unit mass of air and has units of $J.kg^{-1}$ or Gy.

By measuring air KERMA, it becomes possible to estimate the dose absorbed by air using a calibration factor. With a knowledge of relative stopping power, the absorbed dose to air can be converted to absorbed dose for other materials. In this way, the absorbed dose to tissue can be calculated (measured in Gy, where 1 Gy = 1 J/kg).

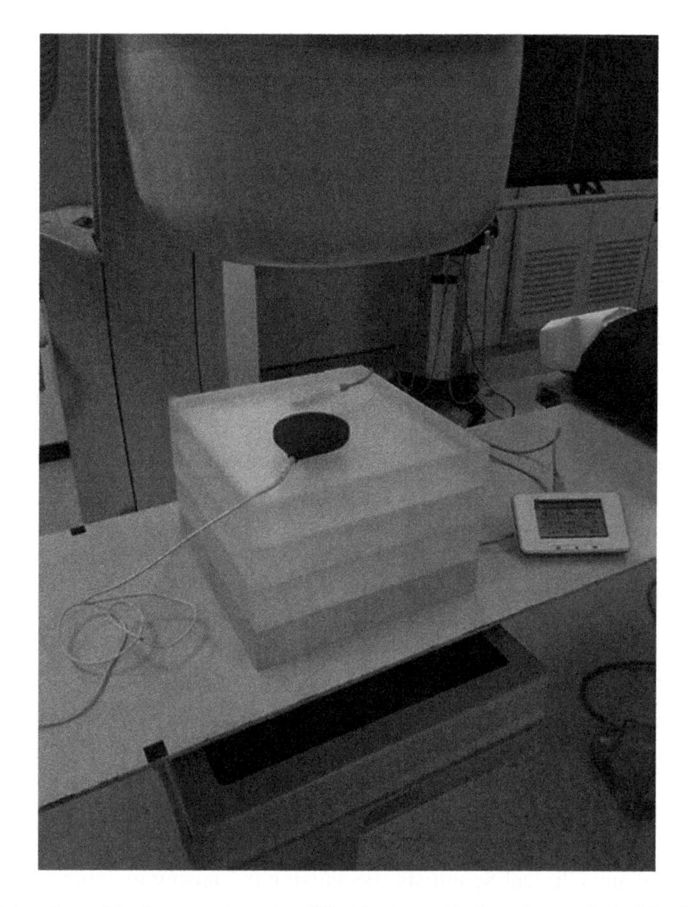

FIGURE 4.14 Setup for a skin dose measurement. The detector is below the couch and the X-ray tube is above the couch. Note that the opposite orientation is usually used clinically to reduce scatter back towards the eye.

The absorbed dose to an organ indicates the amount of energy deposited in that organ, which is very useful for providing an indication of the potential severity of deterministic effects. If the aim is to measure skin dose, for example, in fluoroscopy, the chamber/detector should be placed on a material that creates backscatter (Figure 4.14). However, if you are measuring standard output as described in the performance measurement sections, it is more usual to carry out this measurement without backscatter. This can be achieved by measuring in air or using a lead-backed solid-state chamber.

Another way to measure skin dose is using a tissue equivalent material that has a measurable response to radiation. Thermoluminescent dosimeters (TLD) are used in personal dosimetry worn by staff. Although it is not practical to directly measure the dose to the skin with TLDs for every patient, these are also occasionally used to measure patient dose as described in the IPSM report 53 on patient dosimetry techniques in Diagnostic Radiology (IPSM, 1988).

Equipment used for fluoroscopy and planar X-rays is often fitted with ionisation chambers, or devices that measure a quantity called Dose Area Product (DAP) expressed in units of Gy.cm^2 This is sometimes expressed as KERMA Air Product or KAP. These are used to estimate the air KERMA multiplied by the area of the beam, which gives a good indication of the energy imparted to the patient, but not necessarily where that energy was imparted. Many fluoroscopy systems also measure the dose (in Gy) at a particular reference point and some provide dose maps over the skin of a model patient.

The number of different types of dose and methods of measurement can lead to some confusion and so are summarised in Table 4.4 - Commonly used measures of dose for different X-ray modalities.

If a patient receives a high dose of radiation to the skin this can lead to skin reddening (called erythema). In severe cases, radiation can lead to ulceration requiring skin grafts. Erythema is unlikely to occur for doses lower than 2 Gy, although the actual threshold will vary depending on the individual and severity will increase with dose as described by the International Commission on Radiological Protection (ICRP 2012). As these levels of dose are possible with fluoroscopy, it is important that patients are fully aware of possible side-effects and provide informed consent. If a high enough dose accumulates, which could lead to erythema, patients are provided with extra information. Doses can be estimated retrospectively from an understanding of the equipment settings and doses delivered, reference dose, or a combination of both. There are also some systems that display skin dose maps that account for the position of the beam in relation to the body throughout the procedure.

TABLE 4.4
Commonly Used Measures of Dose for Different X-ray Modalities

Modality	Quantity	Example Units
Plain imaging	Entrance surface dose	mGy
	Kerma Area Product (KAP) or Dose Area Product (DAP)	mGy.cm^2
Fluoroscopy	Kerma Area Product (KAP) or Dose Area Product (DAP)	mGy.cm^2
	Reference dose	mGy
	Fluoroscopy irradiation time	Minutes
CT	Computed tomography dose index (CTDI) air	mGy
	$CTDI_{vol}$ [1]	mGy
	Dose length product (DLP)	mGy.cm
Dental	Patient entrance air KERMA ($K_{a,e}$)	mGy
	Kerma Area Product (KAP) or Dose Area Product (DAP)	mGy.cm^2
Mammography	Mean Glandular Dose (MGD)	mGy

Note
[1] Kalendar (2001)

MPEs may be asked to provide advice on the threshold doses at which skin erythema should be followed up. The MPE will need to listen to clinical colleagues, discuss the threshold that is appropriate for a given situation and translate this into the dose units provided to the operator. To understand the difference in the thresholds, see Table 4.5 - Dose quantities and units commonly used in fluoroscopic and angiographic procedures.

It is interesting to note that the absorbed dose threshold for skin erythema is of the order of 2–3 Gy, whereas the effective dose for angiography is of the order of 10 mSv. The reason that these two values are so different is not because the patient has been exposed to a higher dose of radiation in one situation compared to another, but due to the differences in how absorbed dose and effective dose are calculated. Another area where absorbed dose may be used directly for patient dose estimates is in the assessment of dose to a foetus. This calculation may be requested prospectively, for example, if a clinician needs the information to justify the choice of examination, or retrospectively, as can occur if a patient has received a scan and later found out that they were pregnant. In early pregnancy calculations, it is common to use absorbed dose to the uterus as a surrogate for the foetal dose.

4.4.3 Stochastic Risk

While absorbed dose can provide a good indication of the severity of deterministic effects, this does not give an indication of stochastic risks associated with an exposure in terms of the increased risk of cancer. Absorbed dose is converted to equivalent dose by using a radiation weighting factor (1 for X-rays). The concept of effective dose takes the equivalent dose to each organ, the proportion of that organ irradiated and sums these to give a whole body effective dose. This allows risks to be compared between different types of scans and between organs, for example, comparing a chest X-ray to an X-ray of the pelvis. As we cannot measure the dose absorbed within the organ of an individual, how do we estimate effective dose? We estimate it using computerised models such as those described in the sections below.

TABLE 4.5
Dose Quantities and Units Commonly Used in Fluoroscopic and Angiographic Procedures

Parameter	Example Follow-up Threshold Dose	Reasoning
Peak Skin Dose (PSD)	3000 mGy	A skin dose map is produced and the point at which the skin dose is the highest inclduing dose backscattered radiation weightingis used to estimate PSD.
Kerma to air at a reference point (Kar)	5000 mGy	The reference point is fixed and defined by the equipment. This dose measurement does not consider motion of the equipment which would spread the dose over a larger area requiring a higher follow-up threshold.
Kerma Area Product (KAP)	500 Gy cm^2 [1]	This assumes an area of 10 cm by 10 cm is exposed, which is a common dimension for cardiac catheter labs. If the area is reduced the limit would have to be lower. Again, this is fixed with respect to the equipment and does not account for the movement of the equipment.
Fluoroscopy time	60 minutes	This is not closely linked to the actual skin dose and should only be used as a supplementary indicator.

Note
1 Stecker et al. (2009)

4.4.4 Monte Carlo Simulations

In addition to being based in hospitals, Medical Physicists are also based in:

- Private enterprises
- Public Health England
- The National Physics Laboratory
- Academia

These organisations have worked together to develop phantoms that can be combined with mathematical models, in the form of Monte Carlo simulations, to estimate the absorbed dose for individual organs. The Oak Ridge phantom is an example of a mathematical phantom that was originally developed to estimate doses following exposure to atomic weapons but has since been used for medical dosimetry. Phantoms come in a range of sizes so that doses can be modelled more accurately. More recently, CT scans of cadavers have been used to generate anatomically realistic phantoms, with separate phantoms for men and women. As these models consider interactions with photons at diagnostic energies, they are viewed as giving more realistic dose estimates, as described by Shrimpton et al. (2016).

Monte Carlo phantoms, such as the Oak Ridge phantom, possess both a prostate and breasts, termed hermaphrodite phantoms. Phantoms do not always reflect the typical size and shape of patients and all individuals are different in terms of their distribution of fat and the positioning and size of their organs. Therefore, effective dose is based on several assumptions that we know are not likely to be accurate for individuals patients. Since we cannot measure dose practically, we must be cognisant of the potential pitfalls of using these phantoms for individual dose assessments. There are several different phantoms used currently, and therefore the phantom used for any calculations must be clearly specified so that calculations can be replicated.

CASE STUDY: A PLANAR DOSE ASSESSMENT

A thoracic spine examination of a female patient was performed; however, the intended examination should have been for the abdomen. Therefore, as well as investigating the circumstances that led to the error to prevent a similar incident from occurring again, the Clinical Scientist decided to perform a dose assessment.

One method of estimating patient dose is to use commercial Monte Carlo software such as PCXMC to estimate radiation dose using a mathematical anthropomorphic phantom. This requires geometric information about the examination, radiation output during the examination, and details regarding the patient's age and size.

Details of the examination included the total DAP of 140 $cGy.cm^2$, two projections: anterio-posterior and lateral, an X-ray energy of 80 kV_p, and a focus-to-detector distance of 110 cm. This information was introduced to the PCXMC software, which computed an estimated additional effective dose of 0.2 mSv, with the anterio-posterior projection contributing approximately ¾ of this additional dose.

A similar approach was used to estimate the intended effective dose for an abdomen examination, which PCXMC computed to be 0.6 mSv. Therefore, the total dose of both these examinations would be 0.8 mSv. The ratio of the total dose to the intended dose is 1.3.

Following this incident, the patient contacted the department to confirm she was pregnant at the time of the examination, which she did not know at the time, and was worried about the radiation dose to her unborn child. This is a typical question that a Clinical Scientist might be involved with answering. In this instance, rather than using PCXMC to estimate the equivalent dose to the foetus, another reference could be used to reassure the patient of the relatively low risk to the foetus. From the RCR "Protection of pregnant patients during diagnostic medical exposures to ionising radiation", the risk of additional childhood cancer due to additional dose received during a thoracic spine examination would be less than 1 in 1,000,000. It is also possible to use the uterine dose from PCXMC to calculate a surrogate for the foetal dose in early pregnancy.

However, it should be remembered that these dose calculators and reference documents provide estimates and risks based on generalised populations.

Dose calculations for fluoroscopy resemble planar radiology dose assessments, except that it is also necessary to account for motion of the equipment.

CASE STUDY: CT DOSE ASSESSMENT

There are Monte Carlo simulators for CT such as the ImPACT CT dosimetry programme that can be obtained free of charge. This can be extremely useful when first exploring the impact of different protocol factors such as kV on the dose to the patient. Below is an example dose assessment performed using the ImPACT dose calculator. The scan range is included to show the position on the mathematical model (Figure 4.15).

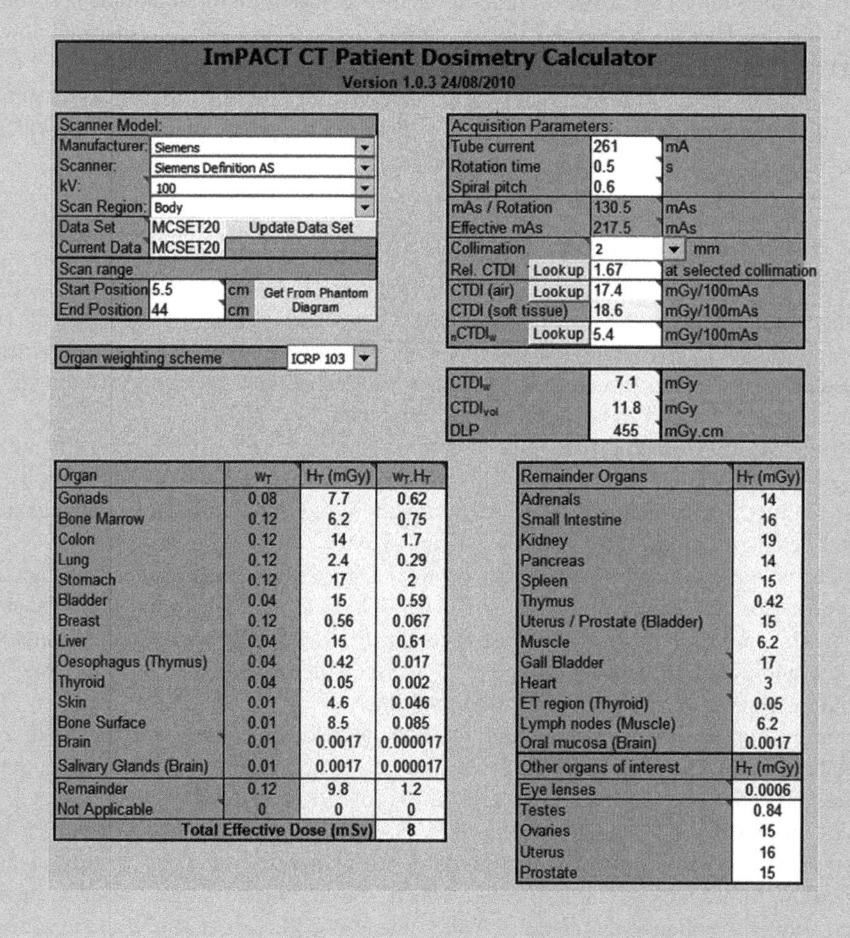

ImPACT CT Patient Dosimetry Calculator
Version 1.0.3 24/08/2010

Scanner Model:		
Manufacturer:	Siemens	
Scanner:	Siemens Definition AS	
kV:	100	
Scan Region:	Body	
Data Set	MCSET20	Update Data Set
Current Data	MCSET20	
Scan range		
Start Position	5.5	cm
End Position	44	cm

Get From Phantom Diagram

Organ weighting scheme	ICRP 103

Acquisition Parameters:			
Tube current		261	mA
Rotation time		0.5	s
Spiral pitch		0.6	
mAs / Rotation		130.5	mAs
Effective mAs		217.5	mAs
Collimation		2	mm
Rel. CTDI	Lookup	1.67	at selected collimation
CTDI (air)	Lookup	17.4	mGy/100mAs
CTDI (soft tissue)		18.6	mGy/100mAs
$_n$CTDI$_w$	Lookup	5.4	mGy/100mAs

CTDI$_w$	7.1	mGy
CTDI$_{vol}$	11.8	mGy
DLP	455	mGy.cm

Organ	w_T	H_T (mGy)	$w_T.H_T$
Gonads	0.08	7.7	0.62
Bone Marrow	0.12	6.2	0.75
Colon	0.12	14	1.7
Lung	0.12	2.4	0.29
Stomach	0.12	17	2
Bladder	0.04	15	0.59
Breast	0.12	0.56	0.067
Liver	0.04	15	0.61
Oesophagus (Thymus)	0.04	0.42	0.017
Thyroid	0.04	0.05	0.002
Skin	0.01	4.6	0.046
Bone Surface	0.01	8.5	0.085
Brain	0.01	0.0017	0.000017
Salivary Glands (Brain)	0.01	0.0017	0.000017
Remainder	0.12	9.8	1.2
Not Applicable	0	0	0
		Total Effective Dose (mSv)	8

Remainder Organs	H_T (mGy)
Adrenals	14
Small Intestine	16
Kidney	19
Pancreas	14
Spleen	15
Thymus	0.42
Uterus / Prostate (Bladder)	15
Muscle	6.2
Gall Bladder	17
Heart	3
ET region (Thyroid)	0.05
Lymph nodes (Muscle)	6.2
Oral mucosa (Brain)	0.0017

Other organs of interest	H_T (mGy)
Eye lenses	0.0006
Testes	0.84
Ovaries	15
Uterus	16
Prostate	15

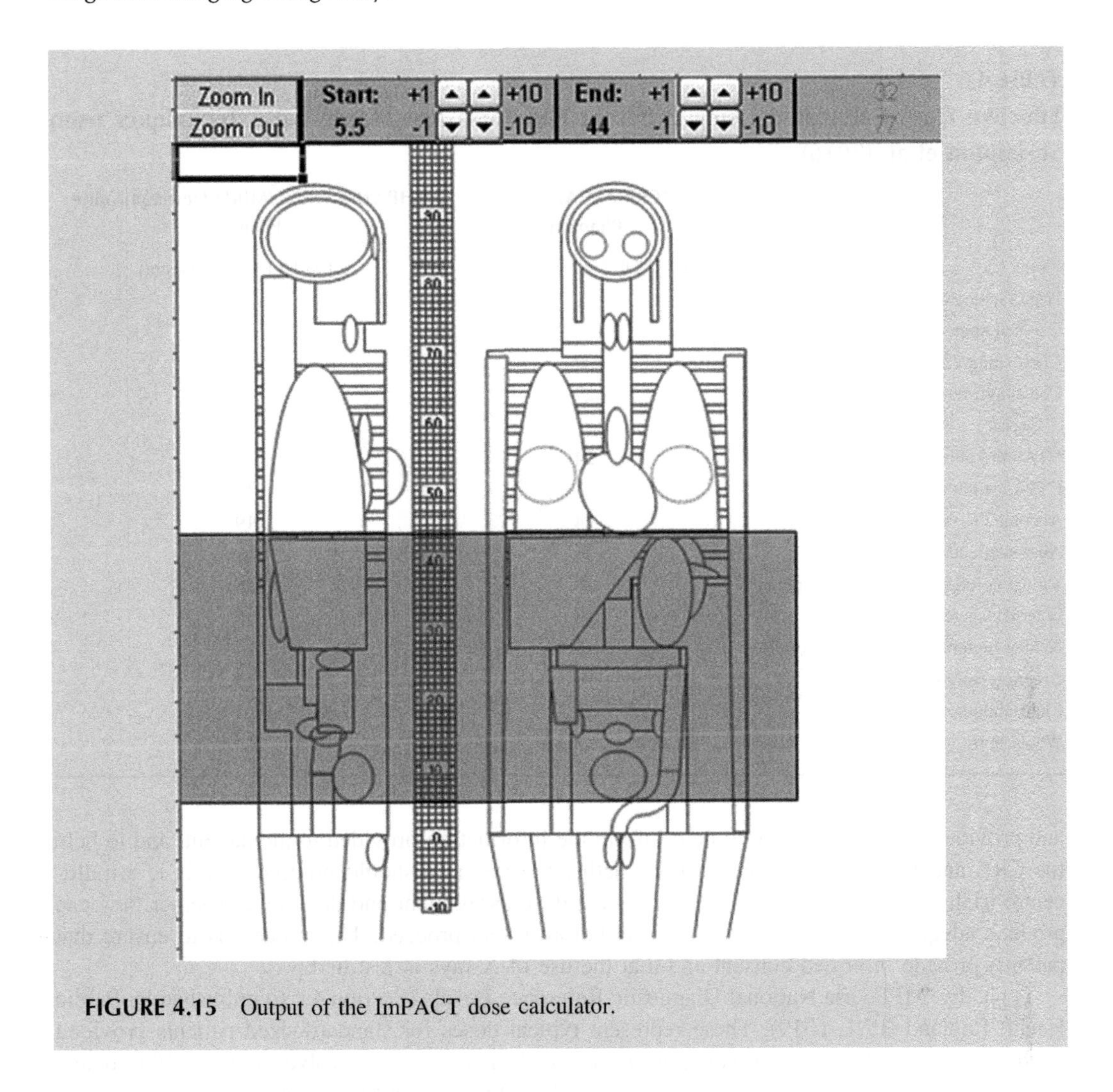

FIGURE 4.15 Output of the ImPACT dose calculator.

4.4.5 Standard Dose Assessments

Most dose assessments that Medical Physicists undertake on a day-to-day basis are to determine whether an incident is reportable or not. Under IR(ME)R, all incidents leading to clinically Significant Accidental and Unintended Exposures (SAUE) should be reported to the CQC (2020). Standard assessments using published data are often used so the use of Monte Carlo simulation programs is not necessary.

Shrimpton et al. (2016) have produced several factors based on Monte Carlo simulations that can be multiplied by the dose length product (DLP) displayed by the CT scanner to estimate effective dose. These "k factors" are summarised in Table 4.6. The DLP is effectively the volumetric equivalent of the DAP in planar imaging. It is instructive to examine these and relate them back to the tissue weighting factors described previously. You will notice that the chest examination has the highest dose per DLP. This is because breast tissue is sensitive to radiation.

It is also instructive to look at the differences between the two different phantoms. Differences are most pronounced for the chest. One situation where differences between phantoms can cause issues is in research applications. When research involves radiation, an Integrated Research Application System (IRAS) form must be completed by the investigator for review by an MPE and a Clinical Research Expert (CRE), often a radiologist. The Medical Physicist will review the form

TABLE 4.6
Effective Dose Calculated Using Different Phantoms and Monte-Carlo Techniques from Shrimpton et al. (2016)

	ICRP AM/AF Voxel Phantom	HPA18+ ORNL/MIRD Hermaphrodite Phantom
Exam	E103/DLP (μSv/mGycm)	E103/DLP (μSv/mGycm)
Head (acute stroke) 16 cm	2	2.2
Cervical spine (fracture) 16 cm	5.7	6
Chest (lung cancer)*	27	19
Chest-high resolution (interstitial lung disease)	27	18
CTA (abdominal aorta/blood vessels)	24	19
CTPA (pulmonary embolism)	27	18
Abdomen (liver metastases)	24	19
Abdomen and pelvis (abscess)	20	16
Virtual colonoscopy (polyps/tumour)	20	16
Enteroclysis (Crohn's disease)	20	16
Kidney-ureters-bladder (stones/colic)	18	16
Urogram (tumour/stones/colic)	18	16
Chest-abdomen-pelvis (cancer)	21	17
Whole body	9.3	8.5

and provide a dose calculation to help inform the information provided to the patient, and to help the CRE and Ethics Committee decide whether the research should proceed. If this is a multi-centre trial, MPEs at each site must review the dose assessment and determine whether they can produce adequate images at this dose level before it can proceed. This process is to ensure that patients provide informed consent and that the use of X-rays is justified.

Typically, MPEs use National Diagnostic Reference Levels (Section 4.4.6) published by Public Health England (PHE 2019). These represent typical doses for standard-sized patients provided good practice is followed. For example, if a research investigation involves a chest CT to detect lung cancer, the National DRL is 610 mGy.cm. The k-factor from Shrimpton using the ICRP AM/AF voxel phantom is 27 μSv/mGycm, whereas the same factor using the HPA18+ ORNL/MIRD hermaphrodite phantom is 19 μSv/mGycm. Using one phantom, the dose would be 16 mSv and the other phantom would be 12 mSv. The patient has received the same dose. There is no wrong method, just different methods of estimation. This emphasises the level of uncertainty involved in patient dose calculations. It is important that Medical Physicists appreciate the conceptual nature of effective dose and the potential errors associated with estimating it.

4.4.6 Diagnostic Reference Levels

One of the main principles of radiation safety is that all exposures to radiation should be optimised (Chapter 7). Therefore the dose to the patient in diagnostic radiology should be kept as low as reasonably practicable (ALARP). For guidance, national benchmarks have been established in the form of diagnostic reference levels (DRLs). It is important first to state that a DRL is <u>not</u> intended to be used on an individual patient basis, which at first may seem counterintuitive – how can you practically determine an individual's patient dose is ALARP if DRLs are not intended for use in individual patients? The answer in part to this question is the concept of *Why* DRLs exist and *How* these values are determined.

4.4.6.1 Why DRLS?

The concept of a DRL is used to evaluate different clinical procedures or examinations, under certain conditions for a patient cohort, or population, to test whether exposure to ionising radiation is ALARP. Within each patient population, there will be variations in patient sizes, habitus, anatomical structure and physiological function. Optimisation means that the optimum dose should be used for each patient. Therefore, there will be some patients that exceed the DRL, even where the exposure has been fully optimised.

For UK organisations, there is a legal requirement under IR(ME)R17 to produce and use DRL values and to document this process in each employer's procedures. There are also other statements set out in this legislation (regulations 6 and 13), where an organisation is required, upon request from the secretary of state, to provide doses to enable population dose estimates to be made.

There is the international ICRP Report 135 (ICRP 2017) and national guidance in the form of IPEM Report 88 (IPEM 2004) around the establishment and the use of DRLs. It is worth noting that, even though there may not be national DRLs for examinations or procedures, it would be acceptable to apply a regional DRL if this was available. Additionally, although the definition states "Diagnostic", a DRL can be applied to interventional procedures, and as part of hybrid imaging systems. This includes image-guided procedures for angiography, cardiovascular procedures, radiotherapy, PET-CT and SPECT-CT.

4.4.6.2 How Are DRLs Produced?

There is currently some debate around how DRLs should be produced. National guidance on the subject states that the mean for average-sized patient (weighing 50–90 kg) should be used. In the past, patient data were collected retrospectively by Radiographers. This was time-consuming and led to small sample sizes. With the advent of dose management systems such as OpenREM (open source, freeware) and other proprietary software, large quantities of patient dose information can be collated. However, the data tend not to include information on the weight of the patient. Therefore, although the national guidance has not yet caught up, some scientists have moved to using the median of a large set of non-weight selected data, as recommended by the ICRP (2017). Patient dose data typically follow a log-normal distribution, associated with a "long" tail of higher dosimetry values, due to larger patients and the use of more complex procedures. Thus, if the arithmetic mean is used, the DRL value would be skewed towards a small number of higher doses. The median is more robust to outliers within the tail of the distribution.

There should be a process for the establishment of DRLs, which clearly states how this process is undertaken and how sites with multiple rooms are processed. The DRL process should also indicate how frequently patient dose samples are compared against DRL values for specific examinations (e.g. every 3 years). Where a local DRL quantity exceeds a specific DRL value (local, regional or national), it can be considered that patient examinations or procedures are consistently greater than expected and therefore may not be ALARP.

4.4.6.3 Practical Considerations

It is recommended that a minimum of 20 patient dose data points are used in a distribution to calculate a median to compare against a DRL value. Where this is not achievable, or the data are not available, this could suggest that the examination or procedure is not typical for that centre and does not warrant producing a local DRL value. Where possible, automated dosimetric data gathering is preferred; however, the advantages and disadvantages of different methods are summarised in Table 4.7.

TABLE 4.7
Advantages and Disadvantages of Different Methods of Data Collection for the Establishment of DRLs

	Advantages	Disadvantages
Automated data collection using a dose management system (DMS)	Takes patient dosimetry information directly from modality No human transcript errors Able to pool large data sets Able to define collection period Able to collate other dosimetric information that can be used in addition to further analysis of DRL quantity median values Allows for patient dose audit to be completed more regularly Doesn't require local clinical staff (Radiographers, Radiology Department Assistants) time to manually collate data Universally compatible with RDSR from DICOM images Can be integrated into other software packages that automatically analyse and pool the data, making the analysis of DRL quantity data more automated	Requires IT infrastructure to set up and local expertise and support Requires input from vendors to set up systems, that may be at additional cost, possibly per connection or per annum Examination protocol selected or indicated on system may not relate to the real clinical examination Several different examination protocols may exist for the same clinical examination or protocol, between which the system may not be able to differentiate Is limited to certain modalities, such as direct digital radiography (DDR), CT, *in situ* fluoroscopy systems, would exclude CR and retrofitted DDR portable detectors
Extracting data from a Radiology Information System	Able to export from computer system into electronic format allowing easier analysis Able to pool large data sets Able to define collection period Allows for patient dose audit to be completed more regularly Part of patient management system, less chance of losing data	Relies on user input can lead to incorrect data being added May not distinguish if single or multiple series or projections have been used, such an accumulated dose rather than single exposure doses Requires support and access to database or system, may not be possible or require other service within a centre Analysis of the data is less automated, and would require further processing and analysis of DRL quantity
Manual collection	Low tech methodology does not rely on computer systems to collate datails accessible to any person collating the data, could use a central data collection sheet as simple as a paper record There is no significant technology or software outlay costs, making this a robust and cheap solution Able to record data for all patient examinations and procedures	Relies on user input can lead to incorrect data being added Including transcript errors in correct dose metric conversion Cumulative dose instead of individual exposure events Difficult to pool large data sets, limited size Limited to prospective and not retrospective data collection Possibility of losing patient dose information May have missing or incomplete information, specifically patient weight or age

CASE STUDY: CHEST X-RAY DRLs

Data have been gathered for adult plain film chest imaging. A histogram of the data is shown in Figure 4.16. The typical log-normal distribution can be seen with a low peak dosimetry value and a long tail. It is a large dataset taken from a dose management system. The mean can be seen to be skewed by the presence of a long tail. The median of 7 cGy.cm^2 is well below the current national DRL for this exam.

Comparing the distribution of DRL quantity data against an expected log normal distribution can be useful for assessing how close this pool of data is to expected values. When any differences occur, this may need further investigation into the possible causes. Higher or lower calculated DRL, and non-typical distributions such as bimodal distributions with two peaks, may occur for the following reasons:

- Inclusion of different numbers of radiation events
- Inconsistent dosimetry value units, that is, μGy.cm^2 and cGy.cm^2
- Inclusion of different examinations
- Patient weight range outside of specified tolerance
- Variable exposure technique for the same examination
- Inclusion of different patient ages, for example, paediatric for adult examinations
- Dosimetry measurement device outside of working tolerance or faulty

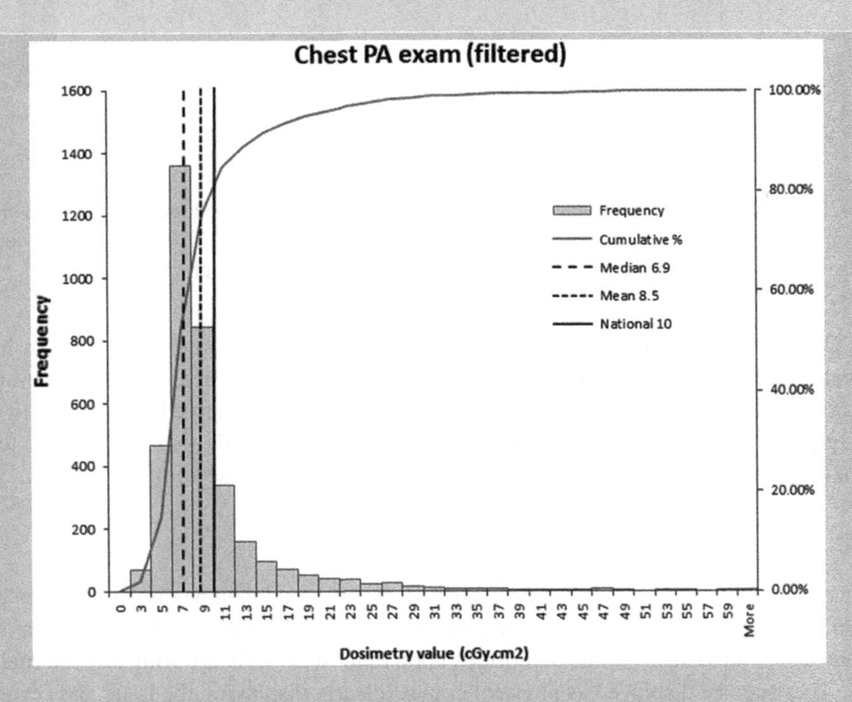

FIGURE 4.16 Histogram of adult PA chest doses.

CASE STUDY: ABDOMEN ANTERIOR-POSTERIOR (AP) DRLS

It is sometimes useful to present DRL values of an examination for different rooms or areas, for comparison with indicated national or regional DRL values. An example for a number of rooms carrying out AP abdominal X-rays can be seen in Figure 4.17, for comparison with the LDRL (mean) and NDRL value (from HPA-CRCE-034, 2010 review). The majority of the Room DAP values fall below the NDRL, with the exception of Room F (Figure 4.17).

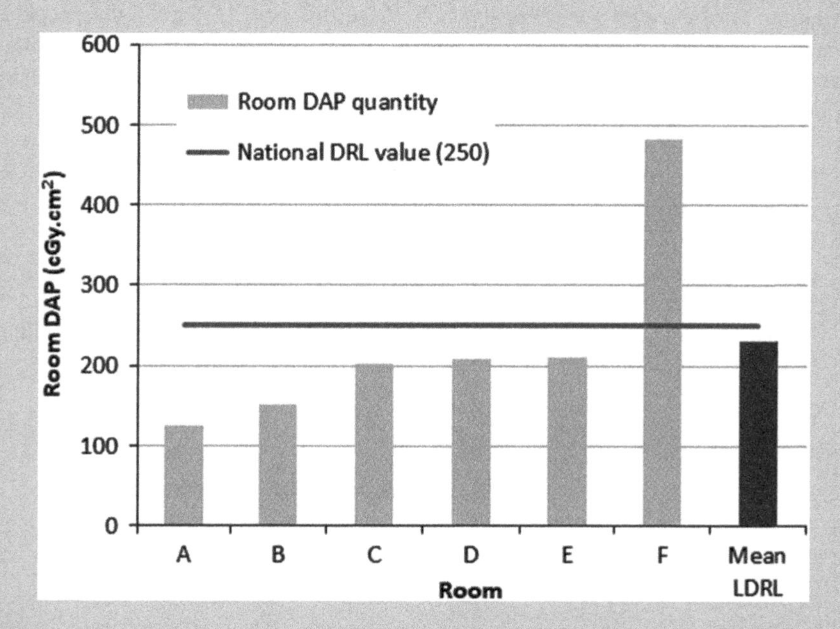

FIGURE 4.17 Dose distribution for several rooms conducting abdominal X-ray examinations, for comparison with the LDRL (mean) and NDRL value (from HPA-CRCE-034, 2010 review). The majority of the Room DAP values fall below the NDRL, with the exception of Room F.

Comparison with the DRL can be used to highlight any rooms with high or low dose levels. In this example, there are six different room DRL values, which are combined to to create a local DRL value. It is obvious when the data are viewed in this way that Room F requires optimisation. Feedback from clinicians should also be sought to verify whether image quality is adequate in some of the lower dose rooms.

Once local DRL values have been established, these should then be disseminated to the relevant clinical areas for which examinations or procedures are performed for reference. Data may need to be made available to inspectorate authorities upon request, as well as during any internal legislative audits. These can be displayed as physical copies (clearly displaying the issue and review date), as well as in electronic format. Staff should be reminded that DRLs are not intended to be applied to individual patients.

For new installations and systems, the use of a local or other DRL value can be useful for evaluating whether commissioned clinical protocols or practices are appropriate. It is recommended that patient dose data is routinely collated and analysed once the system has been in clinical use for enough time to provide a representative sample. Thereafter, this equipment would be included in the DRL process review cycle.

This strategy of dose calculation and review to derive national and local DRLs has resulted in large drops in dose in all modalities. However, it is also known that lower doses can result in poorer image quality. The subjective assessment of what constitutes "adequate" image quality, consistent with the desired clinical purpose, is one of the biggest challenges facing scientists in this field. Patients come in all shapes and sizes, with different levels of fat, and even prostheses, which all impact dose and image quality.

4.4.7 Optimisation

Determining whether the amount of ionising radiation used for imaging is sufficient to achieve reliable clinical information represents a fundamental challenge for ensuring patient exposure is ALARP. DRL audits provide a good indication of whether the amount of radiation used is suitable. This forms the basis of optimisation. Optimisation is not solely concerned with the quantity of radiation, but also ensuring that clinical information is adequate for the justified task. The presentation and quality of radiographic images is just as important. In principle, optimisation could result in an increase in the amount of ionisation radiation used for a clinical task, as the image quality used currently may currently be sub-optimal.

DRL audits will generally provide an initial trigger for optimisation of clinical tasks that are greater or lower than other DRL reference values but should not be the only route for starting any investigations or projects. If DRL dose audits are planned every 3 years, there may be a significant period in between, where the system may not be appropriately optimised in terms of patient dose or clinical performance. Having a forum for clinicians to report *ad hoc* concerns regarding image quality or being able to carry out tasks will allow management of any optimisation. It is recommended that multidisciplinary (MDT) teams are created that include a Radiologist or Consultant for the specialism, Radiographers and Clinical Scientists. These groups are sometimes called Imaging Optimisation Teams (IOT) or Medical Exposure Committees (MEC). They can help facilitate any optimisation work, as well as providing a structured approach and framework for the management of records. This concept has been considered by a report published by the Committee on Medical Aspects of Radiation in the Environment (COMARE 2014). Although the COMARE report focused on CT examinations, these recommendations can be extended to other modalities. The success of any optimisation work is to establish close working relationships with each of the different disciplines. This needs specialist expertise from each area to form strategies and assess the efficacy of any changes.

4.4.7.1 Optimisation Strategy

Determining whether an imaging or clinical task using ionising radiation is optimised is a difficult question, as it cannot be assumed there is a linear relationship between the quantity of radiation and the clinical benefit. Questions you might consider include:

- Does the clinical task need a specific image quality relating to image noise, spatial resolution, or image contrast?
- Is there a minimum acceptable image quality, or are there different acceptable image quality levels (to perform an examination or procedure)?
- Will the task be used for a specific patient cohort, is the detrimental radiation risk greater or lower, or more acceptable, in this cohort?
- Can the task only be performed on or with certain equipment due to practical or technological limitations?
- How significant are patient management and prognosis? What would the impact be of misdiagnosis or an unsuccessful procedure?

- Based on population size, is there an impact on screening programs that would be detrimental to patients by increasing false negatives (or false positives)?
- Is there consistency in image interpretation within and between observers? Do clinicians who perform the examination or procedure always give the same results?

These are some of the considerations, but this list is not exhaustive. It is important to set out a question related to the specific optimisation; for example, just asking "Are chest radiographic X-ray examinations dose optimised?" is not specific enough. A better question might be to add the following qualification to the original question "for trolley bound patients undergoing AP projection examinations for clinical indication of pneumonia infection chest radiographic examinations".

Once a question has been established, the process of optimisation (including standardisation) can be carried out, which ideally should include input from Radiologists (or Consultants), Radiographers and Medical Physics staff (inclusive of Radiotherapy, Nuclear Medicine and Radiation Safety). Any previous data or information will form the basis of any work, which will then guide how to carry out such an optimisation task. This can be as simple as interrogating

Date	All doses below the NDRL (Y/N)	Any other concerns? (Y/N)	Please list/summarise any actions taken so far	Comments	Signed
12/09/18	No	Yes	Clinical Scientist presented a DRL comparison paper for all DR units across organisation. Room A exposures for all exams were consistently high and above the Local and National DRL	Room Aissues were discussed and agreed that an applications specialist needed to return to re-set the baseline/protocols as it was felt that the protocols had been re-set to the factory settings possibly following an engineer visit.	Radiographer
19/11/18			Applications specialists attended, there were several inconsistencies found throughout the data base. The decision was made to copy the database from Room B and copy this on to room A as the software versions are the same. The exposure factors have now been set the same/similar to the exposure factors in room B	One thing which may be a contributing factor to the high DRL's aside from the pre-set exposures, is that if the grid is left in for an examinations which does not require a grid, the system will still allow an exposure to commence despite a warning message been displayed. The system will adjust the exposure accordingly to give a grid exposure hence higher exposure factors. The majority of examinations undertaken in room Bare extremities therefore a grid should not be needed for most examinations. Radiation safety team to review dose data early 2019.	Superintendent Radiographer
15/04/19			Clinical Scientisthas supplied the revised DRLs from Room A–still concerns especially with regarding abdomen, Foot L/R and thoracic spine which are well above the local and NDRL		Radiographer Radiation Safety Lead
18/04/19			Discussed at MEC meeting on 18/04/19. Applications Specialist to be asked to visit and re-look at exam protocols. Radiographerto organise.		Radiographer Radiation Safety Lead
28/05/19			Manufacturer's engineer arrived instead of applications specialist as requested. Engineer unable to change exposures.	Engineer to contact applications to find out if they are aware of the work order.	Superintendent Radiographer
21/06/19			Manufacturer's applications specialist attended to adjust exposures with clinical scientist in attendance.	• Foot mAs reduced from 3mAs to 2mAs • Lateral hand reduced from 3mAs to 1.5mAs • PA Hand reduced from 2mAs to 1.5mAs • C-Spine lateral reduced from8mAs to 6mAs • Protocol for Abdomen EVAR Stent with fixed exposures of 77kV 20mAs for AP and 80kV 40mAs for Lateralshas been created.	Superintendent Radiographer

FIGURE 4.18 Example of a log of changes implemented as part of continuing optimisation of a plain imaging examination X-ray room, with this being managed as part of an MDT. This gives an insight into the time frame and involvement of different people including external engineers and application specialists, with the latest description of adjustments made to exposure parameters.

clinical equipment protocol settings and comparing with other similar systems to more complex additional image analysis, or dose analysis, using other methods and tools. This process will involve suggested changes or adaptions to the clinical task.

The next phase would be to evaluate the efficacy of any changes prior to these being implemented. This can be performed through retrospective clinical audits and patient DRL quantity audits, or simply asking for feedback on image quality or procedures from clinicians. Once any proposed changes have been agreed involving the MDT, then its effectiveness should also be evaluated through clinical audit, DRL audit and retrospective observer studies.

It is important that the optimisation process is managed through MDT groups such as an MEC or IOT. Part of this management is to document any changes made, as well as assessing the result of any changes by follow up patient dose audits or clinical audits. An example of a log of changes and progress regarding optimisation of plain imaging chest PA projection examination is shown in Figure 4.18.

CASE STUDY: OPTIMISATION FOR ENDOVASCULAR AORTIC REPAIR (EVAR) PROCEDURES

Previously, it was seen that Room F was greater than the national DRL values and other Room DRL quantity values. On investigation, the cause was due to the original patient dosimetry data record on the Radiology Information System (RIS) including post-endovascular repair (EVAR) stent examinations as abdomen examinations which are a higher dose examination. The action was to create a new specific examination protocol for post-EVAR examinations, so these could be separately identified from standard abdomen examinations. As a result, a follow-up patient dose audit for abdominal examinations resulted in a lower DRL value, below the national DRL value, as well lowering the local DRL, as shown in Figure 4.19.

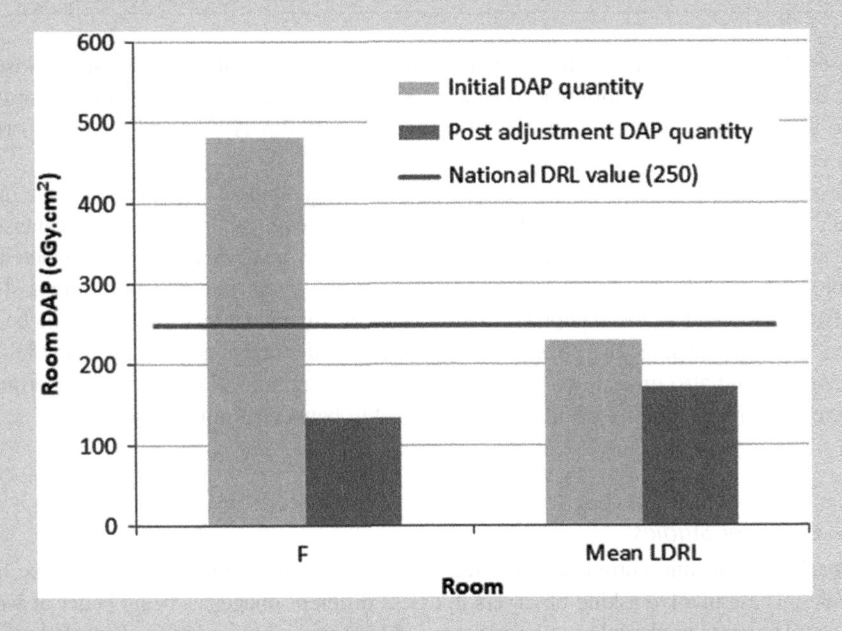

FIGURE 4.19 Doses pre- and post-optimisation.

CASE STUDY: A SIMPLE PRACTICAL EXAMPLE OF OPTIMISATION

The example below shows what could be a typical exposure chart for plain imaging – with comparison between two rooms. When dose collection and DRL analysis has identified differences between rooms a review of exposure charts may help to identify possible solutions. For example, looking at a thoracic spine AP projection, the exposure settings are as follows:

Site A: 75 kVp, 25 mAs set at 110 cm distance with grid
Site B: 90 kVp, 13 mAs set at 110 cm distance with grid

The exposure factors are different, whereas the geometry is the same. As discussed in previous chapters, higher X-ray energy will result in lower image contrast, although as the energy is more penetrating the X-ray flux can be reduced, which is evident in this example. Knowing the relative X-ray tube yield of each system and applying these protocol settings, it would be possible to estimate both a theoretical DAP value and estimated effective dose using PCXMC, with the following example:

Output yield of X-rays *(at protocol X-ray energy and 75 cm)*	**Room A** – 76.8 μGy.mAs^{-1}	**Room B** – 89.2 μGy.mAs^{-1}
Assumed collimated area for thoracic spine AP		24 cm × 43 cm at imaging receptor
DAP *(national DRL value of 100 cGy.cm^2)*	**Room A** – 92 cGy.cm^2	**Room B** – 56 cGy.cm^2
Effective dose	**Room A** – 220 μSv	**Room B** – 160 μSv

Based on this, it could be inferred that Room A is operating at a higher patient dose than Room B. Assuming the equipment, operator technique, image observers and patient cohort are the same, a simple conclusion would be to match Room A protocol settings to that of Room B – standardisation.

"*Two sides of the coin*" – this only considers dosimetry without considering image quality. Although the indicated dose in Room B may be lower, the image quality for the clinical task may not be acceptable. Thus, after checking with clinicians, it may be more appropriate to increase the exposure factors of Room B, or to match that of Room A as part of optimisation. This is particularly important if the equipment varies between manufacturers, as there will be often different image processing, and physical characteristics of the image acquisition chain. Different technologies could also impact image quality; actually, the optimised protocol, considering both dosimetry and image quality, could differ considerably between Rooms A and B.

4.4.7.1.1 Observer Studies

One approach to evaluating current and adapted clinical techniques and protocols could be using observer studies. These involve asking observers to assess different images as being better or worse than other images. The same methodology can also be used to assess intra- and inter-observer agreement. A disadvantage of observer studies is that they can be resource intensive, and as there is currently a national shortage of Radiologists, they can be difficult to implement within busy departments.

Therefore, Clinical Scientists often prefer to analyse images from test objects and phantoms as described in Webb's Physics for Medical Imaging (Flower 2012). Patient images can also be analysed to assess changes in equipment performance and differences between disease characteristics. The multi-factorial nature of optimisation is why MDTs are important. MPEs should oversee the process at all stages and optimisation is a key part of the training of any Clinical Scientist specialising in imaging with ionising radiation.

4.5 THE FUTURE OF X-RAY IMAGING PHYSICS

Recent developments in terms of dose monitoring mean that it is now possible to collate information on the doses received by every patient within a hospital Trust and monitor for any significant deviations from the standard. This means that the establishment of DRLs and identification of errors has become more efficient. It also means that national dose registries have become a possibility and it is foreseeable that dose league tables comparing different Trusts may soon be possible.

Further developments in computing may mean that it becomes more practical to produce more personalised phantom-based dose estimates. While work towards this has begun with the AAPMs Size Specific Dose Estimates it is likely that in the future this will be superseded by automated segmentation of CTs to allow individual dose estimates. The theory that underpins the concept of effective dose relies on the "Linear No Threshold model" described in Chapter 7. If this is proved to be incorrect, mechanisms for calculating effective dose will need to be adjusted accordingly. Other possibilities for the future are the development of more sensitive tests for DNA damage caused by radiation. This may allow personalised radiation doses, rather than using theoretical constructs that introduce large degrees of error.

As technology develops, it is anticipated that artificial intelligence (AI) will have an increasing impact on the work of Medical Physicists and Radiologists. Computer-aided detection (CAD) of abnormalities is already available in several areas and is likely to expand in years to come. How we ensure that this does not negatively impact on clinical diagnosis is likely to form a part of the role of the MPE going forward. In addition, AI is starting to be used for image processing in equipment already on the market. It is likely that AI could come to impact every part of the imaging chain from referral to image optimisation, diagnosis and treatment.

The future is exciting and unpredictable, and methods for calculating patient dose are not fixed or founded on indisputable science, so the potential for future scientists to contribute to improvements in the way that patient dose is assessed is substantial.

REFERENCES

Allisy-Roberts PJ and Williams J (2007). *Farr's Physics for Medical Imaging*, 2nd edition. Saunders (WB) Co Ltd.

COMARE (2014). Committee on Medical Aspects of Radiation in the Environment "Patient Radiation Dose Issues Resulting from the Use of CT in the UK". OCOMARE. online available from https://www.gov.uk/government/publications/review-of-radiation-dose-issues-from-the-use-of-ct-in-the-uk [accessed 10 March 2021].

CQC (2020). Care Quality Commission "Significant Accidental and Unintended Exposures under IR(ME)R". Online Available from: https://www.cqc.org.uk/sites/default/files/20200826_saue_guidance_updated_aug20.pdf [accessed 13 September 2020].

Dendy PP and Heaton B (2012). *Physics for Diagnostic Radiology*, 3rd edition. CRC Press, London.

Flower, MA (2012). *Webb's Physics of Medical Imaging*. CRC Press, London.

HPA (1976). *Hospital Physicists Association "The Physics of Radiodiagnosis"*. HPA, York.

ICRP (2012). International Commission on Radiological Protection "ICRP Statement on Tissue Reactions and Early and Late Effects of Radiation in Normal Tissues and Organs – Threshold Doses for Tissue Reactions in a Radiation Protection Context". *Ann ICRP* **41**(1–2): 1–322.

ICRP (2017). International Commission on Radiological Protection "Diagnostic Reference Levels in Medical Imaging. ICRP Publication 135". *Ann ICRP ICRP* **46**(1).

IPEM (1995). *The Institute of Physics and Engineering in Medicine "Measurement of the Performance Characteristics of Diagnostic X-ray Systems. Report 32, Part I: X-ray Tubes and Generators)"*. IPEM, York.

IPEM (1996). *The Institute of Physics and Engineering in Medicine "Measurement of the Performance Characteristics of Diagnostic X-ray Systems. IPEM Report 32, Part II: Image Intensifier TV Systems"*. IPEM, York.

IPEM (1998). *The Institute of Physics and Engineering in Medicine "Measurement of the Performance Characteristics of Diagnostic X-ray Systems. IPEM Report 32, Part VI: Image Intensifier Fluorography Systems"*. IPEM, York.

IPEM (2002). *The Institute of Physics and Engineering in Medicine "The Medical and Dental Guidance Notes; A Good Practice Guide to Implement Ionising Radiation Protection Legislation in the Clinical Environment"*. IPEM, York.

IPEM (2004). *The Institute of Physics and Engineering in Medicine "Guidance on the Establishment and Use of Diagnostic Reference levels for Medical X-ray Examinations (Report 88)"*. IPEM, York.

IPEM (2005a). *The Institute of Physics and Engineering in Medicine "The Commissioning and Routine Testing of Mammographic X-ray Systems. IPEM Report 89"*. IPEM, York.

IPEM (2005b). *The Institute of Physics and Engineering in Medicine "Recommended Standards for the Routine Performance Testing of Diagnostic X-ray Systems. IPEM Report 91"*. IPEM, York.

IPEM (2010). *The Institute of Physics and Engineering in Medicine "Measurement of the Performance Characteristics of Diagnostic X-ray Systems. Part VII: Digital Imaging Systems"*. IPEM, York.

IPEM (2012). *The Institute of Physics and Engineering in Medicine "The Critical Examination of X-ray Generating Equipment in Diagnostic Radiology. IPEM Report 107"*. IPEM, York.

IPSM (1988). "Patient Dosimetry Techniques in Diagnostic Radiology, IPSM Report No 53." IPSM, York.

IR(ME)R (2017). *The Ionising Radiations (Medical Exposure) Regulations*. SI 2000/1059. The Stationery Office, London, England.

IRR17 (2017). "Ionising Radiations Regulations". SI. 1075. Online Available from: http://www.legislation.gov.uk/uksi/2017/1075/contents/made [accessed 13 September 2020].

Kalendar WA (2001). *Computed Tomography: Fundamentals, System Technology, Image Quality, Applications*, 3rd edition. Wiley Erlangen.

NHS England (2019). "Diagnostic Imaging Dataset 12 Months to March 2019". Online Available from: https://www.england.nhs.uk/statistics/wp-content/uploads/sites/2/2019/07/Provisional-Monthly-Diagnostic-Imaging-Dataset-Statistics-2019-07-18.pdf [accessed 13 September 2020].

PHE (2019). "Public Health England "National Diagnostic Reference Levels (NDRLs) from 19th August 2019". Online Available from: https://www.gov.uk/government/publications/diagnostic-radiology-national-diagnostic-reference-levels-ndrls/ndrl [accessed 1 October 2020].

RCR (2009). Royal College of Radiologists "Protection of Pregnant Patients during Diagnostic Medical Exposures to Ionising Radiation". Online Available from: https://www.rcr.ac.uk/system/files/publication/field_publication_files/HPA_preg_2nd.pdf [accessed 1 October 2020].

RCR (2020). Royal College of Radiologists "iRefer: Making the Best Use of Clinical Radiology". Online Available from: https://www.rcr.ac.uk/clinical-radiology/being-consultant/rcr-referral-guidelines/about-irefer [accessed 13 September 2020].

Sedentexct (2020). "Cone Beam CT". Online Available from: http://www.sedentexct.eu/index.htm [accessed 13 September 2020].

Shrimpton PC, Jansen JT and Harrison JD (2016). "Updated Estimates of Typical Effective Doses for Common CT Examinations in the UK Following the 2011 National Review". *Brit J Radiol* **89**(1057): 20150346.

Stecker MS, et al. (2009). "Guidelines for Patient Radiation Dose Management". *J Vasc Interv Radiol* **20**: S263–S273.

Waites E and Drage N (2013). *Essentials of Dental Radiography and Radiology*, 5th edition. Churchill Livingstone eBook.

Wong KM, Tan BS, Taneja M, et al. (2011). "Cone Beam Computed Tomography for Vascular Interventional Radiology Procedures: Early Experience". *Ann Acad Med Singapore* **40**(7): 308–314.

5 Nuclear Medicine Imaging and Therapy

David Towey, Lisa Rowley and Debbie Peet

CONTENTS

5.1 INTRODUCTION

Nuclear Medicine involves the diagnostic and therapeutic use of radioactive materials. In the NHS, approximately 0.6 million Nuclear Medicine diagnostic tests, including positron emission tomography (PET)/computed tomography (CT), are conducted in England each year (NHS England 2019). Historically, scientists were encouraged to specialise in either Nuclear Medicine or Diagnostic Radiology using X-rays. However, with increasing adoption of hybrid imaging techniques, the boundary between disciplines has become blurred. Within the Clinical Scientist training programme, this specialism is referred to more generally as "Imaging with Ionising radiation". This chapter describes the role of the Clinical Scientist working in Nuclear Medicine.

In Nuclear Medicine (sometimes also called molecular imaging or molecular radiotherapy), ionising radiation is administered in the form of a radionuclide, which is chemically tagged to a pharmaceutical. For diagnostic imaging in Nuclear Medicine, this radiopharmaceutical is usually administered intravenously (injected into a vein) but sometimes orally, in food or drink, or inhaled as a gas. Depending on how each pharmaceutical interacts with the body, it can be absorbed or concentrated within an organ or target tissue of interest (e.g. a cancer tumour or cardiac muscle) and the tagged radionuclide is then detected using an external camera to form an image. Gamma rays emitted by the radionuclide pass out of the body and are collected using a gamma camera. An example of an image obtained following intravenous injection of a radiopharmaceutical that has uptake in the body in areas of bone growth is provided in Figure 5.1. There will be high uptake of the radiopharmaceutical where bones are growing or healing, e.g. in children, fractures, or metabolically active areas, such as cancerous disease. Imaging reveals these areas of high uptake as hot spots where the pharmaceutical, and therefore radiation, have localised within the image. This information can then be used by the patient's medical team to diagnose and better understand the stage of disease spread – such as metastatic spread of cancer, to assess the patient's suitability for treatment. Alternatively, samples of blood or urine can be analysed using a gamma counter to assess the concentration of radioactive material as an indicator of organ function. Alpha and beta-emitting radiopharmaceuticals may also be used for therapeutic Nuclear Medicine applications, such as the treatment of hyperthyroidism or localised radiotherapy treatment of cancer.

Scientists working in this specialism will have studied nuclear physics, radiation interactions, and imaging physics, including CT technology in the context of hybrid single-photon emission CT (SPECT) and PET-CT imaging. They will be experts in the equipment used for detection of radiation and radiation safety. They will also be familiar with relevant human anatomy and physiology.

5.2 PATIENT FLOW THROUGH NUCLEAR MEDICINE

Requests for Nuclear Medicine scans and procedures are received from referring healthcare professionals and then need to be independently authorised by a clinician with an IR(ME)R Practitioner license, awarded by the Administration of Radioactive Substances Advisory Committee (ARSAC). For a summary of key professional groups that Clinical Scientists in Nuclear Medicine work with, including their official roles under IR(ME)R, please see Table 5.1.

The variety of tests performed, and the need to consider the timing of administration of radionuclides for later imaging or sample collection, means that scheduling of patients in Nuclear Medicine is logistically complicated. When the patient arrives in the department, their details are

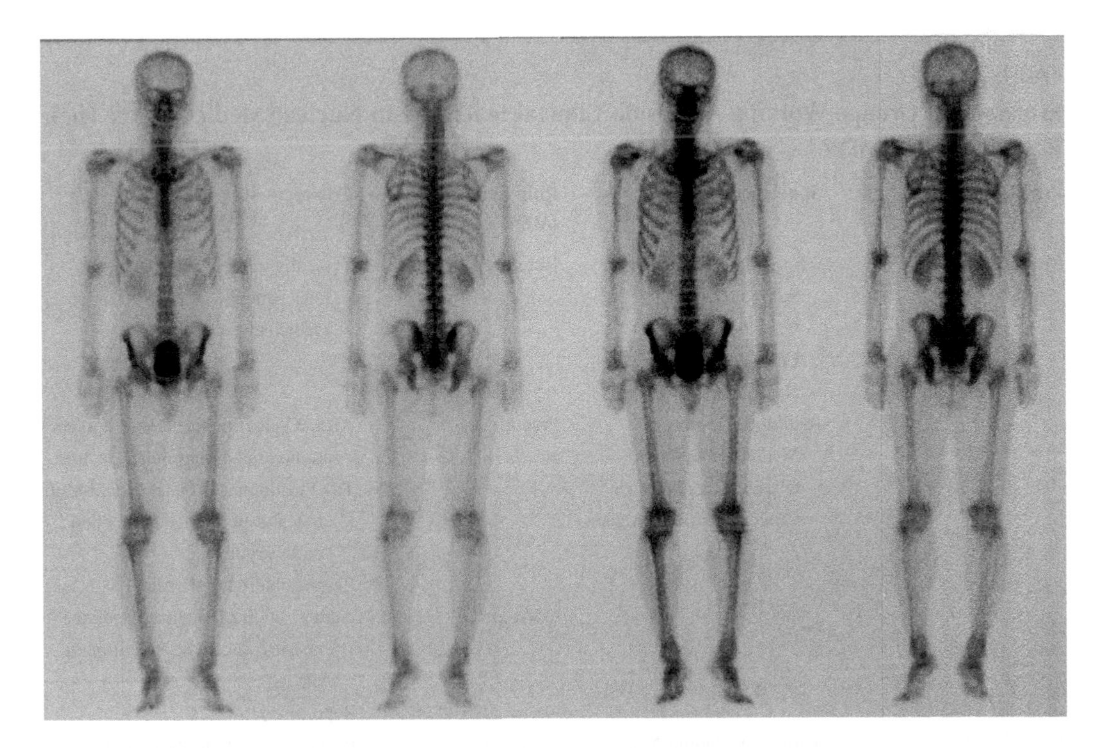

FIGURE 5.1 Nuclear Medicine bone scan showing uptake of radionuclide in the bones.

carefully checked, including ID and medical history, to confirm they are the correct patient (Figure 5.2). They are then administered with radioactive material. Behind the scenes, Scientists are involved in preparing the radiopharmaceutical and ensuring it is administered safely. Generally, after a period of uptake of the radiopharmaceutical, imaging takes place on a gamma camera or PET scanner. Imaging protocols and settings of the scanner are a key responsibility for Scientists in the department.

In Nuclear Medicine, images often need processing using specialised software before being interpreted and reported by a Radiologist or Nuclear Medicine Physician. Scientists are involved with the processing and analysis of data, and for some tests are also responsible for reporting the results.

Clinical Scientists are also involved in setting up local IT networks to enable PET/CT scans obtained from different locations in the hospital to be accessed from the Nuclear Medicine department. Analysed images are then stored within the hospital's "Picture Archiving and Communications Systems" (PACS), as described in Chapter 1.

Once the equipment has been installed, the layout of Nuclear Medicine facilities is rarely altered. For new installations, Scientists will be involved in deciding the shielding design and layout of new facilities.

5.3 THE PROFESSIONAL ROLE OF A CLINICAL SCIENTIST IN NUCLEAR MEDICINE

Nuclear Medicine is a multi-disciplinary service, including Radiopharmacists, Healthcare Practitioners, Clinical Technologists and Radiographers, Nurses, Play Specialists (for paediatric patients), Administrative Staff, Radiologists and Clinical Scientists. These roles are summarised in Table 5.1. Clinical Scientists work closely with all Nuclear Medicine staff, as well as referring medical teams, ward staff, domestic staff, Trust Health and Safety boards, managers and the patients themselves. Clinical Scientists set QC schedules for equipment

TABLE 5.1
Professional Groups Working Alongside Clinical Scientists in Nuclear Medicine and Their Roles Under IR(ME)R

Profession	Specialism	Role Under IR(ME)R	Comment
Medical Doctors	All specialisms including: Surgeon, Oncologist, Cardiologist, Neurologist	Referrer	Typically, only specialist doctors refer patients for Nuclear Medicine procedures.
	Radiologist, Nuclear Medicine Physician, Clinical Oncologist or Clinical Endocrinologist	Practitioner	ARSAC practitioner license holders are normally Consultant Doctors. The Practitioner is responsible for all clinical aspects of radionuclide administration. Responsible for reporting.
		Operator	Authorisation of Nuclear Medicine exposure under the Practitioners ARSAC Reporting
Nursing staff	Various, including Nuclear Medicine, Cardiology and Play Specialists	Operator	Patient care, administration of imaging contrast agents and radiopharmaceuticals
Radiopharmacy staff	Various	Operator	Operators are responsible for the production of pharmaceuticals. This impacts the patient's exposure. Operators also have specific roles under manufacturing regulations.
Healthcare Science Practitioners Clinical Technologists/ Radiographers		Operators	QC, administration, imaging, blood samples, and image processing
Clinical Scientists	Nuclear Medicine, Diagnostic Radiology or Radiotherapy	Operators, and Medical Physics Experts (MPEs)	QC, calibration, protocol setup, optimisation of images, image processing, reporting
Administration/Booking Clerks			Administrative support to all other roles.
Equipment Manufacturer/ Service Engineers		Operators	Since they can directly influence the exposure of a patient, service engineers are listed as Operators.
Employer (e.g. the Chief Executive of a hospital)		Employer	Overall responsibility for ensuring the regulations are followed.

checks, advise Radiologists and Physicians on clinical protocols, and provide training in radiation protection, including advice regarding overexposures and incident reporting as shown alongside the diagram showing the patient pathway with steps that do not involve radioactive material in blue and those that do in pink(Figure 5.2). Clinical Scientists also support research

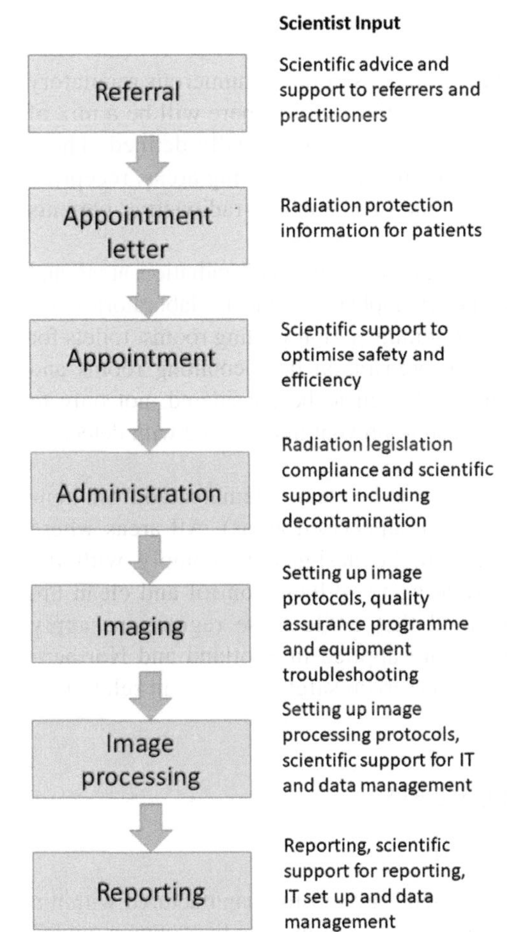

FIGURE 5.2 Patient flow through a Nuclear Medicine department requires input from Clinical Scientists at various stages.

by providing expert advice for research trials as well as leading service innovation within their specialism.

5.4 REFERRAL GUIDELINES

Referral guidelines for Nuclear Medicine procedures and guidelines for clinical protocols are available from the British Nuclear Medicine Society (BNMS) on their website (https://www.bnms.org.uk/; BNMS 2020). PET-CT referral guidelines BFCR(16)3 (2016) are available from the Royal College of Radiologists. There are also clinical guidelines available on the European Associate of Nuclear Medicine (EANM) website (https://www.eanm.org/; EANM 2020a) and some available from the Society of Nuclear Medicine and Molecular Imaging (SNMMI) website (http://www.snmmi.org/) (SNMMI 2020). US organisations generally allow much higher administered activities than European countries, including the UK. The Royal College of Radiologists also have guidelines covering Nuclear Medicine procedures on their "irefer" system (RCR 2019).

Processes and procedures are well documented in Nuclear Medicine departments. These are often managed within a Quality Management System (QMS). Clinical Scientists, and specifically Medical Physics Experts (MPEs), ensure that protocols are in line with national guidance, particularly in relation to diagnostic reference levels (DRLs) and other aspects of dosimetry described in the ARSAC guidance notes (ARSAC 2020).

5.5 FACILITY DESIGN

The design of a Nuclear Medicine facility needs to take into consideration numerous regulatory requirements as well as allowing good clinical working practices. Since there will be a mix of patients, visitors and staff entering the department, various zones are usually defined. These zones include the following non-radioactive areas: pre-administration waiting areas, reception and administrative offices, stores, toilets and changing facilities for non-radioactive patients and visitors.

Other zones are typically designated as "supervised" and/or "controlled" radiation areas and may include storage areas for radioactive materials and radiopharmaceuticals, laboratories, injection rooms, cardiac stressing rooms, uptake areas, post-administration waiting rooms, toilets for radioactive patients, scanning rooms, radioactive sample preparation labs, counting rooms and stores for radioactive waste. Shielding of these rooms will need to be considered, not only to reduce the radiation exposure of staff and the public but also to prevent interference with detection equipment, such as very sensitive radiation counters.

Radiation areas need appropriate local rules in place and to be clearly demarcated from non-radiation areas (see Section 7.7 regarding controlled and supervised areas). All areas where radioactive materials are stored or manipulated need to be designed to comply with the Environmental Permitting Regulations (EPR 2016) relating to spillage control and clean up, and protection from fire and flood. It should be noted that while these regulations apply specifically to England and Wales the same principles are applied in Scotland and Northern Ireland. Departments may also need input from counter terrorism safety advisors in relation to source security.

5.6 PRODUCTION OF RADIOPHARMACEUTICALS

5.6.1 Radiopharmacy Management and Regulations

The radioactive agents (radiopharmaceuticals) given to patients are mostly manufactured within a hospital Radiopharmacy – an aseptic unit conforming to standards outlined by the "orange guide" published by the Medicines and Health Regulatory Agency (MHRA 2017). The Chief Pharmacist within the hospital usually has responsibility for the safety of products manufactured by the Radiopharmacy but may not be involved with the day-to-day running of the unit, which is often led by a Production Manager. The UK Radiopharmacy Group (UKRPG 2020) provides guidance on the manufacture of radiopharmaceuticals for administration to patients, including the provision of radiopharmacy services (UKRPG 2017), proper aseptic processes, radiopharmacy design, product QC and release (UKRPG 2016).

The Radiopharmacy must either have a license to operate issued under the Human Medicines Regulations (2012) or work under a "section 10 exemption" under the Medicines Act. Licensed Radiopharmacies must work to Good Manufacturing Practice (GMP) guidelines (EC 2010; EudraLex 2011), detailed in the Human Medicines Regulations (2012) and outlined in "the orange guide" (MHRA 2017). The license itself requires named people acting as, Production Manager, the person responsible for QC, and the point of contact for the license.

5.6.2 Technetium-99m Generators and Kits

Technetium-99m (^{99m}Tc) is the main radionuclide used in Nuclear Medicine. This isotope is low cost, easy to produce, primarily a gamma emitter (generating photons within an appropriate energy range), and has a suitable half-life. The emission energy (140 keV) provides a good compromise between being high enough to escape the body, which limits absorption, but still low enough to be easily detected by scintillation crystals. ^{99m}Tc is eluted from a molybdenum-99 (^{99}Mo) generator (Figure 5.3) in the form of pertechnetate. The generator contains an aluminium column, with

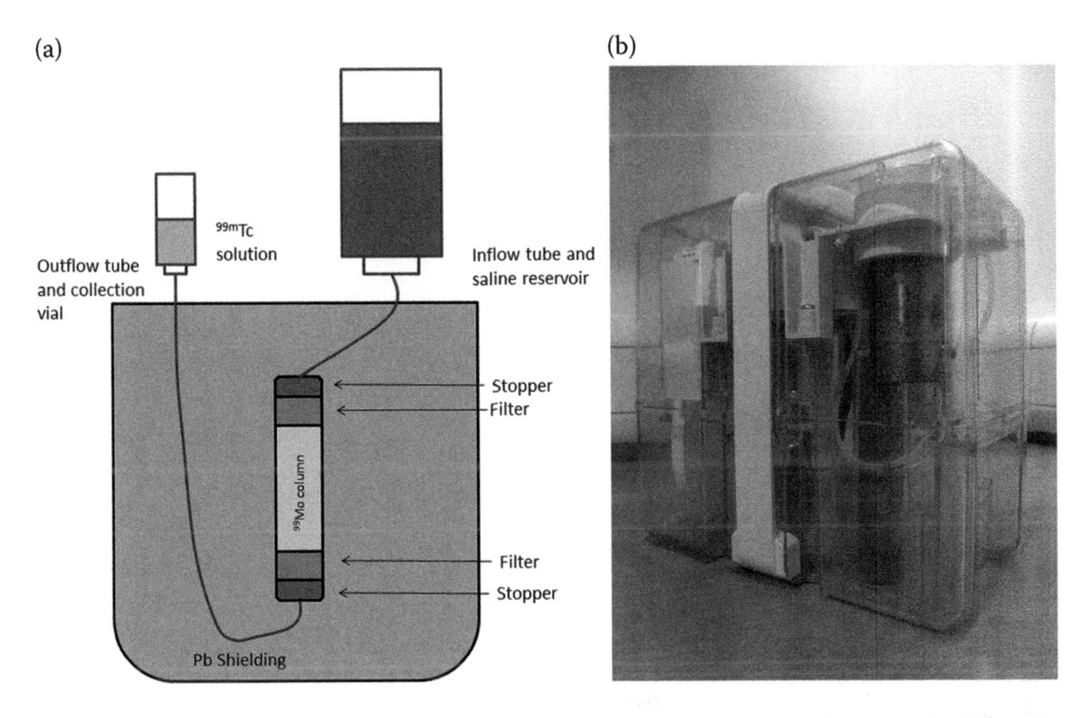

FIGURE 5.3 (a) Schematic of the Mo-99/Tc-99m generator showing the saline reservoir. When the evacuated Tc-99m vial is placed on the outport of the generator, the saline is drawn across the column and the daughter Tc-99m dissolves in the saline and is collected in the vial. (b) Photograph of a generator with the molybdenum column shown behind the lead shield used to shield the Tc-99m vial.

absorbed ^{99}Mo. An evacuated vial is placed at one end of the generator fed by a saline solution at the other. Saline is forced through the column, reacting with the ^{99m}Tc to form sodium pertechnetate, which is then collected in the evacuated vial. This is known as eluting the generator, with the resultant pertechnetate being the eluate.

The ^{99}Mo decays with a half life of 66 hours to ^{99m}Tc and so without elution, the amount of ^{99m}Tc within the generator is in dynamic equilibrium. In normal use, the generator is eluted daily. The ^{99}Mo is usually replaced on a weekly or fortnightly basis as the eluted activity falls (Figure 5.4). Some ^{99}Mo is found in the eluate, and Quality Control (QC) checks are required to ensure there is minimal "breakthrough" of the ^{99}Mo and any aluminium. These checks are usually performed on the first elution of a generator, or if the generator is moved.

^{99}Mo is associated with high-energy Gamma emissions (740 keV and 778 keV). Breakthrough of the eluate (or "radionuclide purity") is determined by measuring the eluate within a dose calibrator "with and without" a lead pot. Lead significantly attenuates the ^{99m}Tc gamma emissions but will have less effect on emissions from ^{99}Mo. The percentage breakthrough is calculated from these measurements, with a tolerance of 0.1% (1 kBq ^{99}Mo per MBq ^{99m}Tc).

Aluminium breakthrough (chemical purity) is measured by combining a drop of the eluate and a drop of known aluminium solvent on test paper. The resultant colour of the paper provides an indication of the aluminium concentration, with a typical pass rate of under 5 μg/ml aluminium.

Once the pertechnetate has been eluted, radiopharmaceuticals can then be made by mixing the pertechnetate with pre-prepared freeze-dried kits. Some kits require multiple steps, or special conditions to allow the radio-labelling process to proceed. MHRA licensed products should always be preferred over unlicensed products. If unlicensed products are to be used, the referrer must be informed, with appropriate QC carried out prior to their release from the radiopharmacy. The MHRA holds copies of the "Summary of Product Characteristics" (SPCs) that are descriptions of

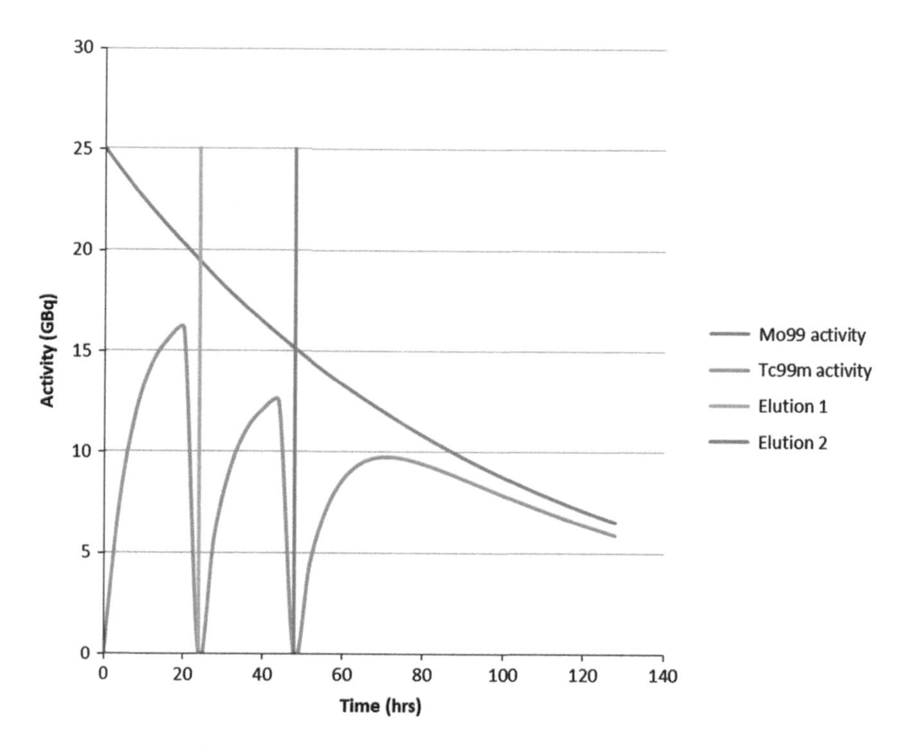

FIGURE 5.4 Activity of ^{99}Mo and ^{99m}Tc in a generator eluted over time.

medicinal products including the kits used to make radiopharmaceuticals. The SPCs are a useful source of information listing the licensed applications of radiopharmaceuticals, their pharmacokinetic and pharmacodynamics characteristics, and expected doses for standard-sized patients. Information on QC is documented in the UKRPG guidance document "Quality Assurance of Radiopharmaceuticals" (UKRPG 2016), with tests including radionuclide purity, radiochemical purity, particle size (for particulate radiopharmaceuticals) and sterility.

The manufacturing process takes place within units that conform to GMP. This includes measures to ensure that all manipulations are performed in an aseptic environment; within an isolator or laminar airflow cabinet. To reduce the risk of microbial ingress, there are strict requirements for passing items into the Radiopharmacy, handwashing, and changing into sterile clothes and shoes. The rooms within the Radiopharmacy are kept at a positive pressure, with isolators kept at negative pressures to ensure that any radiation contamination is contained within the cabinet. A summary of the key information sources relating to radiopharamceutical production is given in Table 5.2.

The Radiopharmacy, and Nuclear Medicine as a whole, must also ensure compliance with The Health and Safety (Sharp Instruments in Healthcare) Regulations (Health and Safety Sharp Instruments in Healthcare Regulations 2013). There must be consideration of risks, and a risk assessment in place for re-sheathing needles. Re-sheathing of needles is required, as during manufacture and administration of radiopharmaceuticals, the activity (amount of radiation) must be measured prior to administration to ensure the patient is receiving the intended radiation dose.

Clinical Scientists specialising in Radiopharmacy can manage or take the quality role under the MHRA license. Scientists working in Nuclear Medicine can also take one of these roles once they have enough relevant experience, but more commonly will work with Radiopharmacy staff, sometimes producing radiopharmaceuticals, or more commonly, being involved in QC checks of kits. All staff must adhere to strict infection control procedures to ensure that the

TABLE 5.2
Key Information Sources Related to Radiopharmaceutical Production

Reference	Description
Quality Assurance of Aseptic Preparation Services: Standards Handbook (RPS 2016)	Chapter 7 in this document from the Royal Pharmaceutical Society describes the requirements for facilities and equipment
Good Manufacturing Practice (GMP) Guidelines (EC 2010)	The European Commission provides a useful chapter in Part I – Chapter 2 on premises and equipment.
The MHRA Orange Guide: Rules and Guidance for Pharmaceutical Manufacturers and Distributions (MHRA 2017)	This is a very important document for production of radiopharmaceuticals – laying down all the requirements in the UJ
Sampson's Textbook of Radiopharmacy (Theobold 2010)	This book provides a helpful description of the requirements
Guidance on Current Good Radiopharmacy Practice (cGRPP) for the Small-Scale Preparation of Radiopharmaceuticals (EANM 2007)	This article on the EANM website is another useful resource
UKRPG Guidance Notes (UKRG 2020)	Contains links to useful documents relating to radiopharmacies
Operational Guidance on Hospital Radiopharmacy (IAEA 2008)	The IAEA has produced a section on the requirements for facilities in Chapter 5 of this resource
Health Building Note 14-01: Pharmacy and Radiopharmacy Facilities (HBN 2018)	Health Building Notes are used by architects and designers in the UK for all health facilities
Annex 1 Guidelines on Good Manufacturing Practices for Radiopharmaceutical Products (WHO 2003)	The World Health Organisation has guidelines that are used across the world
European Commission Pharmaceutical Guidelines for Medicinal Products for Human Use (EC 2010)	European legislation is described in the EC document

resultant medicinal product is not an infection risk. Other QC checks involve recording key environmental parameters within the Radiopharmacy as some pharmaceuticals need to be stored under specific conditions.

5.6.3 Radionuclide Calibrators

Radionuclide calibrators are used to measure radioactivity during the preparation of radiopharmaceutical kits and prior to administration to patients. Clinical Scientists play an important role in the quality assurance of dose calibrators used to measure the actual activity of radioactive material administered to patients. Dose calibrators usually include a gas-filled detector capable of measuring gamma emitters with activities ranging from 1 MBq up to 100's of GBq. The National Physics Laboratory (NPL 2006) good practice guide gives an overview of their properties. The guide sets out the traceability of measurements back to national primary standards and gives guidance on QA frameworks for QC testing.

Daily checks of the calibrator are performed using a long-lived source, such as ^{137}Cs, to confirm that equipment readings are stable and consistent over time, prior to clinical use. A background check is also performed to ensure that there is no contamination, or high activity sources in the vicinity, that may affect any readings. Calibrators have inbuilt QC programmes for daily QC. More detailed checks of the calibrator are carried out on a monthly and annual basis, as per NPL (2006). An example test is described in the following case study.

CASE STUDY: DOSE CALIBRATOR LINEARITY MEASUREMENT

The accuracy of a dose calibrator's reading (measured in Bq) is checked against a secondary standard supplied by a national standards laboratory (e.g. the UK National Physics Laboratory) to ensure appropriate calibration factors are used. However, this accuracy is normally only checked for a single activity; hence, the linearity of response must also be checked (Figure 5.5).

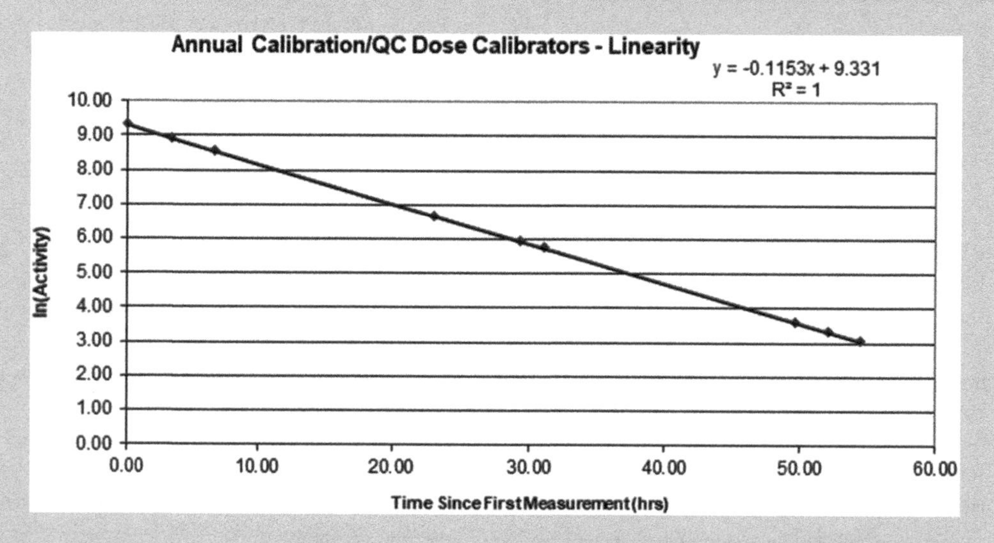

FIGURE 5.5 Logarithmic plot of activity measured against time.

The linearity of a dose calibrator's response is important as the calibrator will be used to measure radioactivity across a large range of activities used clinically. One way of assessing this response is to measure a single source at many time points as it decays. In this example, a ^{99m}Tc source was measured nine times over 3 days (Table 5.3). Since the half-life of ^{99m}Tc is known, the expected activity can be calculated by extrapolating back from the low activity measurements. The error from the expected reading can then be calculated.

TABLE 5.3
Activity of the Same ^{99m}Tc Source Measured Over Time in a Dose Calibrator

Date and Time	Time Difference	Activity Reading (MBq)	Expected Activity (MBq)	% Difference
27/01/2020 08:41	0.00	11310.00	11282.41	0.24
27/01/2020 12:07	3.43	7590.00	7594.17	−0.05
27/01/2020 15:21	6.67	5210.00	5230.87	−0.40
28/01/2020 07:38	22.95	801.00	800.20	0.10
28/01/2020 13:59	29.30	385.00	384.79	0.05
28/01/2020 15:45	31.07	315.00	313.88	0.36
29/01/2020 10:17	49.60	36.90	37.04	−0.39
29/01/2020 12:46	52.08	27.70	27.82	−0.43

The tolerance between expected and measured activity is 1% for a reference instrument, and 5% for a field instrument (NPL 2006).

5.7 ADMINISTRATION OF RADIOPHARMACEUTICALS

Intravenous administration of radiopharmaceuticals is generally carried out by Healthcare Science Practitioners (Clinical Technologists), Radiographers or nurses rather than Clinical Scientists, but the Clinical Scientist may verify (or "double-check") the activity of the radiopharmaceutical prior to administration. Verifying the dose is a precautionary measure to help eliminate the possibility of incorrect administration resulting in an incorrect dose being delivered to the patient.

Clinical Scientists frequently administer therapy agents such as Iodine-131 (^{131}I). Clinical Scientists often oversee the introduction of new processes and services. Once a process has been embedded into routine service, these duties are then carried out by Healthcare Science practitioners or nurses. In the event of an incident, a Clinical Scientist will be involved to investigate the incident and advise on decontamination.

5.8 RADIONUCLIDE IMAGING EQUIPMENT

5.8.1 Gamma Cameras

Clinical Scientists specialising in Nuclear Medicine are expected to have a detailed understanding of the design and function of the gamma camera used to detect emitted gamma photons.

The basic components of a gamma camera are labelled in Figure 5.6. Most modern gamma cameras use scintillation crystals (Sodium Iodide) and photomultipliers tubes to detect gamma photons to produce an image. Some newer designs are based on solid-state detectors. As with diagnostic X-ray imaging, Clinical Scientists need to work with equipment manufacturers, Radiologists, Healthcare Science Practitioners, Technologists and Radiographers, to optimise the sensitivity and resolution of images for different clinical applications. The basic components of an Anger gamma camera are outlined in Figure 5.7; however, the reader is referred to Lawson (2013) for a more comprehensive explanation.

Collimators: These are used to accept photons that are perpendicular to the collimator face. There is a trade-off between resolution and sensitivity; therefore, it is important to ensure that the correct collimator is in use for the intended scan. For example, bone scans require high resolution, whereas renal scans for drainage require high sensitivity. The thickness of the lead septa can be adapted to accept photons of the desired energy range.

Scintillation Crystal: It is a Sodium Iodide crystal, doped with thallium. Incident gamma photons excite electrons in the crystal, causing scintillation. Scintillation photons are then

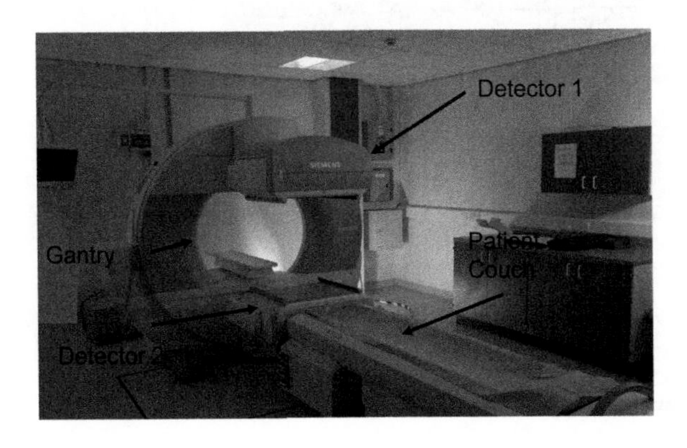

FIGURE 5.6 A gamma camera.

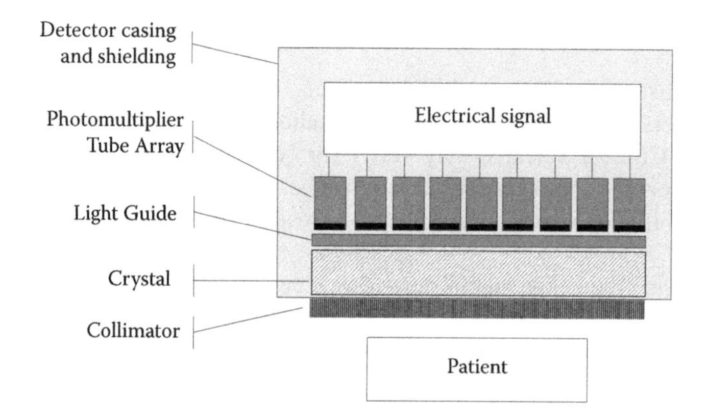

FIGURE 5.7 Gamma camera schematic including details of component parts.

detected and amplified by photomultiplier tubes (PMTs). Solid-state cameras are becoming increasingly common but are generally more expensive than scintillation detectors.

PMTs: Light produced in the scintillation crystal is incident on the photocathode of the PMT, producing photoelectrons. These are then accelerated along the PMT, through a series of dynodes, amplifying the original signal and converting the incident photon into an electrical signal. A PMT is shown in Figure 5.8.

Electronics: The signal from the PMT describes the *x* and *y* coordinates of the signal, with an intensity value (*z*), which is proportional to the incident energy. The *z*-signal is passed through a pulse height analyser, to accept energies within a defined range, i.e. set at 140 keV for ^{99m}Tc. This helps to reduce the scatter accepted, as scatter will add noise to the resultant image.

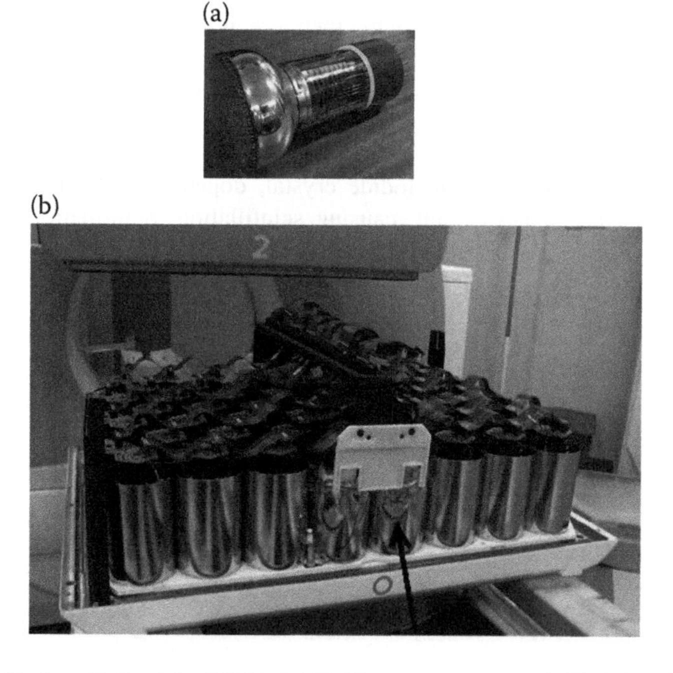

FIGURE 5.8 (a) Photomultiplier tube (PMT); (b) PMT array on a camera behind the scintillation crystal.

Image correction and storage: The signals from the detector electronics are further processed to correct for image artefacts, including energy, linearity and uniformity corrections. Data are then stored as one or more image matrices.

Acquiring static images with a gamma camera requires the following parameters to be defined: collimator choice, energy acceptance window(s), image matrix size, zoom and stop conditions (either time or count-based). Changing these settings will affect the image resolution, image noise and acquisition time. The patient/phantom position relative to the scanner and the activity being imaged will also impact image quality.

5.8.1.1 Image Parameters and Image Quality

The Williams phantom is often used to compare the effects of changing gamma camera acquisition parameters on the resultant image. This phantom is constructed of four layers of Perspex, as shown in Figure 5.9(a). The top and bottom layers are uniform sheets. The second layer has a D-shaped cut-out, within which are four solid discs of Perspex. The third layer is solid except for four disc-shaped holes. When the resulting empty cavity is filled with a radioactive solution, this allows visualisation of "hot" and "cold" circles of differing diameter against a "warm" background, as shown in Figure 5.9(b).

Figure 5.10 demonstrates how choice of collimator impacts the resulting image.

Figure 5.11 demonstrates the trade-off between resolution and sensitivity with varying matrix dimensions. The larger the matrix, the smaller the pixel size, and the better the resolution, but the counts per pixel will also reduce.

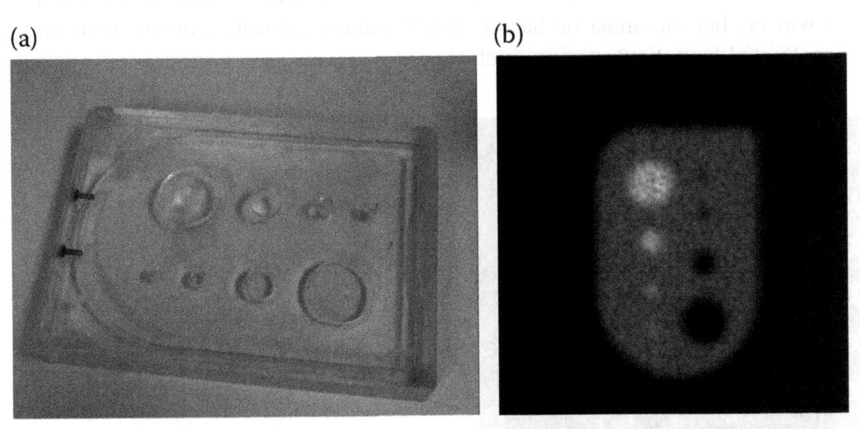

FIGURE 5.9 (a) The Williams phantom; (b) Williams phantom, acquired with Low Energy High Resolution (LEHR) collimator, 256 × 256 matrix, 1.0 zoom, 500,000 counts using ^{99m}Tc on a Siemens gamma camera.

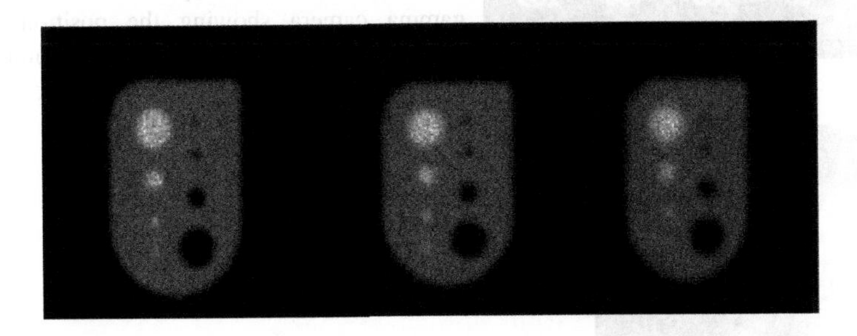

FIGURE 5.10 Effect of changing collimator on the Williams phantom. (Left to right) Low Energy High Resolution (LEHR), Low Energy All (General) Purpose (LEAP), Medium Energy. All images acquired with the same parameters except collimator.

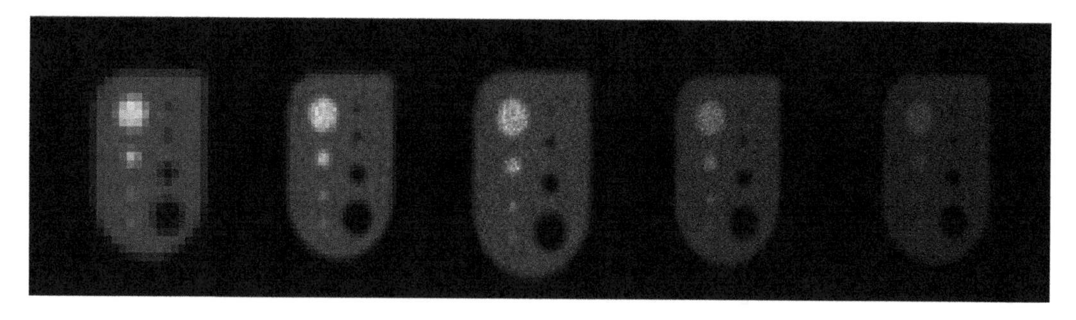

FIGURE 5.11 (Left to right) Matrix sizes of 64 × 64, 128 × 128, 256 × 256, 512 × 512 and 1024 × 1024, with all other parameters held constant. Resolution increases with larger matrix size, but this is accompanied by a reduction in contrast due to the lower number of counts per pixel.

The gamma camera will detect photons with the selected energy, plus or minus a percentage of that energy, e.g. for ^{99m}Tc, 140 keV ± 15%. If the wrong photopeak is chosen, there may be more noise in the image, and it will take longer to acquire an adequate number of counts (Figures 5.12 and 5.13).

As the distance from the camera is increased (Figure 5.14, left to right), the resolution also decreases, due to the wider solid angle.

The zoom function alters both the size and resolution of the image (Figure 5.15).

Increasing the number of counts contributing to the image reduces noise but requires longer acquisition times (Figure 5.16). Clinically, this means the longer the patient is scanned, the better the images will be, but this must be balanced with patient comfort. Longer scans are also more likely to be affected by patient motion artefacts.

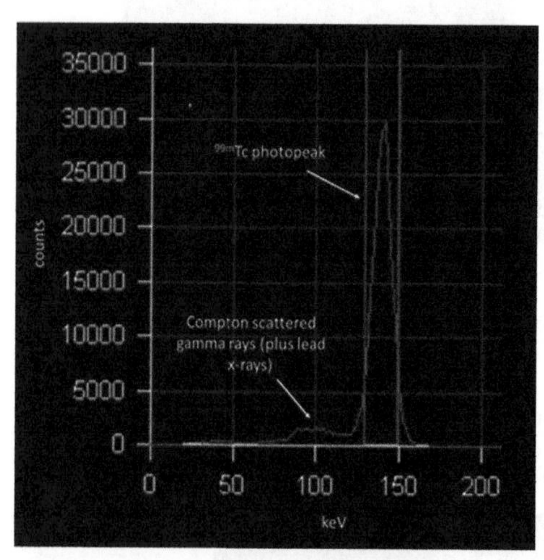

FIGURE 5.12 ^{99m}Tc Spectra obtained from a Siemens gamma camera showing the position of the energy acceptance window centred on the photopeak.

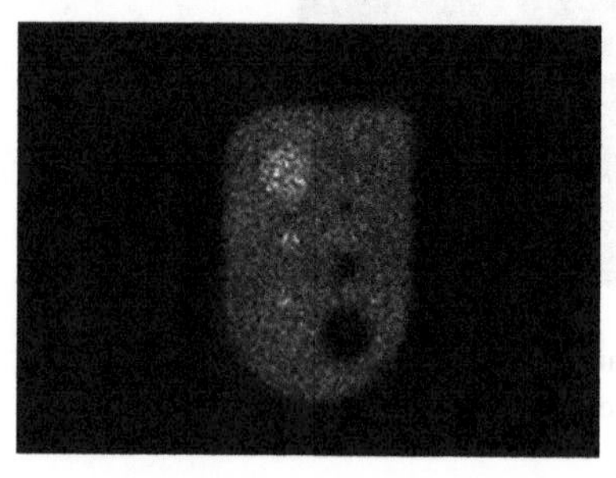

FIGURE 5.13 ^{123}I photopeak for a ^{99m}Tc-filled Williams phantom.

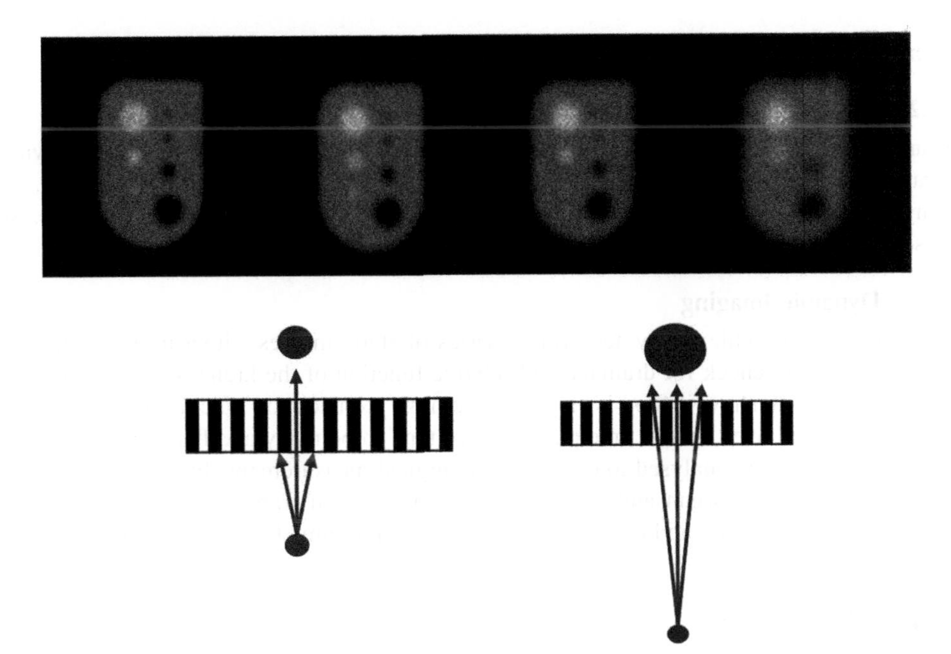

FIGURE 5.14 The distance from the camera impacts image resolution.

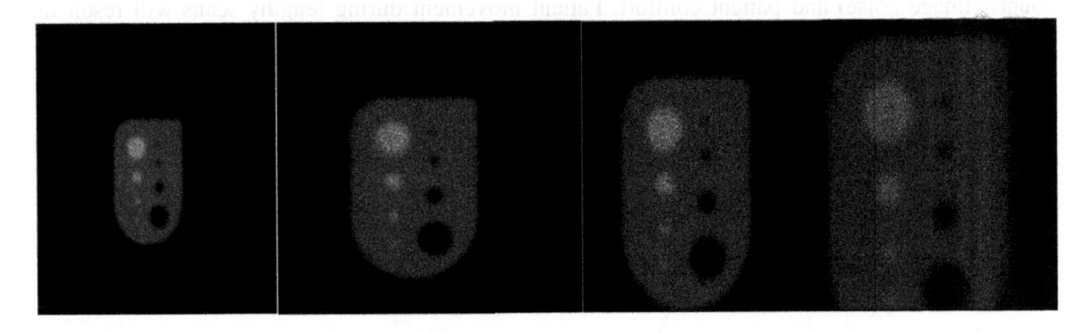

FIGURE 5.15 Illustration of the impact of zoom settings on the Williams liver phantom.

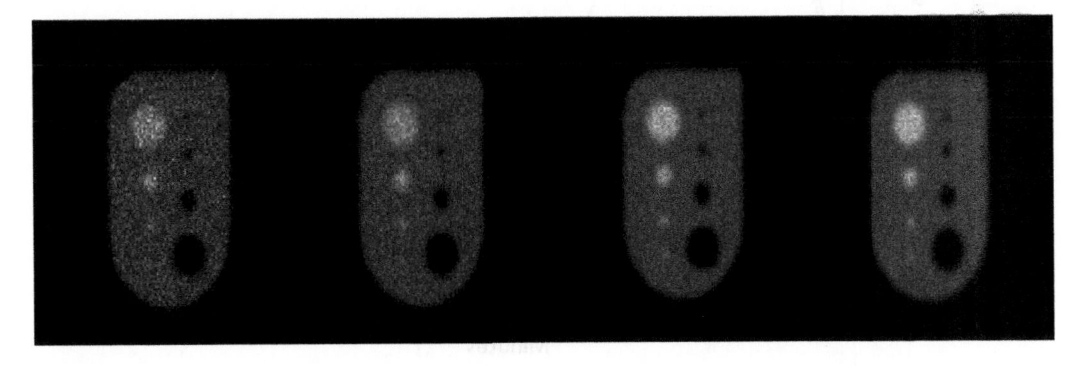

FIGURE 5.16 The impact of increasing the number of counts is demonstrated from left to right.

There are four basic types of image acquisition performed with gamma cameras.

5.8.1.2 Static (Planar) Gamma Camera Imaging

The camera is focused on the area of interest. For example, an image might be obtained with the camera focused only on the kidneys, or on the entire body for a whole-body bone scan. Static imaging is used where a high number of counts is required to see detail within the image. Static images are usually 5–20 minutes in duration.

5.8.1.3 Dynamic Imaging

Dynamic imaging is achieved by acquiring a series of static images, either in quick succession (e.g. renal images to check for drainage and relative function of the kidneys), or over time (e.g. gastric emptying – to show if food is emptying from the stomach too quickly or too slowly). As images are acquired over a short time period, these tend to be noisy and fine detail may be lost. Dynamic images can be analysed to extract physiological measurements by placing a Region of Interest (ROI) over the organ and plotting the counts in each frame over time. A frequently carried out test for children and cancer patients is the "renogram" kidney function test shown in Figure 5.17.

5.8.1.4 SPECT Imaging

This provides 3D images by acquiring a series of static images as the detector heads rotate around an area of interest (Figure 5.18). Images are then reconstructed to provide a 3D dataset. As images are acquired at several time points, a compromise needs to be reached between the number of counts (image noise) and patient comfort. Patient movement during lengthy scans will result in image artefacts. Imaging times are usually around 30 minutes, up to a maximum of 1 hour for patients who can tolerate a longer scan duration.

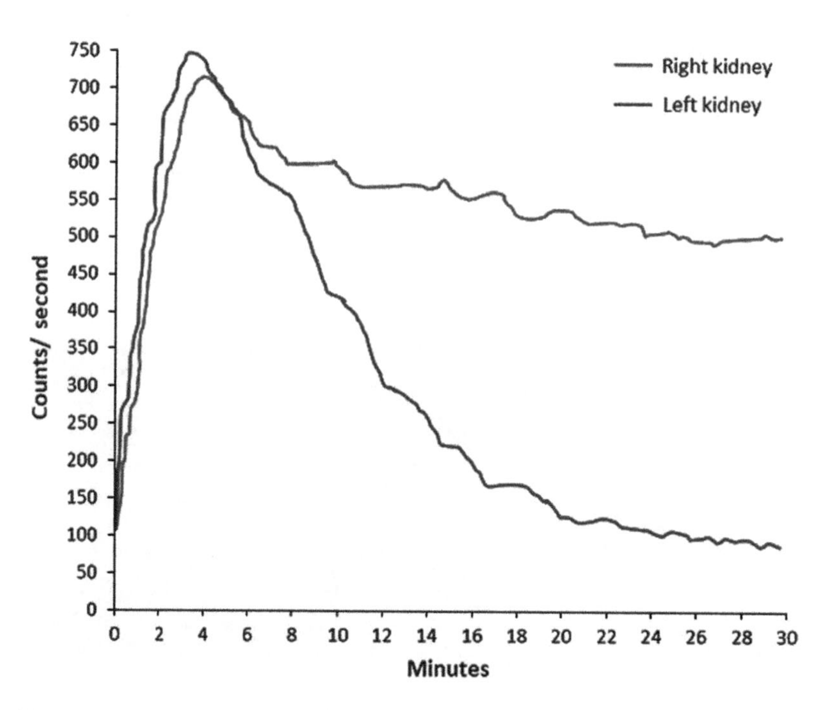

FIGURE 5.17 A renogram showing good clearance from the left kidney but much slower clearance from the right kidney.

FIGURE 5.18 Detectors rotate around the patient to generate a 3D image.

5.8.1.5 Gated Imaging

Images, in either planar or SPECT form, can be obtained by performing an ECG during scanning and then separated out and combined images for different phases of the cardiac cycle to form a composite dynamic image. This type of scan is used to measure motion and blood flow within the heart.

5.8.2 PET-CT

PET imaging utilises the properties of positron-emitting radionuclides. A commonly used radionuclide for PET is Fluorine-18 (^{18}F), which has a half-life of 110 minutes. ^{18}F is produced in a cyclotron (see IPEM (2011b) and IAEA (2012) for more information). A detailed explanation of PET imaging can be found in "PET Physics, Instrumentation and Scanners" (Phelps 2006) or "Physics in Nuclear Medicine" (Cherry et al. 2012). Positrons travel a short distance within the body before encountering an electron and being annihilated. As a result of the annihilation process, energy is released as a pair of two 511 keV photons moving in opposite directions. These photons are then detected by a ring of detectors surrounding the patient (Figure 5.19). To be accepted, both photons must be detected at the expected energy and within a specified narrow time frame. Any other photons will be assumed to originate from scatter and be rejected (Figure 5.20). Due to a combination of factors involved in PET imaging, staff doses are generally much higher and strict radiation protection measures need to be in place, such as the additional use of lead shielding.

The most common application for PET-CT within the UK is ^{18}F – fluorodeoxyglucose (FDG) imaging for cancer diagnosis, staging, management and assessing treatment response. FDG is a glucose analogue, taken up by metabolically active tumours. As the heart and brain are also highly

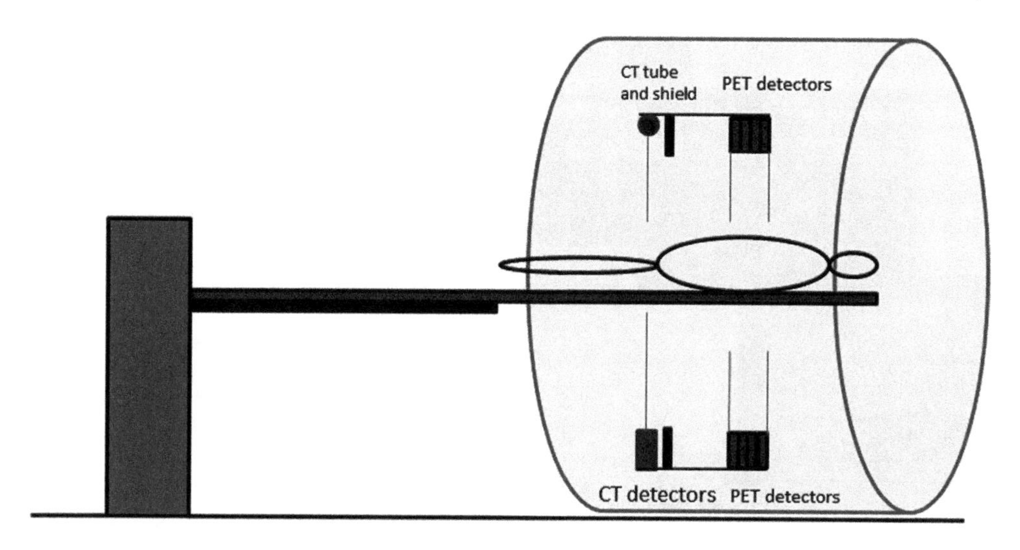

FIGURE 5.19 PET configuration showing the table that moves through the gantry to acquire CT images, resting in a number of positions to image the PET emissions.

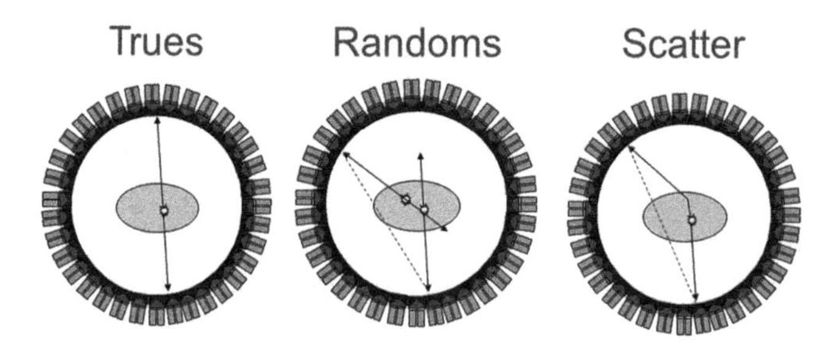

FIGURE 5.20 PET distinguishes photons from true single annihilation events from other scattered photons by analysing the timing and energy of the detected photons.

metabolically active, the patient is asked to rest in a warm room following administration of FDG to prevent brown fat and muscle uptake and enable metabolically active tumours to be clearly visualised.

PET services are commissioned by NHS England, with the private sector and NHS hospitals both eligible to submit bids. Therefore, some PET units are run by NHS providers, while the majority are run through private healthcare companies. At all centres, the local medical physics team MPE, Radiation Protection Advisor (RPA) and Radioactive Waste Advisors (RWAs) provide advice, and the IR(ME)R Practitioner licence holder and the MPE work together to optimise image protocols.

PET detectors typically comprise Bismuth or Lutetium-based scintillation crystals coupled to PMTs. Detectors are arranged in multiple rings circling the patient. The multiple rings allow the scanner to acquire data over a large volume of space (typically around 20 cm long). For whole-body scans, the patient is scanned in one position for a set amount of time, before the bed moves a set distance and another volume is acquired. These multiple "bed positions" are then combined during the image reconstruction stage.

Most modern scanners run in 3D mode. This allows coincidence events to be recorded between different detector rings, increasing the sensitivity in the central rings significantly. However, because of the geometry, the very edge of the outer rings can only run in 2D mode. This means the

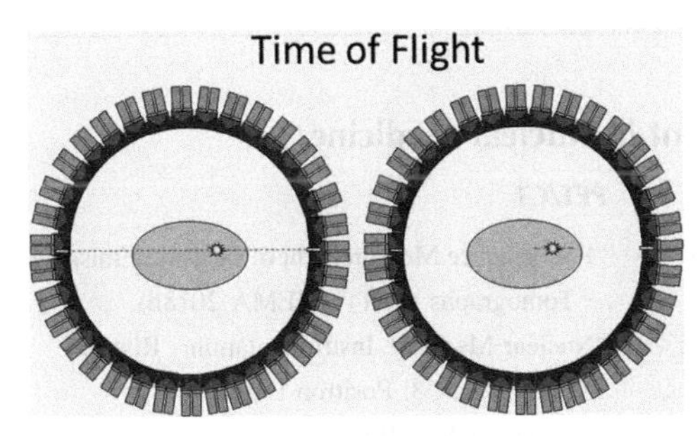

FIGURE 5.21 Illustration of Time of Flight (ToF) analysis within a PET scanner. In the right-hand image, the origin of the annihilation photons can be narrowed down to a small region.

sensitivity varies across the field of view. Because of this, scans normally involve some overlap between bed positions.

The detectors are normally arranged in blocks of detectors covering multiple rings. This modular arrangement means sections can be easily replaced if there are faults.

Time of Flight (ToF) information is used to estimate the origin of the annihilation event (Figure 5.21). Since the timing accuracy of the crystals and electronics can only localise each annihilation event to a few centimetres, analysis of the ToF does not improve image resolution, but will improve the noise characteristics of the resultant image by reducing the number of erroneously placed counts.

5.8.3 Hybrid Imaging

Hybrid imaging combines the gamma camera or PET scanner with CT or MR, linking functional imaging with anatomical detail. This can be used in the form of very low dose CT to correct for photon attenuation (which is important in PET imaging). Attenuation correction uses a look-up table to convert the CT image in Hounsfield units to an attenuation coefficient at the PET or SPECT imaging energy.

Low-to-medium dose CT can also provide localisation information for interpreting the PET or SPECT image, as well as the potential to use full diagnostic CT imaging (medium-to-high dose), potentially offering a "one stop shop".

5.9 EQUIPMENT MANAGEMENT OF GAMMA CAMERAS AND PET/CT SCANNERS

Clinical Scientists in Nuclear Medicine work closely with a range of equipment and are expected to test, adjust and calibrate gamma cameras and PET scanners, as well as identifying faults if issues arise. If faults are identified, this may require replacement parts and repair or calibration by a specialist service engineer.

Scientists are involved over the entire equipment life cycle, as described in Section 1.7. This will include liaising with other professionals when selecting new equipment, acceptance testing (where the scanner is compared to the specification outlined in the purchase), and commissioning (to determine baseline values, cross-calibrations and normal performance thresholds). The cost and complexity of equipment means that most scientists will only be involved in a handful of new installations over the lifetime of their careers. While in use, the system will require regular QC tests. These can be split into three groups of tests: detector performance, mechanical performance and software performance. Scientists will also be involved in disposal of the equipment at the end of its useful life.

Guidelines for performance measurements and checks on gamma cameras and PET/CT scanners are described in several documents (Table 5.4). NEMA standards (National Electrical Manufacturers Association) have been developed and published to support the design and

TABLE 5.4
Standards for Equipment Testing and Measurement in Nuclear Medicine

Organisation	Gamma Cameras	PET/CT
NEMA	"Performance Measurement of Gamma Cameras" (NEMA 2018a)	"Performance Measurement of Positron Emission Tomographs (PET)" (NEMA 2018b)
IEC	"Radionuclide Imaging Devices – Characteristics and Test Conditions – Part 2: Gamma Cameras for Planar, Whole Body, and SPECT Imaging" (IEC 2015)	"Nuclear Medicine Instrumentation - Routine Tests – Part 3: Positron Emission Tomographs" (IEC 2018)
IAEA	"Quality Control Atlas for Scintillation Camera Systems" (IAEA 2003)	
IPEM	"Quality Control of Gamma Cameras and Nuclear Medicine Computer Systems" (IPEM 2013a)	"Quality Assurance of PET and PET/CT Systems" (IPEM 2013b)
EANM	"Routine Quality Control Recommendations for Nuclear Medicine Instrumentation" (EANM 2010)	"Routine Quality Control Recommendations for Nuclear Medicine Instrumentation" (EANM 2010)
ACR/AAPM	"Technical Standard for Nuclear Medicine Physics Performance Monitoring of Gamma Cameras" (ACR – AAPM 2018a)	"Technical Standard for Nuclear Medicine Physics Performance Monitoring of PET/CT Imaging Equipment" (ACR – AAPM 2018b)

testing of equipment. Their standards are written for a wide range of equipment (even outside of healthcare) but the standards have been widely used by scientists in Nuclear Medicine to establish methodologies and to understand the results of measurements on gamma cameras, SPECT systems, CT hybrid systems and PET/CT scanners. There are also IEC standards, and reports by professional bodies, such as the UK-based Institute of Physics and Engineering in Medicine (IPEM) and the American Association of Physicists in Medicine (AAPM) based in the USA.

Many tests require the use of point radioactive sources, either provided as a vial or syringe filled with liquid, or as a phantom. Nuclear Medicine phantoms typically contain hollow vessels that can be filled with an aqueous solution containing a radionuclide. Imaging of the phantom provides well-defined hot and cold spots in the image (as demonstrated by the William's liver phantom presented earlier). Other phantoms utilise rods and different sized spheres to determine image resolution, as shown in Figure 5.22. Filling phantoms with open liquid sources should be done with extreme care, especially when the collimators have been removed from the camera. Where practical, absorbent material (such as Benchkote) is used to reduce the risk of contamination of the camera in the event of a spill or leakage. It is also useful to plan to perform all tests involving the phantom on the same day if possible. Isotope availability and deliveries should also be considered when creating a timetable for carrying out performance measurements.

5.9.1 Acceptance Testing and Commissioning

One of the first tasks performed on new items of equipment is acceptance testing to confirm that the performance of the system matches that quoted by the manufacturer. Often, commissioning is performed in the same session to determine baseline values and set calibration factors.

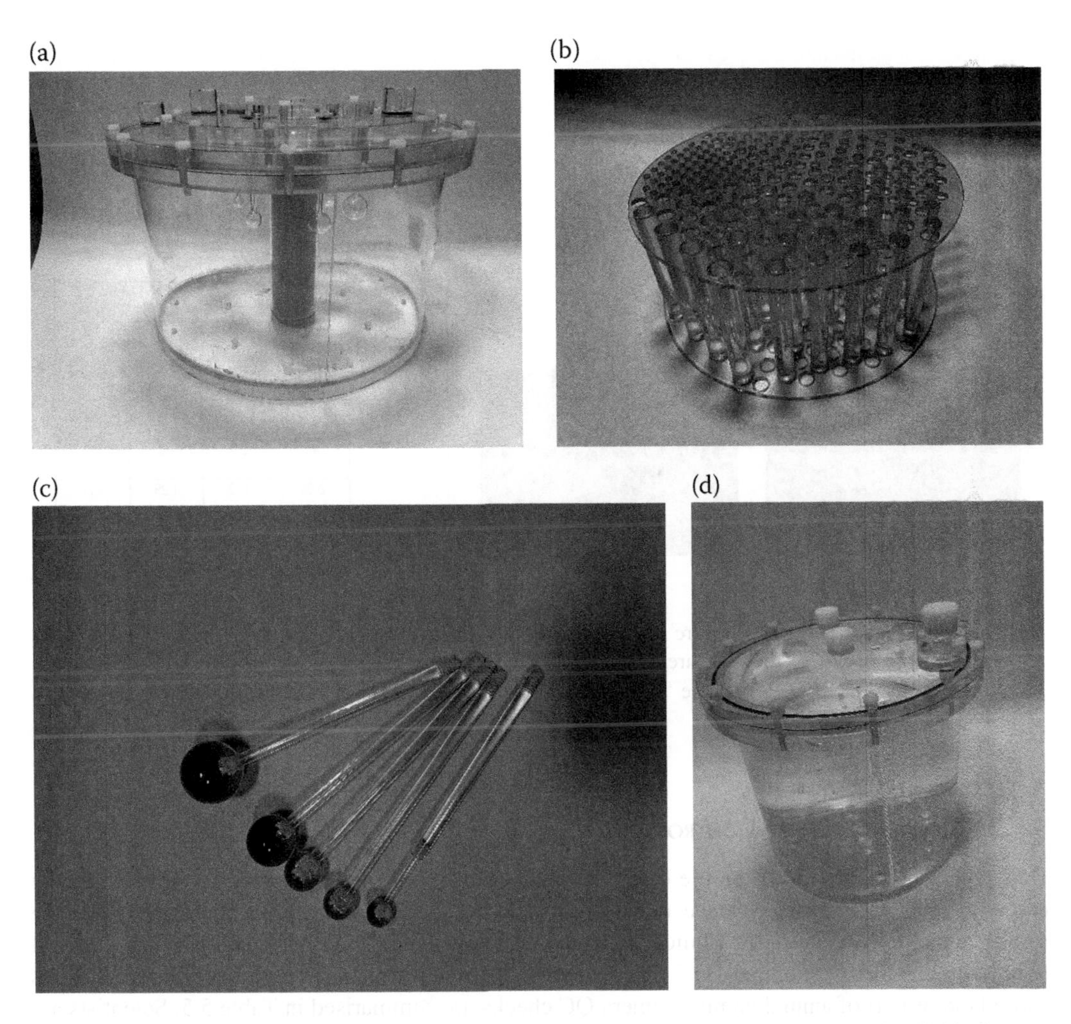

FIGURE 5.22 Fillable phantoms used to assess image contrast and resolution. (a) An IEC NEMA phantom with spheres to assess contrast recovery curves in PET and SPECT, (b) Jaszczack rod inserts to measure resolution, (c) spheres, and (d) the Jaszczack phantom itself.

CASE STUDY: COMMISSIONING OF A GAMMA CAMERA

As part of the commissioning process, a Clinical Scientist performs checks of a dual-headed gamma camera. These checks include an intrinsic flood measurement, which involves collecting over 60 million counts using a small point source positioned on the detector axis a large distance away. The large distance (over five times the size of the useful field of view [UFOV]) is used to ensure the flux of photons is uniform across the detector (Figure 5.23). Regions of interest are placed over the useful field of view (UFOV) as defined in the standard, and over the central field of view (CFOV). Results were compared with the manufacturers' specification and were within tolerance.

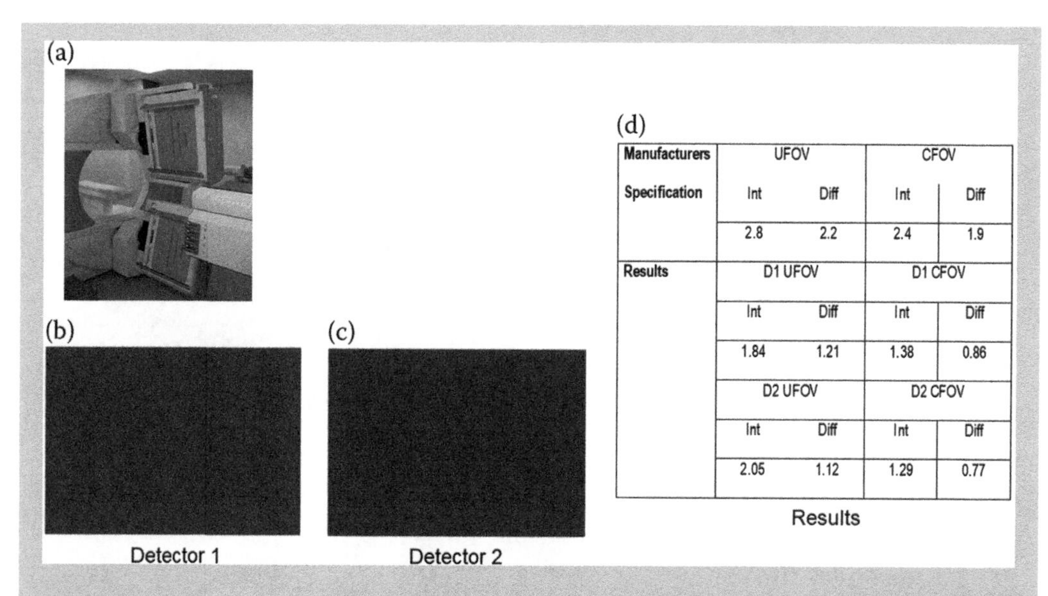

Manufacturers Specification	UFOV		CFOV	
	Int	Diff	Int	Diff
	2.8	2.2	2.4	1.9
Results	D1 UFOV		D1 CFOV	
	Int	Diff	Int	Diff
	1.84	1.21	1.38	0.86
	D2 UFOV		D2 CFOV	
	Int	Diff	Int	Diff
	2.05	1.12	1.29	0.77

FIGURE 5.23 (a) Setup to measure intrinsic uniformity, resultant images from (b) detector 1 and (c) detector 2. The results of the measurements are shown in (d) for the useful field of view (UFOV) and central field of view (CFOV), where "Int" and "Diff" denote the integral and differential uniformities, respectively.

5.9.2 Routine Quality Control

Daily checks are carried out on the equipment to ensure it is operating correctly. This is usually carried out by Healthcare Science (HCS) practitioners, Technicians or Radiographers, but the results are often reviewed by Clinical Scientists. The Scientist may also perform less frequent (quarterly, biannually or annual) QC checks involving more time and specialised knowledge. Tests carried out as part of annual gamma camera QC checks are summarised in Table 5.5. Scientists are involved in setting up protocols and work instructions for each of these tests. A report from the tests should be generated to clearly identify which tests have passed or failed and any remedial action required.

Detector performance:

- Sensitivity
- Uniformity (of sensitivity)
- Spatial Resolution
- Energy Resolution
- Count rate performance
- Multi-energy spatial registration
- Pixel size calibration

Physical and mechanical performance:

- Centre of rotation
- Bed movement and bed movement corrections
- SPECT to CT alignment
- Detector shielding

TABLE 5.5
Annual Gamma Camera and SPECT Checks

QC Test	Recommended Phantom/Source	Radionuclide and Activity
System sensitivity	Locally produced sensitivity phantom (to meet recommendations in IPEM report)	^{99m}Tc (50–70 MBq)
Spatial linearity/qualitative resolution	Flood source and ortho-hole lead transmission mask	^{57}Co flood source
Energy resolution	Point source	^{99m}Tc (1 MBq) and ^{57}Co (marker source)
Multiple window spatial registration	Collimated point source	^{67}Ga (1 MBq)
WB system non-uniformity	Flood source	^{57}Co flood source
WB exposure time correction	Flood source	^{57}Co flood source
SPECT performance – qualitative uniformity, resolution, SPECT/CT registration	Jaszczak phantom	^{99m}Tc (400 MBq)
Off-peak flood	Daily QC source	^{99m}Tc (1 MBq)
Intrinsic spatial resolution	Brass block with 0.5-mm wide slit, capillary tubing (and lead sheeting to mask face of camera)	^{99m}Tc (0.5 ml of eluate in 1 ml syringe)
System spatial resolution	Perspex block with 2 x 1 mm internal diameter cannula	^{99m}Tc (700 MBq/ml – 0.5 ml eluate added to cannula)
System alignment	3-point sources less than 2 mm diameter (NEMA method)	^{99m}Tc (activity of the 3-point sources should vary by less than 10 %)
SPECT spatial resolution	SPECT line phantom	^{99m}Tc (0.5 ml eluate)
WB system spatial resolution	Perspex block with 2 x 1 mm internal diameter cannula	^{99m}Tc (0.5 ml eluate added to cannula)
Non-99mTc spectra and uniformity	Point source in regular QC form	^{67}Ga, ^{123}I, ^{131}I, ^{111}In (1 MBq)
Count rate performance	1 or 2 x Point source in regular QC form	^{99m}Tc (1 at 200 MBq and 2 at 120 MBq)
Performance variation with gantry angle	Point source in regular QC form	^{99m}Tc (1 MBq)
Pixel size	Locally produced hole phantom	^{99m}Tc (drop of eluate in each hole)
Measurement of table speed	Point source	^{99m}Tc (1 MBq)
Detector Shield Leakage	Point source in locally produced shielded pot with 5 mm hole	^{99m}Tc (50–70 MBq), 131I (550 MBq)

- Angular sensitivity (EM shielding of electronics)
- Touch sensors, emergency stops, contour mechanisms

Software performance:

- Attenuation correction systems
- Reconstruction algorithms
- Calibration and tuning

5.10 OPTIMISATION OF IMAGING PARAMETERS IN NUCLEAR MEDICINE

An important role of the Clinical Scientist in Nuclear Medicine is to work with Radiologists and scanning staff to optimise acquisition parameters. This is based on guidelines, but also needs to consider the imaging equipment used, available image processing software, activity administered and patient group or referral criteria.

Various phantoms are available to measure and optimise scanning performance. Some phantoms have standardised features (e.g. spheres of known size) which can be used to measure and compare performance; whereas others are more anthropomorphic so can mimic organs of interest, such as the lungs, heart and brain. Most phantoms are made of acrylic and have one or more fillable sections. Some have solid and/or air-filled sections to mimic higher or lower density tissues. A selection of common phantoms and their uses are described as follows.

5.10.1 The NEMA/IEC NU2 Image Quality Phantom

This is used in PET for determining contrast recovery, as detailed by NEMA (2018b). This includes a semi-anthropomorphic outline, low-density central section and a series of fillable spheres of different sizes (Figure 5.22(a)). An example of the contrast recovery curve as measured with this phantom is shown in Section 5.11.1. The non-active, low-density section in the centre allows checks to be made of attenuation and scatter correction algorithms.

5.10.2 The Jaszczak Phantom

This cylindrical phantom, named after the scientist who designed it, is commonly used to measure SPECT performance and can also be used for PET scanners (Figure 5.22(d)). It is used for several performance tests defined in the IPEM and ACR QA protocols (see Section 5.9.2 for QA). There are also a few variations on this standard cylindrical design. It has a range of inserts available including:

- Series of solid acrylic rods – Figure 5.22(b)
- Series of solid acrylic spheres of different sizes – Figure 5.22(c)
- Series of fillable acrylic spheres of different sizes
- Point source or line source holders – Figure 5.22(d)

Figure 5.22(d) shows the phantom with point sources in position against an otherwise uniform background. This phantom is used to measure SPECT or PET imaging performance against baseline or to compare different scanners. The spheres can be used to measure contrast recovery. The rod sections can be used for qualitative resolution. The line or point sources are used to measure reconstructed resolution, and the uniform liquid filled section for noise measurements. The uniform activity sections are also useful for identifying non-uniformities in planar detector performance.

5.10.3 Anthropomorphic Torso Phantom

The torso phantom contains lung, liver, spine and cardiac sections in realistic shapes and sizes. The spinal insert is made of increased density material to mimic bone. The lung volumes are filled with polystyrene balls and water to mimic the lower density of lung tissue. The liver and cardiac inserts can be filled with varying activity concentrations, and the cardiac model can have defects added to the myocardium. Figure 5.24(a) shows the phantom being imaged on a solid-state dedicated cardiac scanner. Figure 5.24(b) provides a close-up of the cardiac insert used to mimic activity in the myocardium of the left vertical, and/or inside the left ventricle itself. The resulting phantom images are useful for optimising image processing.

Because of the non-homogeneity of attenuation, scatter, and activity concentration, this phantom is particularly useful for checking reconstruction algorithms and image correction techniques.

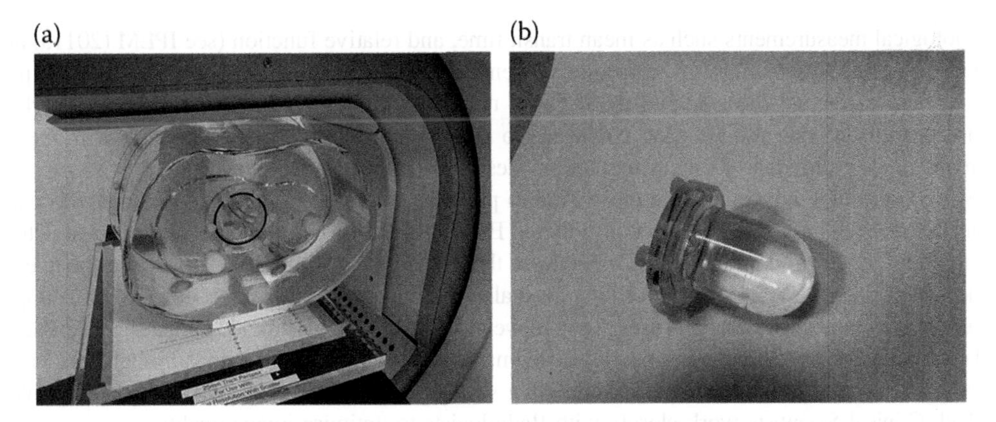

FIGURE 5.24 Examples of (a) an anthropomorphic phantom used for SPECT and PET QC checks and image optimisation and (b) the cardiac insert to mimic the heart.

5.10.4 Brain Phantoms

A brain perfusion phantom is used to simulate SPECT and PET brain imaging and is often referred to by the trademarked name "Hoffman" phantom. It consists of a series of sheets of acrylic with cut-out sections. These are stacked to provide a continuous cavity that mimics a brain volume. The acrylic sheets are thinner than the scanning slice thickness, so by varying the cut-outs between slices, the phantom can mimic differences in uptake within grey and white matter due to the partial volume effect.

Another brain phantom commonly used in both PET and SPECT is the striatum phantom. This includes a uniform brain volume, with four internal fillable structures that mimic the size and shape of the four sections of the striata; the left and right caudate, and left and right putamen. This phantom is used to simulate ioflupane, IBZM and other pharmaceuticals which bind to the striata. The brain volume is mounted in a head and neck anthropomorphic model.

5.10.5 Protocol Optimisation

Protocol optimisation is a collaborative process involving the MPE and clinician and will often also include the Technologist and Radiopharmacist. The optimisation process covers the entire process from patient selection/referral criteria all the way through to reporting.

The acquisition optimisation is a balancing act involving several factors such as image quality, patient dose and speed of acquisition. Since the speed of the scan if often dictated by the bio-kinetics of the radiopharmaceutical, or for more fixed distributions, the length of time a patient can stay still; this usually ends up being the lowest dose that can be administered while still obtaining the minimum clinical information required.

In SPECT imaging, this will involve the choice of acquisition parameters (collimator, camera positioning, camera movement, acquisition time, image matrix, timing of dynamic phases and/or gating).

The image processing steps may include spatial and/or temporal filtering, region or volume of interest measurements, and 3D/4D reconstruction. Table 5.6 shows the parameters identified as optimal for whole body and SPECT imaging of the bones.

5.11 IMAGE PROCESSING

Once images have been acquired, some processing of the images is necessary prior to interpretation and reporting by the Radiologist. For planar or static images, this may require labelling of images, for example, "left", "right", "anterior" or "posterior". Some studies, for example, renograms and gastric emptying studies, will require analysis of activity within specific ROIs to estimate uptake in the organs of interest. Further computational processing of activity as a function of time can give physiological measurements such as mean transit time, and relative function (see IPEM (2011a) and Cherry et al. (2012)). Camera manufacturers often provide commercial software packages for processing and extracting the required data. These medical software packages must be CE-marked. In-house software can also be used but needs to undergo rigorous testing to comply with the requirements of the Medical Devices regulations described in Section 1.7.

SPECT and PET images require processing to produce 3D images. There are two methods commonly used for 3D image reconstruction: Filtered Back Projection (FBP) and iterative reconstruction. For more information on image reconstruction, see the IPEM report 100 on mathematical techniques in nuclear medicine (IPEM 2011a). Processing must also be optimised for the intended use. For example, increasing the number of iterations used for iterative reconstruction will provide a higher level of detail and quantitative accuracy of the images, but such images tend to be very noisy and unsuitable for visual reporting. Conversely, images for visual reporting will have fewer iterations and/or smoothing filters applied. Clinical Scientists work closely with Radiologists to optimise image quality.

TABLE 5.6
Optimal Parameters for Whole Body and SPECT Imaging of Bones

Whole Body	
Workflow:	Bone whole body
Orientation:	Feet first, supine. Optional knee support.
Zoom:	x1
Collimator:	Low Energy High Resolution (LEHR)
Energy window:	140 (±7.5% energy window)
Matrix size:	256 x 1024
Scan speed:	10 cm/min
Views:	Anterior and Posterior views
Distance to detector:	auto contour
SPECT	
Orientation:	Feet first, supine. Detectors over area of interest. *Head first, supine. Must use CT compatible head rest.* Optional knee support.
SPECT – 1 bed	
Zoom:	x1
Collimator:	Low Energy High Resolution (LEHR)
Energy window:	140 (±10% energy window)
Matrix size:	128 x 128
Rotation direction:	CW, 0° starting angle, detector configuration 180°
Stop condition:	7 seconds per view.
Views:	128 views (64 per detector) over 360°.
Distance to detector:	Auto contour (ensure patient positioning allows detector to get as close as possible to patient)
SPECT –2 beds	
Zoom:	x1
Collimator:	Low Energy High Resolution (LEHR)
Energy window:	140 (±10% energy window)
Matrix size:	128 x 128
Rotation direction:	CW, 0° starting angle, detector configuration 180°
Stop condition:	7 seconds per view.
Views:	128 views (64 per detector) over 360°.
Distance to detector:	Auto contour (ensure patient positioning allows detector to get as close as possible to patient)

Iterative reconstruction is more computationally intensive than FPB. To address this, a variation on the standard iterative process called Ordered Subsets Expectation Maximisation (OSEM) is often used. In this method, only a subset of projections is used for the first iteration. A second iteration uses a different subset of projections, and so on. In this way, the activity distribution estimate is updated many more times than in standard iterative reconstruction techniques, and the image converges more quickly. Reconstructing with five iterations using ten subsets has been shown to give similar results as 50 iterations and one subset (i.e. standard iterative reconstruction).

There are also a range of image corrections that can be applied during the iterative reconstruction process. These allow some image degrading processes to be modelled and compensated for within the reconstructed image, as summarised in Table 5.7. Caution should be taken with resolution recovery packages, as these may introduce artefacts into the image, providing false positives. Likewise, filtering and noise suppression software may mask smaller defects giving false negatives.

TABLE 5.7
Physical Interactions, Impact on Images, and Corrections Applied to Improve Image Quality

Physical Effect	Effect on Images	Correction Method	Practical Implementation
Attenuation of gamma photons	Reduction in counts within reconstructed image. Tends to be non-uniform throughout the body. This can lead to areas of uniform uptake being visualised as non-uniform. Quantification inaccurate.	Estimated (Chang correction) Measured (for SPECT and PET) Calculated (for PET-MR)	All commercial packages allow Chang correction for SPECT Either a transmission source or more commonly a CT scanner. Most current PET scanners include a CT scanner Using optimised MR sequence and lookup table
Scattered photons	Scattered photons are imaged in the "wrong" position, adding to image noise. Most clearly seen as body parts with low or zero uptake still showing activity in the reconstructed image.	This can be estimated using scatter window measurements. Can also be modelled using CT data.	Most SPECT systems allow dual or triple energy window scatter correction in SPECT imaging. One manufacturer uses a Monte Carlo model to estimate the scatter contribution. PET scanners tend to model the scatter function.
Geometric distortions	SPECT collimators give varying resolution with distance from the detector. The depth of interaction in PET will lead to drop off in resolution towards the edge of the FOV.	The geometric properties of the SPECT collimators and PET gantry & crystals are fixed and can therefore be modelled in the forward projection of the reconstruction algorithm.	In SPECT imaging this is called "resolution recovery" or "3D-OSEM". Despite the name, the main effect of resolution recovery software is to increase contrast and reduce noise. In SPECT noise reduction has been used to reduce radiation exposure and/or scan times. In PET, this is a Point Spread Function (PSF) reconstruction. The application of PSF corrections in PET produces higher resolution, high contrast, images, with uniform resolution.

(*Continued*)

TABLE 5.7 (Continued)

Physical Effect	Effect on Images	Correction Method	Practical Implementation
Image noise	All Nuclear Medicine acquired data is noisy, and this noise can propagate through the reconstruction process.	Noise metrics can be included in the reconstruction steps to stop the iterative process over-converging.	Bayesian statistics have been used in both SPECT and PET research. The QClear software from GE uses this approach.

5.11.1 Image Quantification in Nuclear Medicine

The standardised uptake value (SUV) is used in PET (and more recently SPECT) for diagnosis, staging and measurement of treatment response. It is a semi-quantitative uptake measure. The tracer uptake in the volume of interest (e.g. a tumour) is measured from the SPECT or PET images in kBq/ml. This is divided by the injected activity to give the fraction of the injected activity in each ml. This figure is then normalised to body mass (or sometimes lean body mass). Note that a decay correction will be applied to either the tracer uptake figure (taken from the image data) or the injected activity. In the special case of uniform activity distribution with no excretion, in tissue of unit density (1 g/ml), the SUV would be equal to

$$\text{SUV} = \frac{\text{tracer uptake}}{\text{activity per unit weight}}$$

where SUV is generally quoted in g/ml, tracer uptake in kBq.ml activity in MBq and weight in kg.

SUV values depend on several factors that must be kept constant to allow SUVs to be reliably compared. These include:

- The tissue being measured
- The radiopharmaceutical
- Delay between injection and time of scanning
- The image quality (e.g. resolution and noise, which are dependent on the reconstruction method, and any image correction techniques applied)
- Patient habitus – for obese patients the lean body mass might be used
- Drug or physiological interactions (e.g. circulating glucose can affect FDG uptake)

Numerous studies have examined these effects in PET and many of these can be applied to SPECT quantitation. One area when SUV can be very useful is in monitoring SUV before and after treatment; as most of the above factors are fixed, SUV measurements can be used to assess if a treatment agent is working.

CASE STUDY: SUPPORTING RESEARCH STUDIES

As part of a multi-centre research, study, the scientist arranged to scan the NEMA image quality phantom on the PET/CT scanner. The process for filling the phantom is defined by the lead research site, but will usually follow the clinical protocols described in Section 5.4. The protocols specify not only an activity ratio within the phantom (typically target ratio of 4:1) but also the activity concentrations at the time of the scan (typically 4.7 Bq/ml in the background volume). Since decay of F-18 is rapid (approximately 10% decay every 15 minutes), allowances must be made for the time it takes to prepare and fill the phantom.

The phantom was scanned and reconstructed using the setting prescribed in the research protocol. These parameters will need to be fixed for each project as changes to them will alter image quality and quantitative accuracy. Hence images will need to be reconstructed with and without certain image correction factors (e.g. ToF) to ensure the resulting images are appropriate. The images are analysed for absolute quantitative accuracy, image contrast as a function of object size, and image noise using the figures from volume of interest (Figure 5.25(a) and (b)).

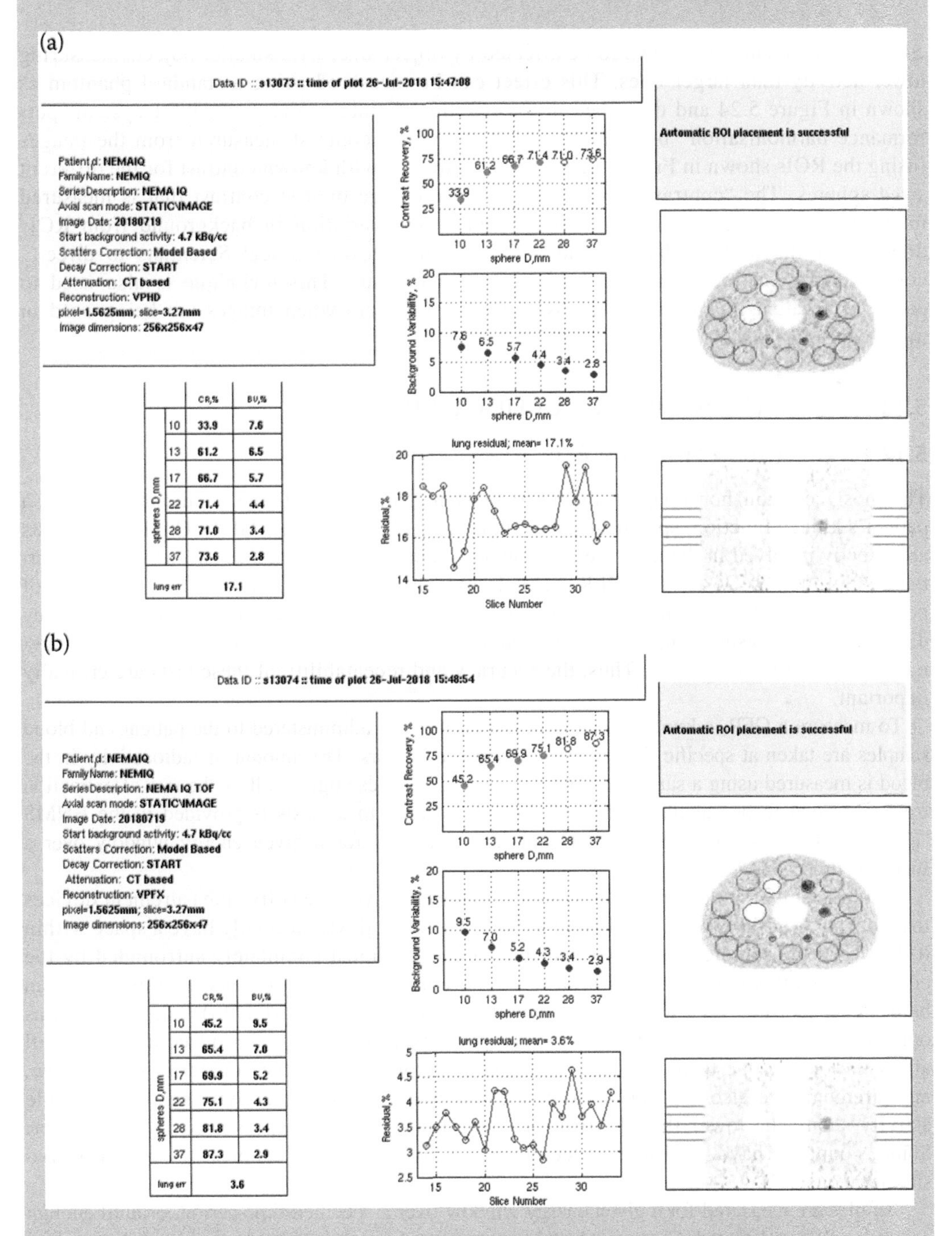

(a)

spheres D,mm	CR,%	BV,%
10	33.9	7.6
13	61.2	6.5
17	66.7	5.7
22	71.4	4.4
28	71.0	3.4
37	73.6	2.8
lung err	17.1	

(b)

spheres D,mm	CR,%	BV,%
10	45.2	9.5
13	65.4	7.0
17	69.9	5.2
22	75.1	4.3
28	81.8	3.4
37	87.3	2.9
lung err	3.6	

FIGURE 5.25 Research study quantitative measurements on the NEMA image quality phantom using (a) standard reconstruction and (b) reconstruction including time of flight information

The images and all the relevant information are sent to the central laboratory coordinating the study to confirm quantitative values across the country and intercomparison of results between centres.

SUV measurements are known to be affected by object size, with smaller objects measuring lower activity than larger ones. This effect can be measured using a standard phantom as shown in Figure 5.24 and described in "Feasibility of state-of-the-art PET/CT systems performance harmonisation" by Kaalep et al. (2020). The contrast measured from the images (using the ROIs shown in Figure 5.25) can be compared with known contrast for the different sized spheres. The "contrast recovery" is the percentage of true contrast that is measured in the images. Image noise can be measured as the variation in background from ROIs drawn over the uniform background activity. This allows Clinical Scientists to directly compare different reconstruction settings and algorithms. This technique is also used to ensure equivalent image quality across different scanners when images are being used in multi-centre trials.

5.12 NON-IMAGING RADIONUCLIDE TESTS

5.12.1 Glomerular Filtration Rate

The most common non-imaging diagnostic test in Nuclear Medicine is measurement of a patient's kidney function, specifically the Glomerular Filtration Rate (GFR). Clinical scientists are directly involved in the measurement and calculation of these studies and will often report the results of these tests. The GFR result is used by clinicians assessing the suitability of potential kidney donors, and for the monitoring of kidney function in patients with kidney disease. GFR measurements are also used by Oncologists to calculate the dose of chemotherapy to give to patients. Thus, the accuracy and repeatability of these tests are critically important.

To measure a GFR, a known quantity of radioactivity is administered to the patient and blood samples are taken at specific time intervals over a few hours. The amount of radioactivity in the blood is measured using a sample counter (Figure 5.26). These figures allow the rate of extraction from the blood to be calculated. More information about these tests is provided in the BNMS "clinical guidelines" (BNMS 2020). Counts are measured for a given energy window over a specified time period.

Gamma counters (also known as sample counters) are very sensitive measurement devices that consist of a NaI crystal attached to a PMT. The crystal will normally have a space within it, in which the sample can sit, so that the sample is almost completely surrounded by the crystal. This allows detection of almost all the emitted gamma photons, so this setup can be used to measure very low activity samples (down to around 100 Bq). Factors, such as other sources in the room, and radioactive patients walking past during counting can all affect the results due to the sensitivity of the detector. Because the counter is so sensitive, measurements are also susceptible to count rate errors for higher activity samples. Sample activity should be lower than 1 MBq, or "dead time" may be experienced in the counter. Some counters have multiple detectors to allow multiple samples to be measured simultaneously.

Samples are measured for a given energy window over a specified time period, or until enough counts have been detected. Care must be taken to minimise sources of background activity, such as other sources in the room, and radioactive patients nearby. The volume of the samples can also influence results as this alters the location of radioactivity relative to the crystal.

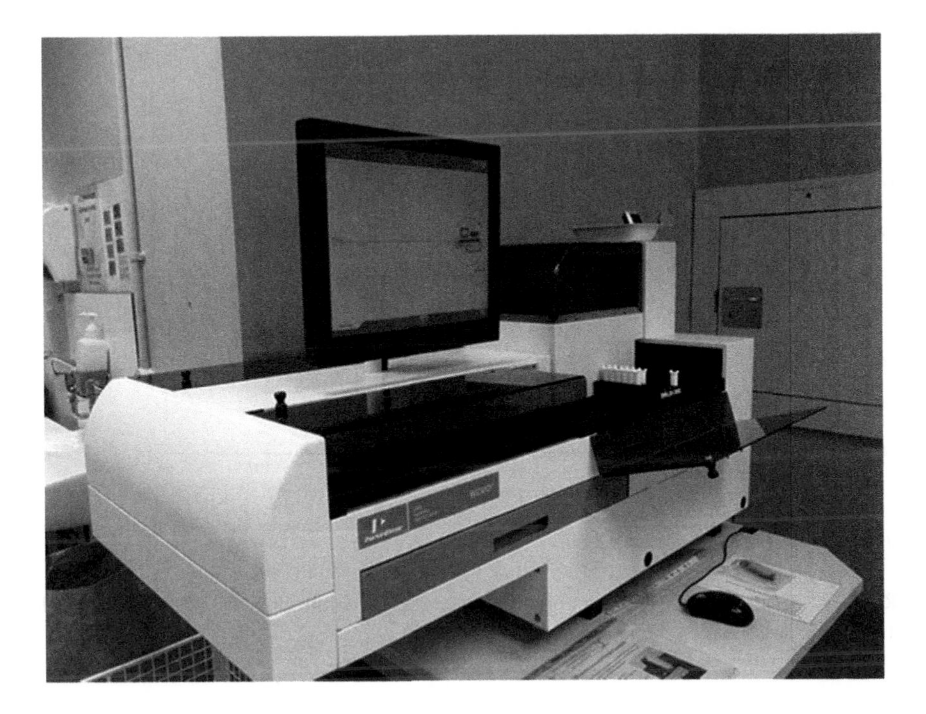

FIGURE 5.26 A sample (gamma) counter.

Sample counters are most often used in nuclear medicine to measure the activity concentration in blood plasma samples. However, they can also be used to measure other low activity samples from patient samples (e.g., blood, urine, faeces, breast milk). This is required when assessing new radiopharmaceuticals in clinical trials, to assess uptake of the pharmaceutical following administration.

The function, use, calibration and QA of samples counters are described in IPEM report 85 "Radioactive Sample Counting – Principles and Practice" (IPEM 2002).

CASE STUDY: GFR MEASUREMENTS

A patient was suspected of impaired kidney function and was referred for a GFR test. The patient was injected with 10 MBq 99m-Tc-DTPA (pentetic acid). Four blood samples were taken from the patient between 2 and 4 hours after the injection. These samples were then spun down in a centrifuge to allow the blood plasma to be extracted. The plasma was pipetted into sample tubes and counted in a sample counter. A calibration source (called a standard) was produced using an extract from the patient's injection vial, to allow the activity in the samples to be expressed as a fraction of the injected activity.

The result was calculated and then checked and reported by a Clinical Scientist. The GFR tests followed BNMS clinical guidance for GFR measurement with 99m-Tc (BNMS 2020). This document details the different methodologies available, including analysis of single or multiple samples, and relevant calculations.

In the example shown in Figure 5.27, the patient's GFR was lower than the normal reference GFR for their age, which implies that their kidney function is impaired.

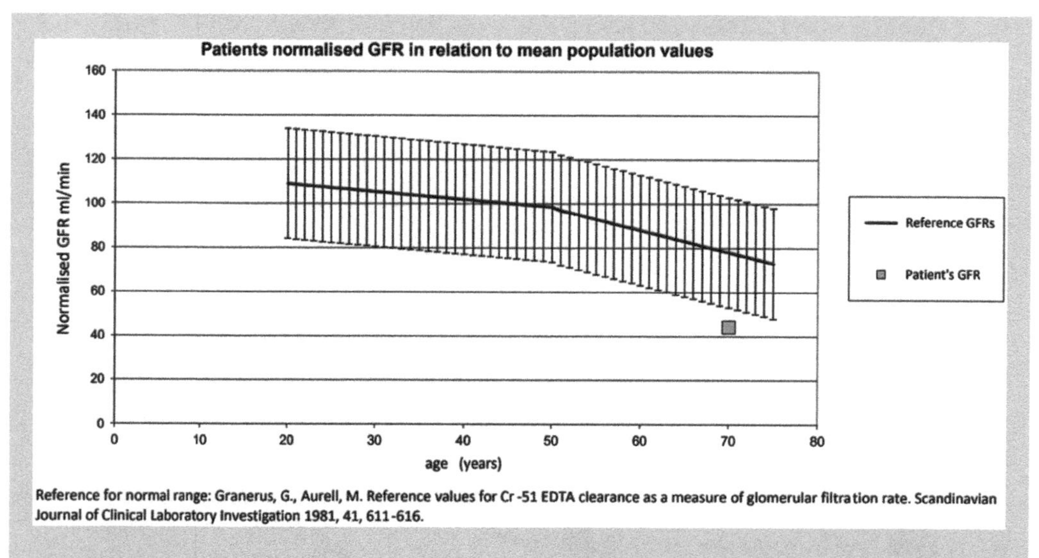

FIGURE 5.27 GFR plotted against normal reference ranges confirming that the GFR was below the normal range as described by Granerus and Aurell (1981).

5.12.2 Bile Acid Malabsorption

Bile acid malabsorption is a chronic condition that affects a patient's digestive system. The process can be measured using a SeHCAT study. SeHCAT (23-seleno-25-homotaurocholic acid, selenium homocholic acid taurine, or tauroselcholic acid) is a drug used to diagnose bile acid malabsorption. The patient is given a low activity ^{75}Se capsule to swallow. The capsule is broken down in the digestive system, and the radiopharmaceutical is actively absorbed by the intestines and re-excreted by the bile duct. Over 7 days, the pharmaceutical is expected to pass around this loop around 35 times, and if malabsorption is present, the level of SeHCAT within the patient will reduce considerably. The amount of SeHCAT in the patient is measured at 4 hours post-administration and again 1 week later.

The difference in counts, corrected for decay, is calculated and quoted in terms of retention (i.e. the percentage of the activity in the patient at 4 hours that is still present on day 7).

Some hospitals have dedicated whole-body counters to measure the total activity within a patient. However, most hospitals use a gamma camera, with the collimators removed to act in a similar way. Figure 5.28 shows how small difference in reabsorption (from 92% to 97%) can have a significant effect on the level retained at day 7.

5.13 RADIATION PROTECTION FOR UNSEALED RADIOACTIVE MATERIAL

Nuclear medicine is heavily regulated because of its use of radioactive material – the Ionising Radiation Regulations and the Health and Safety Executive Approved Code of Practice (HSE 2018) covers protection of staff and the public. They set the principle of ALARP dose limits, local rules, and definitions of relevant staff groups and their roles and responsibilities. These guidelines also detail the requirements around the use and movement of radioactive sources.

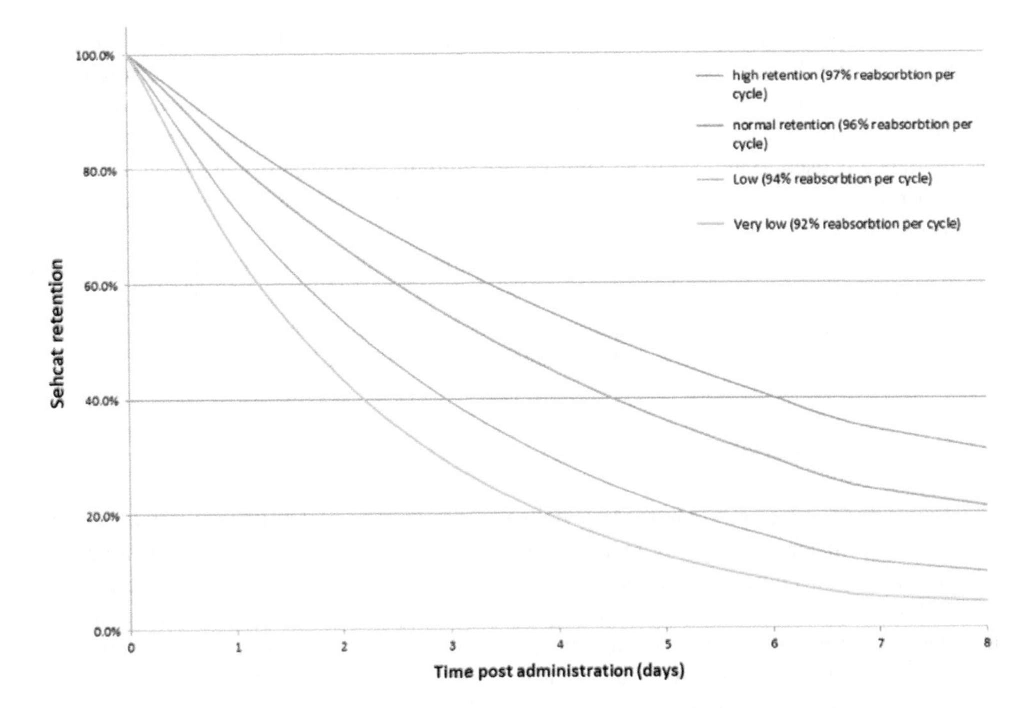

FIGURE 5.28 Results from a SehCAT test.

Scientists in Nuclear Medicine are familiar with working with radioactive sources, including the need to locally shield sources and how to determine the thickness of lead shielding required. The practical skills required to manipulate radioactive material are not easy to acquire. ICRP report 106 annexe E "How to protect your hands in Nuclear medicine Work" (ICRP 2008) gives examples of good technique and a clear explanation of why certain measures are recommended. For example, radioactive sources should never be handled using your fingers – long handled instruments should always be used, as exposure reduces with distance according to the inverse square law. Personal monitoring in Nuclear Medicine includes measurement of whole-body dose, but where appropriate, also dose to the fingers and sometimes to the eyes. Lead glass shields are used to protect the eyes. Dose in PET/CT tends to be higher than those in general Nuclear Medicine and staff fall under the category of classified radiation workers (Chapter 7) as they might receive higher doses.

The unsealed nature of radioactive material in Nuclear Medicine applications means that the risk of contamination must be considered. Gloves will always be worn when handling unsealed material. Drip trays and absorbent material will also be used and disposed of appropriately in the event of any drips or spills.

The Ionising Radiation (Medical Exposures) Regulations (IR(ME)R 2017) protects patients by setting diagnostic reference levels (DRLS) and requiring ARSAC licenses for employers and the practitioners, as well as setting out the role of the MPE, Referrer and Operator.

The Environmental Permitting (England and Wales) Regulations (EPR 2016) cover the management of radioactive sources and waste. Organisations require permits to hold radioactive sources and need to appoint a Radioactive Waste Adviser (RWA) to ensure control over radioactive waste, usually a senior Clinical Scientist in Nuclear Medicine or Radiation Safety. Permits issued are radionuclide specific, site specific, and limit activity and require detailed records to be kept regarding radioactive sources and their disposal. Scientists are often responsible for record-keeping systems, developing or using commercially available spreadsheets and databases to ensure that records are accurate and up to date. Many radionuclides are excreted and discharged to the public sewers. Scientists work together to ensure the latest scientific evidence is used to apportion an appropriate amount form each organisation to the drain. The IPEM produces regular reports on excretion factors (IPEM 2018), which are reviewed by the government.

The Carriage of Dangerous Goods Act 2009 (as amended in 2019) covers transport of radioactive material, and deliveries to and from the Nuclear Medicine Department Radiopharmacy to other sites (CDGR 2019). There is a requirement to train consignors and drivers involved with the delivery of radioactive packages, which is usually carried out by Clinical Scientists.

CASE STUDY: CONTAMINATION EXERCISE

Handling a radioactive spill requires knowledge, understanding and practice. Staff who work with unsealed material every day react instinctively when there is a spill, such as a patient who urinates in a clinic room after an injection of ^{99m}Tc. New staff need to be able to respond as part of a team, to seal off the spill, and use appropriate methods to mop up any liquid and them monitor themselves and the area and will undertake exercises to practise the skills required (Figure 5.29). After any real incident, there is also a period of analysis and reflection, as described in more detail in Chapter 7.

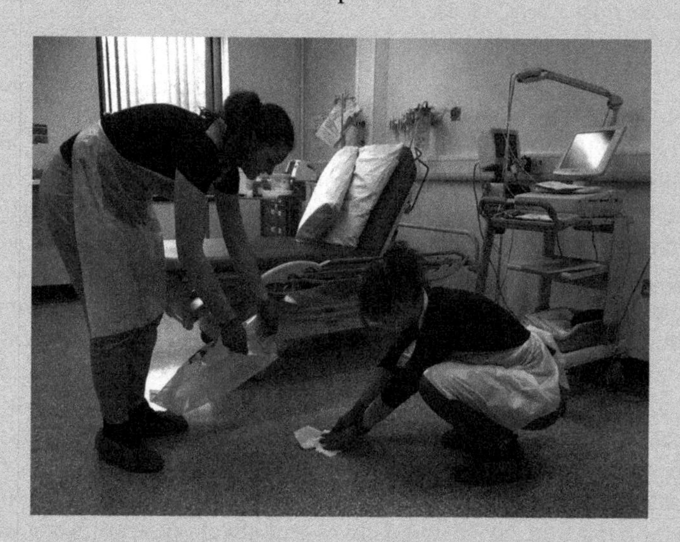

FIGURE 5.29 A decontamination exercise.

5.14 THERAPEUTIC TECHNIQUES IN NUCLEAR MEDICINE

Nuclear Medicine therapies are also known as Molecular Radiotherapy or mRT. The therapeutic effect is produced by the administration of isotopes that emit beta and/or alpha particles. The administration of these radiopharmaceuticals is by ingestion (in the form of a capsule or a drink), or injection (intra-venously, intra-arterially or intra-articularly into the joints).

TABLE 5.8
Common Nuclides in Nuclear Medicine Therapies

Isotope	Half Life	Therapy Emissions	Main Imaging Emissions
Phosphorus-32	14.3 days	Beta 1.7 MeV max (100%)	Bremsstrahlung from Beta
Yttrium-90	2.7 days	Beta 2.3 MeV max (100%)	Bremsstrahlung from Beta, or very rare secondary positron for PET imaging
Lutetium-177		Beta 498 keV max	208 keV (11%)
			113keV (6%)
Iodine-131	8 days	Beta 606 keV max (90%)	365 keV (82%)
		Beta 334 keV max (7%)	
Radium-223	11.4 days	Alpha 5.7 MeV	81.1 keV (15.2%)
		Alpha 6.8 MeV	
		Alpha 7.4 MeV	
		Alpha 6.6 MeV	
		Alpha 7.4 MeV	83.8 keV (25.1%)
		Beta 1.4 MeV max	94.9 keV (11.5%)
		Beta 570 keV max	

5.14.1 Radionuclides

Common radionuclides used for mRT treatment are listed in Tables 5.8 and 5.9. Some of these can also be used for imaging, to allow treatment verification and post-therapy calculation of dose. However, for diagnostic imaging, and pre-therapy dosimetry calculations, other isotopes are often used.

5.14.2 Clinical Scientist Involvement with Therapeutic Techniques

Clinical Scientists are usually directly involved with the introduction of new therapeutic mRT techniques, especially for more complex administrations.

Under IR(ME)R (2017) the prescribed dose for any ionising-radiation-based treatment must be individually planned. This planning must take into consideration the planned dose to the target volume(s) and exposure of non-target organs at risk. See Section 5.15 for a description of dosimetry in Nuclear Medicine and an example of a dose calculations for organs at risk.

Due to the long half-life of many of these isotopes, radiation protection advice will need to be provided to patients to limit exposure to radiation of themselves, as well as family, friends and members of the public. Radiation protection advice has been standardised for some well-understood therapies (e.g. iodine treatment for benign disease). More tailored therapies will need individual dose and risk assessments with patient-specific advice.

Restrictions are stricter for therapies with high activities and significant gamma radiation, such as 131-Iodine, whereas restrictions for radium therapies are less stringent. Restrictions are aimed at limiting the potential for exposure to external radiation, as well as reducing the risk of introducing radiation interally if coming into contact with body fluids. For day-case and in-patient therapies, there will often be considerable contamination hazards to be managed.

TABLE 5.9
Summary of Common mRT Therapies

Therapy	Condition	Targeting Mechanism	Typical Activity	Setting	Notes
131I-NaI	Overactive thyroid (Graves' disease or toxic nodule(s))	Active uptake of iodine by thyroid tissue	200–800 MBq	Outpatient clinic	Treatment for benign disease. High uptake and retention mean long radiation protection precautions
131I Thyroid ablation	Thyroid cancer – post thyroidectomy surgery	Active uptake of iodine by thyroid tissue	1.1 GBq	Day case or in-patient	As an adjunct to surgery. Very low retention
131I-Thyroid cancer treatment	Thyroid cancer- disease remnants and metastases	Active uptake of iodine by thyroid tissue	3.7–7.2 GBq	In-patient typically 2–5 days	Very variable uptake
Xofigo 223Ra-dichloride	Bone metastases from prostate cancer	Preferentially absorbed by fast-growing bone	~3.5 MBq 55 kBq/kg 6 treatments at 4-week intervals	Out-patient	Currently approved for bone metastases from prostate cancer, but could be extended to other bone metastases in future
90Y labelled glass or resin micro-spheres	Metastases in liver, and liver cancer	Direct injection into liver using fluoroscopy-physical fixing in capillary bed	1–5 GBq based on tumour load and patient size	In-patient	A dummy run will be performed using ^{99m}Tc-MAA to evaluate vascular structure of liver and possible shunting to lung
177Lu-peptide and/or 90Y-peptide	Neuroendocrine Tumours	Binding to somatostatin receptors on tumour cells	177Lu 5.5–7.4 GBq 90Y 2.7–7.4 GBq	Day case or in-patient	Slow administration (30–60 minutes). Some centres are experimenting with combined 90Y + 177Lu administrations. Some centres are adjusting subsequent treatments based on dosimetry

131-mIBG	Neuroendocrine tumours	Uptake by neuro-ectodermal cells	3.7–12 GBq	In-patient	Slow administration (1–4 hours). Some centres are adjusting subsequent treatments based on dosimetry
177Lu-PSMA	Prostate cancer	Prostate-Specific Membrane Antigen	~7 GBq 177-Lu with multiple treatments at 4–6 week intervals	In-patient	Exact treatment regimes are still under investigation
32P	Polycythaemia	Absorption of phosphorous by bone marrow	~250 MBq based on patient size		Being replaced by drug-based therapies
90Y colloid	Synovitis, typically knee	Direct injection into joint capsule	~200 MBq per knee	Out-patient	Benign arthritic conditions. Highly skilled administration, often under X-ray guidance

CASE STUDY: ROOM PREPARATION AND RESTRICTIONS ON PATIENT'S ACTIVITIES

An inpatient therapy for thyroid ablation is carried out by administering 3.7 GBq of I^{131}. The room is prepared by nursing staff or others, placing non-slip and waterproof material across the floor and areas of heavy interaction so that any contamination can be removed from the room by taking up the lining (Figure 5.30).

The Clinical Scientist will discuss how the radioactive material will be excreted while in hospital with the patient. They will explain in terms the patient can easily understand that there is a phase in the first couple of days when their urine sweat and saliva are highly radioactive and the practical measures they can take to minimise exposure to staff who care for them. For example, men may be asked to sit on the lavatory to empty their bladder. The patient may be asked to flush the toilet twice and to use copious amounts of water when cleaning their teeth or washing their hands.

The Clinical Scientist is likely to be closely involved during the in-patient stay, monitoring the dose rate at a distance from the patient to assess when the patient can safely go home. This will depend on excretion but also home circumstances. For example,

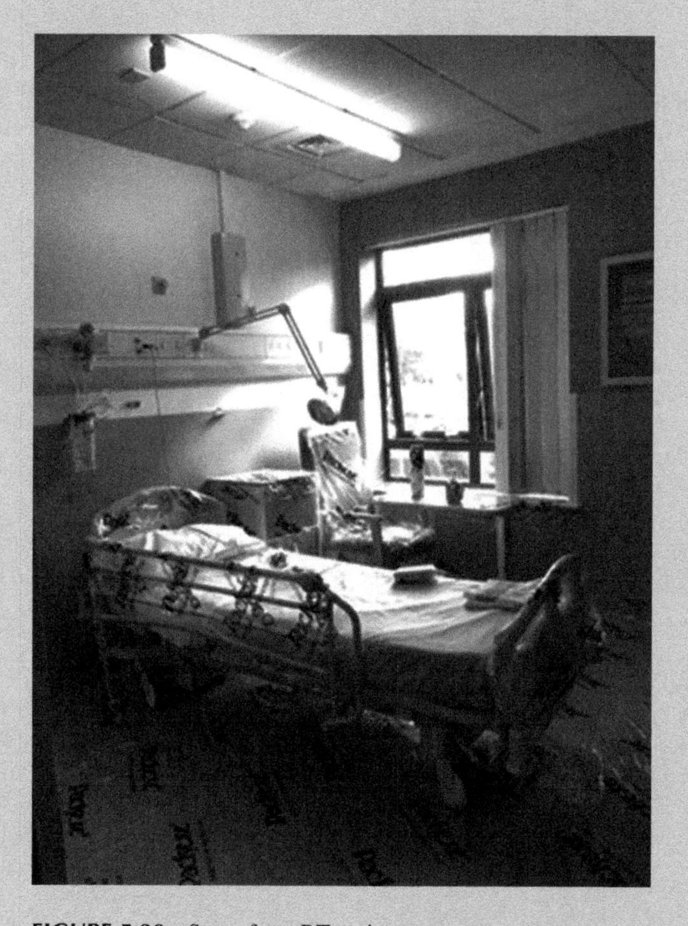

FIGURE 5.30 Setup for mRT patient treatment.

patients who live alone may be able to go home sooner than someone living in with young children. The contacts the patient has are discussed in some detail to ensure no one who might come into contact with the patient will be put at risk, either through potential contamination or the radiation dose they might receive. They are usually advised to drink plenty of water and are given a card to remind them of the restrictions and advice they have been given.

For administration of 1.1 GBq or 3.7 GBq restrictions typically last around 7 days, depending on uptake and patient physiology.

5.15 PATIENT DOSIMETRY IN NUCLEAR MEDICINE

The dose that a patient is exposed to in Nuclear Medicine is complicated by the dynamic nature of many radiopharmaceuticals. Not only will different pharmaceuticals be taken up by different organs at different rates but also their residence time within each organ will vary. If the pharmaceutical is metabolised in some way, some of the resultant metabolites may also be radioactive, and their bio-kinetics will differ from that of the original pharmaceutical. To complicate matters further, the activity in an organ will not only irradiate the immediate vicinity but also more distant tissue and organs. Patient pathology may also alter the bio-kinetics. The dose distribution can be modelled using Monte Carlo techniques where the radiation source itself is modelled in space and time and radiation transport calculations made taking account of biological distribution including excretion and physical half-life. Computer programmes such as MIRDOSE (Stabin 1996) and then OLINDA (Stabin et al. 2005) were written using mathematical phantoms to model the human body and ICRP data, such as that found in ICRP53 to relate the administered activity in Bq to the absorbed dose in organs. A useful summary of the data and methodologies for patient dosimetry in Nuclear medicine is given in ICRP 128 (ICRP 2015a). The International Commission on Radiological Protection (ICRP) are an independent organisation made up of expert volunteers who have published recommendations and guidelines that are implemented into EU directives and radiation legislation internationally such as the dose limits laid out in ICRP publication 103 (ICRP 2007).

Patient dosimetry in Nuclear Medicine is typically divided into two groups of calculations:

1. **Clinical diagnostic and standardised therapeutic procedures:** For common radiopharmaceuticals, average patient exposures have been measured and calculated as effective doses. These figures can then be used as reasonable estimates for patient or research subject exposures.
2. **Patient-specific dosimetry:** It is used for some therapeutic work and for novel diagnostic pharmaceuticals. For some therapies, the kinetics of the pharmaceuticals through organs can vary significantly. For some treatments, patient measurements can be made to tailor the treatment, either prior to treatment, or as an adjustment to later administrations in fractionated treatments.

The ARSAC guidance notes have published standard administered activities (DRLs) for diagnostic Nuclear Medicine procedures (ARSAC 2020). These should not be exceeded for diagnostic procedures, and local DRLs are generally set that account for the equipment and software used at each centre. These standard activities are published alongside the Effective Dose for each radiopharmaceutical, allowing a quick estimate of patient dose to be achieved. The guidance notes also contain information with respect to breastfeeding restrictions and uterus dose (for calculation of foetal dose in pregnancy).

CASE STUDY: ASSESSMENT OF DOSE AND RISK FOR A RESEARCH STUDY

A research study was designed to include nine whole-body PET-CT scans obtained in the same patient over time to assess the impact of treatment on tumour size. An expert Clinical Scientist in Nuclear Medicine (an MPE) assessed the protocol and the doses from the potential scans for each of the study participants. A clinical expert outlined which of the proposed scans would have been carried out as part of routine follow-up for the participants in the study. The MPE liaised with a colleague in diagnostic radiology to assess the effective dose from the CT element of the study and prepared a dose assessment and considered the lifetime risk of developing cancer as a result of that exposure. Using the ARSAC guidance notes (ARSAC 2020), the effective dose was confirmed to be 7.6 mSv per procedure from the administration of the proposed 400 MBq of F-18 FDG. The diagnostic radiology MPE confirmed the dose for whole body attenuation scanning as 6.5 mSv per procedure. A report was written, which was submitted for ethical review by an expert committee outlining the total protocol dose of 127 mSv, which results in an additional risk of cancer induction from radiation exposure of 0.6%. In their report, they put this in context against the UK natural background radiation exposure of 2.3 mSv per annum. They also advised that the natural incidence of cancer in the UK is 50% and included a contextual statement considering each participant's prognosis. A further review was also made of the participant information leaflet (PIL) to ensure any statements describing radiation usage and risk were appropriate and correct.

While there are no limits laid down in legislation for the activity administered to a patient undergoing molecular radiotherapy, as there are for diagnostic nuclear medicine procedures, in many cases standard administered activities are set out in guidance and prescribed, based on the patients' clinical presentation, sometimes being adjusted for patient mass or surface area as described in IPEM report 104 (IPEM 2011b). The British Thyroid Association has published guidelines for the management of thyroid cancer, setting recommendations for administered activities of 1.1 GBq, 3.7 GBq and 5.5 GBq of Iodine-131 (^{131}I) depending on patient risk, the extent and spread of disease, and repeat exposures (Perros et al. 2006). The maximum tolerated dose is established in Phase I trials prior to clinical use, with further phase II and III trials testing effectiveness and safety (IPEM 2011b).

EANM (2020b) and SNMMI (2020) in the USA have published guidance for molecular radiotherapy, which includes ways to perform dosimetric assessments for various therapies, with the EANM guidelines endorsed by the BNMS. EANM guidelines are also endorsed by the UK Internal Dosimetry User Group (IDUG 2015), a group of experts advising on dosimetry. The SPC for radiopharmaceuticals contains tabulated organ and whole-body doses. These figures have been measured during the trial and licensing stages for each radiopharmaceutical and can be useful in assessing the dose to organs that might be at risk when high activities are administered.

CASE STUDY: CALCULATING THE DOSE TO ORGANS AT RISK

In support of an IR(ME)R (2017), practitioner applying for a license a Clinical Scientist was asked to calculate the dose to organs at risk from ^{131}I therapies for benign thyroid disease.

The scientist considered the uptake in the thyroid for a number of benign conditions using data from Leslie et al. (2003), Al-Muqbel and Tashtoush (2010) and Hooper and Caplan (1977). These papers suggest that the uptake ranges for these diseases are around 50% for Graves' disease, 30–35% for toxic multinodular goitre, and 15% for euthyroid (which could be used as a surrogate for mild hyperthyroidism). Hence, maximum doses to organs at risk were calculated for uptake at 15%, 35% and 55%, with up to 800 MBq administered activity.

The dose to various organs from this treatment is quoted in the product characteristics summary for Sodium Iodide (I-131) Capsules, Summary of Product Characteristics (MHRA 2020).

The maximum permitted administered activity of 800 MBq was used to calculate the dose to organs at risk adults, as shown in Table 5.10.

These dose estimates assume normal physiology, especially gastric and renal functions. Renal impairment could significantly alter these estimates

The effect of abnormal focal uptake outside of the thyroid would need to be assessed individually but is not normally expected.

The Scientist concluded that the maximum doses to organs at risk in this therapy are all <1 Gy, and therefore well below the levels of significant early effects as described in the ICRP report 118 (ICRP 2015b) describing the threshold doses for tissue reactions. The findings were then discussed with the clinician to support their understanding.

TABLE 5.10
Dose to Organs at Risk from 800 MBq I-131 for Different Thyroid Uptake

Thyroid Uptake	Maximum Administered Activity (MBq)	Organ at Risk	Dose to Organ at Risk (mGy/MBq)	Maximum Dose to OaR (mGy)
55%	800	Stomach wall	0.46	368
		Bladder wall	0.29	232
35%	800	Stomach wall	0.46	368
		Bladder wall	0.4	320
15%	800	Bladder wall	0.52	416
		Stomach wall	0.46	368
		Small intestine	0.28	224

5.16 FUTURE DEVELOPMENTS

Future developments in Nuclear Medicine include the widespread roll-out of solid-state detectors for imaging in Nuclear Medicine. These systems generally offer increased sensitivity and resolution, which can greatly reduce both patient radiation dose and scan time, leading to greater patient throughput. The limiting factor for such units is cost.

The development of PET-MR, which is currently primarily used for research and neurology applications, is ongoing. At present, this is not funded by NHS England, limiting its use to research and private organisations.

Another trend in Nuclear Medicine is the roll-out of quantitative imaging in SPECT, which is now available at many Nuclear Medicine departments and is being evaluated by Clinical Scientists and Radiologists to identify future clinical applications. There are also new developments in radiopharmaceuticals, with new pharmaceuticals labelled with both diagnostic and therapeutic isotopes. This combination, often referred to as "theranositics", allows better planning and monitoring of mRT, which converges with some of the themes in the next chapter on Radiotherapy Physics.

REFERENCES

ACR – AAPM (2018a) American College of Radiography and American Association of Physicists in Medicine "Technical Standard for Nuclear Medicine Physics Performance Monitoring of Gamma Cameras". Online Available from: https://www.acr.org/-/media/ACR/Files/Practice-Parameters/Gamma-Cam.pdf [accessed 14 September 2020].

ACR – AAPM (2018b) American College of Radiography and American Association of Physicists in Medicine "Technical Standard for Nuclear Medicine Physics Performance Monitoring of PET/CT Imaging Equipment". Online Available from: https://www.acr.org/-/media/ACR/Files/Practice-Parameters/pet-ct-equip.pdf?la=en [accessed 14 September 2020].

Al-Muqbel KM and Tashtoush RM (2010) "Patterns of Thyroid Radioiodine Uptake: Jordanian Experience". *J Nucl Med Technol* **38**(1): 32–36.

ARSAC (2020) Administration of Radioactive Substances Advisory Committee "Notes for Guidance on the Clinical Administration of Radiopharmaceuticals and Use of Sealed Radioactive Sources". Online Available from: https://www.gov.uk/government/publications/arsac-notes-for-guidance [accessed 14 September 2020].

BNMS (2020) British Nuclear Medicine Society "Clinical Guidelines". Online Available from: https://www.bnms.org.uk/page/BNMSClinicalGuidelines [accessed 22 September 2020].

CDGR (2019) "Carriage of Dangerous Goods Regulations". Online Available from: https://www.hse.gov.uk/cdg/regs.htm [accessed 22 September 2020].

Cherry S, Sorenson J and Phelps M (2012) *Physics in Nuclear Medicine*, 4th edition. Imprint Saunders, Elsevier Inc

EANM (2007) European Association of Nuclear Medicine "Guidelines on Current Good Radiopharmacy Practice (cGRPP) in the Preparation of Radiopharmaceuticals". Online Available from: https://www.eanm.org/publications/guidelines/gl_radioph_cgrpp.pdf [accessed 22 September 2020].

EANM (2010) European Association of Nuclear Medicine "Routine Quality Control Recommendations for Nuclear Medicine Instrumentation". *Eur J Nucl Med Mol Imaging* **37**: 662–671

EANM (2020a) European Association of Nuclear Medicine "Clinical Guidelines". Online Available from: https://www.eanm.org/publications/guidelines/ [accessed 22 September 2020].

EANM (2020b) European Association of Nuclear Medicine "Guidelines – Dosimetry". Online available from: https://www.eanm.org/publications/guidelines/dosimetry/ [accessed 22 September 2020].

EC (2010) European Commission "EU Guidelines to Good Manufacturing Practice. Medicinal Products for Human and Veterinary Use". Online Available from: https://ec.europa.eu/health/documents/eudralex/vol-4_en [accessed 22 September 2020].

EPR (2016) *The Environmental Permitting (England and Wales) Regulations.* SI 2010/67. The Stationery Office, London, England.

EudraLex (2011) "Good Manufacturing Practice (GMP)". Online Available from: https://ec.europa.eu/health/documents/eudralex/vol-4_en [accessed 22 September 2020].

Granerus G and Aurell M (1981) "Reference Values for 51Cr-EDTA Clearance as a Measure of Glomerular Filtration Rate". *Scand J Clin Lab Invest* **41**(6): 611–616. doi: 10.3109/00365518109090505.

HBN (2018) Health Building Note 14-01 "Pharmacy and Radiopharmacy Facilities". Online Available from: https://www.gov.uk/government/publications/guidance-on-the-design-and-layout-of-pharmacy-and-radiopharmacy-facilities [accessed 22 September 2020].

Health and Safety (Sharp Instruments in Healthcare) Regulations (2013) *SI 645*. The Stationery Office, London, England.

Hooper PL and Caplan RH (1977) "Thyroid Uptake of Radioactive Iodine in Hyperthyroidism". *JAMA* **238**: 411–413.

HSE (2018) "Work with Ionising Radiation. Ionising Radiation Regulations 2017. Approved Code of Practice and Guidance". *L121*, 2nd edition. Online Available from: https://www.hse.gov.uk/pubns/priced/l121.pdf [accessed 30 October 2020].

Human Medicines Regulations (2012) "Human Medicine Regulations SI 2012 No. 1916". Online Available from: https://www.legislation.gov.uk/uksi/2012/1916/contents/made [accessed 22 September 2020].

IAEA (2003) International Atomic Energy Agency "Quality Control Atlas for Scintillation Camera Systems". Online available from: https://www-pub.iaea.org/MTCD/Publications/PDF/Pub1141_web.pdf [accessed 14 September 2020].

IAEA (2008) *International Atomic Energy Agency "Operational Guidance on Hospital Radiopharmacy".* IAEA, Vienna.

IAEA (2012) International Atomic Energy Agency "Cyclotron Produced Radionuclides: Guidance on Facility Design and Production of [18F] Fluorodeoxyglucose (FDG)". *IAEA Radioisot Radiopharm Ser* (3): 153.

ICRP (2007) International Commission on Radiological Protection "Publication 103: The 2007 Recommendations of the International Commission on Radiological Protection". *Ann ICRP* **37**(2–4).

ICRP (2008) International Commission on Radiological Protection "Radiation Dose to Patients from Radiopharmaceuticals – how to protect your hands in Nuclear Medicine" Annexe 3 to ICRP Publication 53 ICRP Publication 106". *Ann ICRP* **38**: 1–2.

ICRP (2015a) International Commission on Radiological Protection "Radiation Dose to Patients from Radiopharmaceutical: A Compendium of Current Information Related to Frequently Used Substances". *ICRP Publ 128 Ann ICRP* **44**(2S).

ICRP (2015b) International Commission on Radiological Protection "ICRP Statement on Tissue Reactions/ Early and Late Effects of Radiation in Normal Tissues and Organs – Threshold Doses for Tissue Reactions in a Radiation Protection Context". ICRP Publication 118. *Ann ICRP* **41**(1/2).

IDUG (2015) Internal Dosimetry Users Group "Whole Body Dosimetry Guidance". Online Available from: http://www.idug.org.uk/wp-content/uploads/2017/05/IDUGI-131-Whole-Body-Dosimetry-Final.pdf [accessed 24 August 2020].

IEC (2015) International Electrotechnical Commission "Radionuclide imaging devices - Characteristics and Test Conditions - Part 2: Gamma Cameras for Planar, Whole Body, and SPECT Imaging". IEC 61675-2:2015.

IEC (2018) International Electrotechnical Commission "Nuclear medicine instrumentation - Routine Tests - Part 3: Positron Emission Tomographs". IEC TR 61948-3.

IPEM (2002) Institute of Physics and Engineering in Medicine "Radioactive Sample Counting - Principles and Practice". Report 85. IPEM, York.

IPEM (2011a) Institute of Physics and Engineering in Medicine "Mathematical techniques in Nuclear Medicine". Report 100. IPEM, York.

IPEM (2011b) Institute of Physics and Engineering in Medicine "Dosimetry for Radionuclide Therapy". Report 104. IPEM, York.

IPEM (2011c) Institute of Physics and Engineering in Medicine "Medical Cyclotrons (Including PET Radiopharmaceutical Production)". Report 105. IPEM, York.

IPEM (2013a) Institute of Physics and Engineering in Medicine "Quality Control of Gamma Cameras and Nuclear Medicine Computer Systems". Report 111, IPEM, York.

IPEM (2013b) Institute of Physics and Engineering in Medicine "Quality Assurance of PET and PET/CT Systems". IPEM Report 108. IPEM, York.

IPEM (2018) Institute of Physics and Engineering in Medicine "Excretion Factors: the Percentage of Administered Radioactivity Released to Sewer for Routinely Used Radiopharmaceuticals". Online Available from: https://www.ipem.ac.uk/Portals/0/Excretion%20factors%20Sept%202018.pdf?ver=2018-10-03-150031-463 [accessed 20 November 2020].

IR(ME)R (2017) "Ionising Radiations Regulations". SI. 1075. Online available from: http://www.legislation.gov.uk/uksi/2017/1075/contents/made [accessed 13 September 2020].

Kaalep et al. (2020) "Feasibility of State-of-the-Art PET/CT Systems Performance Harmonisation". *Eur J Nucl Med Mol Imaging* **45**(**8**): 1344–1361. doi: 10.1007/s00259-018-3977-4.

Lawson RS (2013) *The Gamma Camera: A Comprehensive Guide*. IPEM, York.

Leslie WD, Ward L, Salamon EA, Ludwig S, Rowe RC and Cowden EA (2003) "A Randomized Comparison of Radioiodine Doses in Graves' Hyperthyroidism". Online Available from: https://doi.org/10.1210/jc.2002-020805 [Accessed online 18th March 2021]

Medicines Act (1968) "Medicines Act". Online Available from: https://www.legislation.gov.uk/ukpga/1968/67/contents [accessed 13 September 2020].

MHRA (2017) Medicines and Healthcare Products Regulatory Agency "Rules and Guidance for Pharmaceutical Manufacturers and Distributors - The Orange Guide". Online Available from: https://mhrainspectorate.blog.gov.uk/2016/12/02/the-2017-orange-and-green-guides/ [accessed 13 September 2020].

MHRA (2020) "Summary Of Product Characteristics Name of the Medicinal Product Theracap131 37 MBq-5.55 GBq Capsules, Hard". https://mhraproducts4853.blob.core.windows.net/docs/a8d7affb68d7672c8015da045bf296aa85d91342 [accessed 19 November 2020].

NEMA (2018a) National Electrical Manufacturers Association "Performance Measurement of Gamma Cameras". Online Available from: https://www.nema.org/standards/view/performance-measurements-of-gamma-cameras [accessed 14 September 2020].

NEMA (2018b) National Electrical Manufacturers Association "Performance Measurement of Positron Emission Tomographs (PET)". Online Available from: https://www.nema.org/standards/view/Performance-Measurements-of-Positron-Emission-Tomographs [accessed 14 September 2020].

NHS England (2019) "Diagnostic Imaging Dataset 12 Months to March 2019". Online Available from: https://www.england.nhs.uk/statistics/wp-content/uploads/sites/2/2019/07/Provisional-Monthly-Diagnostic-Imaging-Dataset-Statistics-2019-07-18.pdf [accessed 13 September 2020].

NPL (2006) National Physical Laboratory "Protocol for Establishing and Maintaining the Calibration of Medical Radionuclide Calibrators and their Quality Control" Measurement Good Practice Guide No. 93". Online Available from: https://www.npl.co.uk/special-pages/guides/establishing-maintaining-calibration-radionuclide.pdf [accessed 13 September 2020].

Perros P et al. (2006) "British Thyroid Association Guidelines for the Management of Thyroid Cancer". *Clin Endocrinol (Oxf)* **81**(Suppl 1): 1–122.

PHE (2016) Public Health England "Ionising Radiation Exposure of the UK Population 2010 Review". Online Available from: https://assets.publishing.service.gov.uk/government/uploads/system/uploads/attachment_data/file/518487/PHE-CRCE-026_-_V1-1.pdf [accessed 19 October 2020].

Phelps M, ed. (2006) *PET Physics, Instrumentation and Scanners,* Springer Verlag, New York Inc.

RCR (2016) Royal College of Radiologists BFCR(16)3 "Evidence-Based Indications for the Use of PET-CT in the United Kingdom". Online Available from: https://www.rcr.ac.uk/publication/evidence-based-indications-use-pet-ct-united-kingdom-2016 [accessed 13 September 2020].

RCR (2019) Royal College of Radiologists "RCR Referral Guidelines". Online available from: https://www.rcr.ac.uk/clinical-radiology/being-consultant/rcr-referral-guidelines/about-irefer [accessed 13 September 2020].

RPS (2016) Royal Pharmaceutical Society "Quality Assurance of Aseptic Preparation Services: Standards". Online Available from: https://www.rpharms.com/Portals/0/RPS%20document%20library/Open%20access/Professional%20standards/Quality%20Assurance%20of%20Aseptic%20Preparation%20Services%20%28QAAPS%29/rps---qaaps-standards-document.pdf [accessed 13 September 2020].

SNMMI (2020) Society of Nuclear Medicine and Molecular Imaging "Clinical Guidelines". Online Available from: https://www.snmmi.org/ClinicalPractice/content.aspx?ItemNumber=10817&navItemNumber=10786 [accessed 13 September 2020].

Stabin MG (1996) "MIRDOSE: Personal Computer Software for Internal Dose Assessment in Nuclear Medicine". *J Nucl Med* **37**(3): 538–546.

Stabin MG, Sparks RB and Crowe E (2005) "OLINDA/EXM: The Second-Generation Personal Computer Software for Internal Dose Assessment in Nuclear Medicine". *J Nucl Med* **46**(6): 1023–1027.

Theobold A (2010) *Sampson's Textbook of Radiopharmacy*, 4th edition. Pharmaceutical Press, London.

UKRPG (2016) UK Radiopharmacy Group "Quality Assurance of Radiopharmaceuticals", 4th edition. Online Available from: https://www.bnms.org.uk/page/UKRGGuidelines [accessed 13th September 2020].

UKRPG (2017) British Nuclear Medicine Society "Guidelines for the Provision of Radiopharmacy Services in the UK". Online Available from: https://cdn.ymaws.com/www.bnms.org.uk/resource/resmgr/guidelines/radiopharmacy_support_docume.pdf [accessed 14 September 2020].

UKRPG (2020) UK Radiopharmacy Group "Guidance Notes". Online Available from: https://www.bnms.org.uk/page/UKRGGuidelines [accessed 19 October 2020].

WHO (2003) World Health Organisation "Good Manufacturing Practices for Radiopharmaceutical Products". Online Available from https://www.who.int/medicines/areas/quality_safety/quality_assurance/GMP-RadiopharmaceuticalProductsTRS908Annex3.pdf?ua=1 [accessed 24th September 2020].

6 Radiotherapy Physics

Andrea Wynn-Jones, Caroline Reddy, John Gittins, Philip Baker, Anna Mason and Greg Jolliffe

CONTENTS

6.1 INTRODUCTION

Cancer is a condition where cells divide and grow uncontrollably. One in two of us will develop a cancer at some point in our lifetimes and it has been estimated that around 1000 cases of cancer are diagnosed each day in the UK as described by CRUK (CRUK 2020). Forty-five percent of patients will undergo surgery, 28% will have chemotherapy and 27% will have radiotherapy at some point during their cancer journey, either for primary management or for symptom control.

Surgery is often used to remove the bulk of the tumour if it is easily accessible and if the patient does not have any co-morbidities such as obesity or heart disease, which would make surgery too risky. Both chemotherapy and radiotherapy are often used following surgery to treat remaining cancer cells and prevent microscopic spread. Medical and clinical oncologists can prescribe a variety of treatments depending on the type and stage of the tumour and considering the patient's suitability for treatment. Precision, or personalised medicine, where knowledge of an individual's genetic make-up is used to decide which treatment(s) would be most effective, is also becoming increasingly important.

Radiotherapy usually involves directing high-energy X-rays from a linear accelerator, (or LINAC), to kill cancer cells, but as radiation will damage normal as well as cancerous cells, beams must be accurately delivered to keep side effects to a minimum. For radiobiological reasons (discussed briefly later), radiotherapy is commonly delivered over several days or weeks to allow healthy tissue time to recover between treatments. This is known as fractionation. There are often historic reasons for the choice of fractionation regimes for certain sites. Treatment regimes in the past were often derived from clinical experience. Nowadays guidance is more established and good practice recommendations support Clinical Oncologists to use evidence-based treatment regimens, for example, those recommended by Royal College of Radiologists (RCR) guidance, on dose fractionation, National Institute for Health and Care Excellence (NICE) guidance such as the NICE guideline for breast radiotherapy (NICE 2018), or from established clinical trials.

Many different professionals are involved in cancer care: radiologists interpret and report diagnostic images, laboratory pathology staff analyse tissue biopsies and blood biomarkers to diagnosis and identify the stage of progression (staging) of the cancer, physiotherapists, occupational therapists, and dieticians help patients to manage their symptoms during treatment and rehabilitation. Radiotherapy delivery is a complex process that involves multiple specialist staff groups, so multidisciplinary working is essential. Clinical Scientists work closely with Clinical Oncologists, who prescribe treatment and take overall clinical responsibility, and Therapeutic Radiographers, who CT scan patients at the pre-treatment stage, deliver treatment and support the patient throughout the radiotherapy process. The main staff groups that Clinical Scientists come into contact within radiotherapy are summarised in Table 6.1.

TABLE 6.1
Professional Groups That Clinical Scientists Will Encounter While Working in Radiotherapy

Profession	Title	Role	Comment
Medical Doctors	Surgeon	Surgery	Primary responsibility for the patient during surgery – removes the tumour and nearby tissue during surgery. Surgeons also perform certain types of biopsy to help diagnose cancer.
	Medical Oncologist	Prescribes chemotherapy and/or immunotherapy	A medical oncologist treats cancer using chemotherapy or other medications, such as immunotherapy.
	Clinical Oncologist	Prescribes radiotherapy	Treats cancer with radiation, working with Clinical Scientists and dosimetrists to optimise patient treatment using radiation.
Nursing Staff	Various	Patient care	Throughout the patient's journey, nurses help to co-ordinate and deliver direct patient care, including managing side effects, cancer symptoms, problems at home, and follow-up appointments.
Therapeutic Radiographers	Various	Radiotherapy treatment delivery	Primarily involved in the practical aspects of safe treatment delivery, but can also include pre-treatment imaging, some treatment planning activities, and pre-treatment checks of patient treatment plans.
Clinical Technologists	Planning Dosimetrists	Immobilisation and Dosimetry	Treatment planning using specialist software and immobilisation of patients ready for treatment.
	Engineering Technicians	Engineering	Engineering technicians (often called engineers) keep radiotherapy equipment functioning.
Various	Radiotherapy Assistants	Patient care	Undertake patient care activities such as taking ECGs and observations and assisting them moving through the department.
	Clerical staff	Admin	Manage the appointments to ensure patients achieve the cancer waiting time target for radiotherapy.

Radiotherapy Physics as a specialism employs the highest number of Clinical Scientists in Medical Physics, although they are largely unnoticed by patients as direct patient contact is rare. A medium-sized hospital department will treat over 100 patients a day, spread over three or more linear accelerators. Approximately half of these patients will be treated with the intent of completely removing the cancer (radical intent), requiring an individualised plan and complex imaging during treatment. Preparatory scans and planning can take several days per patient.

Radiotherapy delivery is highly complex, requiring Clinical Scientists to be involved in many of the more scientific and technical aspects to assure high-quality treatment delivery as well as safety, as the high doses of radiation exposure involved can be lethal if not delivered correctly. Clinical Scientists specialising in Radiotherapy Physics need to have a good understanding of radiation beam production, radiation interactions within human tissues and biological effects. They should understand the concepts of absorbed dose, isodose curves and dose-volume histograms (DVH) involved in treatment planning and be able to use sophisticated treatment planning computers to model dose distribution.

Clinical Scientists working in radiotherapy need to pay great attention to detail to perform ionising radiation checks on treatment equipment and achieve high levels of precision and accuracy. They must also have good communication skills and be logical thinkers in order to formulate and communicate solutions and manage and advise in unusual scenarios. Clinical Scientist duties can include:

- Commissioning of new equipment and techniques to avoid systematic errors
- Producing and checking treatment plans
- Providing expert advice with difficult plans, patient positioning and non-standard scenarios
- Establishing and evaluating clinical trials
- Quality control (QC) and maintenance of equipment
- Providing advice to meet regulatory compliance
- Evaluating and managing risks from ionising radiation
- Scientific and technical research and innovation in radiotherapy
- Radiation safety aspects of radiotherapy to protect patients, staff and visitors

Importantly, current legislation in the UK (IRMER 2017) requires that experienced Clinical Scientists are available to provide Medical Physics Expert (MPE) oversight and scientific input to all aspects of radiotherapy treatment.

Clinical Scientists also work with other professionals in Radiotherapy Physics such as Clinical Technologists, of which there are two main types traditionally referred to as Dosimetrists and Engineering Technicians. Dosimetrists have direct patient contact when preparing immobilisation systems for each patient to help ensure that the patient is in the same position for treatment each day. Using specialised treatment planning computers, dosimetrists also prepare individualised treatment plans that are optimised to treat the tumour while sparing the normal tissue. Engineering Technicians usually receive specialist training from equipment manufacturers to maintain and repair treatment equipment. Having technicians on site aims to ensure that equipment is always functioning, typically achieving 98% "uptime" performance. Although many staff groups are involved in delivering radiotherapy, this chapter focuses on the role of the Clinical Scientist.

6.2 RADIOBIOLOGY

Many of the improvements in cancer treatment achieved over the last couple of decades are the result of rapid developments in radiotherapy treatment, enabled by advances in technology and computing power. Developments include the ability to calculate and deliver beams conforming to the exact shape of tumours (conformal radiotherapy) using Intensity Modulated Radiotherapy (IMRT) techniques, and the use of pre-treatment imaging to ensure treatments are delivered as planned. The ability to deliver precision radiotherapy, however, is only one part of the story. At the end of every clinical radiotherapy plan that is created, there is a patient, and a biological system that could potentially respond to the delivered radiotherapy in a unique way; this is where the scientific disciplines of Radiotherapy Physics and radiobiology overlap.

Radiobiology can inform radiotherapy practice at various levels. On a conceptual basis, modelling mechanisms and processes that explain observed responses of both tumour and normal tissues can be used to propose developments and adjustments to treatment strategies. Examples include either decreasing or increasing the number of treatment sessions (hypo- and hyper-fractionation schedules) or the use of agents to either sensitise cancer cells to radiation effects (cell sensitisers) or protect healthy tissue. A good example of hypo-fractionated radiotherapy is a treatment technique known as Stereotactic Ablative Body Radiotherapy (SABR) where high doses of precisely targeted radiation are used to control small tumours, typically in the lung or for cancer that has only just spread to other areas. In SABR, treatment is delivered over relatively few visits compared to conventional radiotherapy treatments.

The use of high linear energy transfer (LET) radiation, such as proton therapy, is also possible. Proton therapy potentially offers some advantages to using high-energy X-rays, relating to the physics of how the radiation dose is distributed in the target, but is not suitable for all patients. NHS proton therapy facilities in the UK are currently limited to a small number of specialist centres.

In routine daily practice, advice may be sought from a Clinical Scientist for managing unscheduled gaps in radiotherapy treatment (e.g. due to illness, social issues or machine breakdowns). Clinical Scientists provide advice on the rare occasions when treatments are not delivered as planned, for radiation exposure during pregnancy and radiotherapy in children (paediatric exposures). Decision making in such situations may require a detailed understanding of not only the physical dosimetry, including anatomical dose distribution, but also the radiobiological processes involved, and the impact of age, sex, dose rate and time factors.

6.2.1 Radiation Damage

Radiation damage and cancer cell death, while minimising damage to healthy cells, is the main purpose of radiotherapy treatment. Ionising radiation reduces the ability of cells to live and reproduce by damaging their DNA. To further understand the radiobiological ideas underpinning radiotherapy, it is useful to introduce the concepts of lethal damage, potentially lethal damage and sub-lethal damage. In lethal damage, the cellular repair is not possible and cell death becomes inevitable. Potentially, lethal damage could lead to cellular death, but this depends on post-irradiation conditions, whereas sub-lethal damage is damage that could be repaired given time. Sub-lethal damage can result in increased normal cell survival, especially if the total dose is split into fractions with a suitable time interval between fractions. The prescribed dose of radiation is usually divided into multiple smaller doses called fractions. This allows healthy cells to recover between treatments.

As the exposure time for a given total dose increases, there will be a resulting overall reduction in biological effect. Not all tissues have the same capacity for repair; the differences exhibited in repair kinetics between normal cells and cancer cells can be exploited by careful selection of the dose delivered per fraction. This means that the timing between fractions and the number of fractions delivered can be adjusted for therapeutic advantage.

6.2.2 Radiation Modifiers and the "5 Rs" of Radiobiology

The 5 Rs of radiobiology (Figure 6.1) as described by Dale and Jones (2007) provide a framework from which the success or failure of treatment of cancer cells using ionising radiation can be understood. These are now very briefly described.

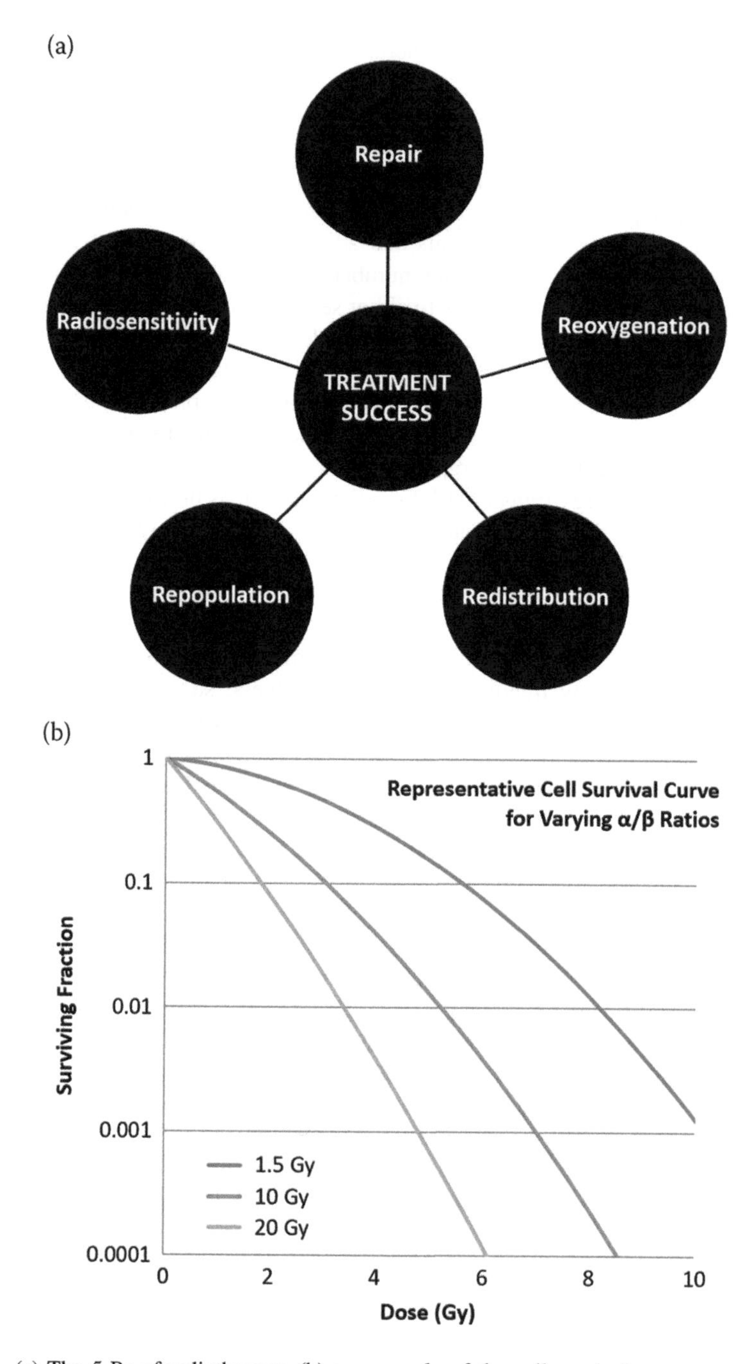

FIGURE 6.1 (a) The 5 Rs of radiotherapy, (b) an example of the cell survival curve.

6.2.2.1 Radiosensitivity and Repair

Cell survival curves, such as that shown in Figure 6.1(b), are used to show how the surviving fraction of cells varies with the total ionising radiation dose delivered. The cell survival curves typically observed for mammalian cells can be qualitatively described in terms of their shape. For low LET radiation and low doses, the survival curve typically starts as a straight line when plotted as a log (survival) vs linear (total dose) graph. At higher doses, a bending of the graph can be observed, which tends to straighten again for higher doses. Various mathematical models have been proposed to relate the shape of the observed curves to biophysical processes that occur within the cells due to exposure to ionising radiation.

The linear-quadratic model, as described by Dale and Jones (2007), is a commonly used model, which assumes two contributions to cell death. The first component is proportional to dose (D), the second component is proportional to the square of the dose, reflecting the observation that lethal damage to cells results from two separate DNA breaks. The form of this model can be expressed as

$$S = e^{-\alpha D - \beta D^2}$$

where S corresponds to the fraction of surviving cells, D the dose (Gy), α and β are parameters that characterise the cell radiosensitivity. The α/β ratio has units of Gy and corresponds to the delivered dose when the linear and quadratic contributions to cell kill are equal.

Different tissues in this model possess differing ratios of α/β. Typically, early responding tissues are characterised by a relatively high α/β ratio, whereas late responding tissues are characterised by low α/β ratios. In general, tumours are characterised by relatively high α/β ratios and therefore behave more similarly to early responding tissues, although current evidence suggests that some types of tumour cells (e.g. prostate adenocarcinoma) exhibit relatively low α/β ratios. The concept of iso-effective total dose can be used to explain the relative effects of radiation on early and late responding tissue from altering the dose per fraction. For late responding tissues, characterised by lower α/β ratios, the iso-effective total dose increases more rapidly with the increasing number of fractions compared to early responding tissue, that is, late responding tissues are more sensitive to a change in dose per fraction.

The shoulder exhibited by cells with a low α/β ratio can be exploited when the total treatment dose is split into daily fractions. Assuming repair of cells can occur between fractions (sub-lethal damage), cell kill effectively remains in the shoulder region, reducing the magnitude of the quadratic component, resulting in significant sparing of cells for a given total dose. This means that fractionation can be exploited to change the relative cell killing in cell populations with different α/β ratios.

Scientists should be careful to ensure that the values used in any radiobiological calculations correspond to the most recent published evidence and match as closely as possible to the irradiation conditions used in the clinical situation.

6.2.2.2 Redistribution and Repopulation

For normal tissue, homeostatic control mechanisms strike a balance between cell proliferation, differentiation and cell loss, whereas in tumours, loss of such control mechanisms leads to uncontrolled proliferation. Human cell populations typically proliferate asynchronously, which means that in any cell population, a certain proportion of cells will exist in a relatively radio-resistant state. Following treatment with radiation, cells that were in a radio-resistant state may have moved to a radio-sensitive phase. This is known as redistribution. These cells could be damaged by subsequent delivery of radiotherapy. Repopulation is the increase in cell division and cell populations occurring after radiation is delivered, in cells that have not been lethally damaged.

6.2.2.3 Reoxygenation

It has been conclusively established that tumour cells that are deprived of oxygen (hypoxic) respond less well to radiotherapy than well-oxygenated cells (e.g. McKeown 2014). Although radio-sensitisation of radiation-resistant hypoxic cells has been achieved in the lab, this has not yet translated to improvements in patient outcome. The state of the tumour vasculature and the relative oxygenation of tumour cells is not known at the time of treatment and will dynamically change over the course of radiotherapy, which may help to partly explain why clinical trials have generated mixed results. Tumour cell hypoxia currently remains a significant barrier to the success of radiotherapy treatment.

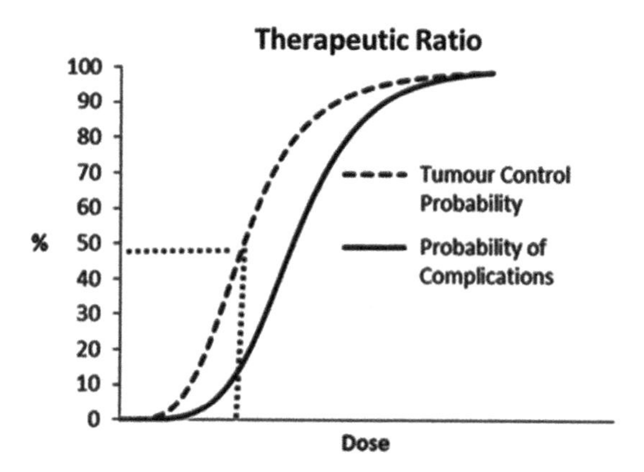

FIGURE 6.2 Therapeutic ratio. A dose used for a given probability of tumour control will result in the corresponding probability of treatment complications.

6.2.2.4 Radiosensitivity

After accounting for the aforementioned factors in the success, or otherwise, of radiotherapy treatments, the concept of intrinsic radiosensitivity was added to the framework to account for an inability to explain, at a mechanistic level, the differences in treatment success for different types of tumour.

6.3 THERAPEUTIC INDEX

The therapeutic index (or window) is a concept that relates the dose response for tumour control (Tumour Control Probability [TCP]) to the dose response curve for normal tissue complications (Normal Tissue Complication Probability [NTCP]). More information can be found in Basic Clinical Radiobiology (Joiner and van der Kogel 2018).

Typically, higher doses are required to increase the probability of tumour control, but this will also increase the level of normal tissue damage leading to complications (Figure 6.2). Manipulation of treatment variables including the use of drugs affecting cell radiosensitivity can increase the separation of theses curves resulting in a wider therapeutic window and more favourable treatment outcomes. In a clinical situation, many factors need to be considered such as early and late effects for normal tissue damage, the emergence of different patterns of toxicity and the practicality of a new regime. Convenience to both staff and patients and potential cost implications should also be considered. Thus, the concept of therapeutic index in a clinical setting must also include some form of mathematical cost model.

6.4 ACCOUNTING FOR GAPS IN TREATMENT

In real life, situations inevitably arise that result in delays or appointment cancellations, such as illness or family-related reasons, bank holidays, or machine breakdowns. A Clinical Scientist is expected to understand the basic principles of radiobiology and should be able to perform calculations to inform clinicians of the dose and fractionation regimes that can be used to compensate for a missed treatment fraction using calculations and RCR guidance on the timely delivery of radical radiotherapy. The relative importance of delays or gaps in treatment schedules depends to some extent on the type of cancer being treated. For example, prolonging treatment for squamous cell head and neck cancers has been shown to be detrimental to overall tumour control and such delays should be avoided wherever possible, whereas prolonging treatment for a relatively slower growing adenocarcinoma, typical of prostate cancer, is not so detrimental. Current guidelines categorise patients as:

Category 1: Patients with rapidly growing tumours such as squamous cell carcinomas. The treatment of such patients should not extend more than 2 days beyond the original prescription.
Category 2: Patients being treated with radical intent but with slower growing tumours such as adenocarcinomas. The treatment period for these patients should not extend more than 5 days beyond the original prescription.
Category 3: Treatment of patients with palliative intent. Prolongation of treatment may not affect the palliative intent.

Wherever possible, changes to the treatment schedule should be made as conservatively as possible considering the aforemenioned categories. Strategies could include treating patients at the weekend to make up for lost fractions, treating twice daily with enough time between fractions, modifying the dose per fraction, or increasing the total dose to compensate for loss of tumour control and giving extra fractions.

If modifications to the treatment regime are proposed, then the linear quadratic model can be manipulated to compare competing or alternative treatment regimens using the concept of Biological Equivalent Dose (BED). BED calculations are also used to indicate previous treatment as an Equivalent Dose in 2 Gy fractions (EQD2 dose). Traditionally, 2 Gy per fraction was common, so organ at risk (OAR) tolerance studies often report the effects equivalent to 2 Gy per fraction regimes. BED calculations are also used to adapt and inform dose constraints for OAR.

$$BED = nd\left[1 + \frac{d}{\alpha/\beta}\right] \text{ for normal tissue}$$

$$BED = nd\left[1 + \frac{d}{\alpha/\beta}\right] - K\left(T - T_{delay}\right) \text{ for tumours}$$

where n is number of fractions, d is daily fraction dose, T is total treatment time and K is daily dose factor accounting for repopulation.

CASE STUDY: ADJUSTING FOR A GAP IN RADIOTHERAPY TREATMENT

A patient is prescribed 55 Gy over 20 fractions (often shortened to #), corresponding to 2.75 Gy per #. The expected treatment duration was 26 days (daily treatment Monday to Friday for 4 weeks, so duration includes three weekends), but due to gaps in treatment, this patient is now due to finish in 43 days. The patient has so far received 10 fractions with 10 remaining. The role of the Clinical Scientist would be to suggest corrections to the treatment regime and calculate the potential impact on the expected biological effect of the originally prescribed dose.

Parameters
Use the linear quadratic model and BED.
For tumour:

$$BED = nd\left[1 + \frac{d}{\alpha/\beta}\right] - K(T - T_{delay})$$

K = 0.42 Gy/day for generic cancer type
T_{delay} = 28 days for generic treatments
Calculate for α/β = 10 Gy (standard tumour value)
For normal tissue:

$$BED = nd\left[1 + \frac{d}{\alpha/\beta}\right]$$

Use α/β = 3 Gy

Prescribed BED
Normal Tissue: BED = 105.4 Gy_3
Tumour: BED = 70.1 Gy_{10}

Options:
1. Make no changes to treatment fractionation
Treatment time is increased by 17 days, with T – T_{delay} = 15 days

Normal Tissue:
BED = 105.4 Gy_3 as prescribed

Tumour:
BED = 63.8 Gy_{10} (–9.0%)

2. Add a fraction on at end
Overall treatment time increased to 44 days

Normal Tissue:
BED = 110.7 Gy_3 (+5.0%)

Tumour:
BED = 66.9 Gy_{10} (–4.6%)

3. Increase dose per fraction to keep same tumour BED
Tumour BED:
Need to account for decrease in BED of 6.3 Gy_{10} – so the required tumour BED = 76.4 Gy_{10}
Already delivered 35.1 Gy_{10} – leaves 41.3 Gy_{10}
Required dose per fraction = 3.1 Gy (+12.7%)

Normal tissue BED:
Already received 10 fractions at 2.75 Gy = 52.7 Gy_3
Would be followed by 10 fractions at 3.1 Gy = 63.0 Gy_3
Overall NT BED = 115.7 Gy_3 (+9.8%). It is generally not advisable to increase the normal tissue BED by more than 5%, so this may not be the best option.

4. Increase dose per fraction by more modest amount
Increase the dose per fraction to 2.9 Gy

Tumour BED:
10 fractions at 2.75 Gy = 35.1 Gy_{10}
10 fractions at 2.9 Gy = 37.4 Gy_{10}
Reduction in BED due to extension = 6.3 Gy_{10}
Overall BED = 66.2 Gy_{10} (–5.6%)

Normal tissue BED:
10 fractions at 2.75 Gy = 52.7 Gy_3
10 fractions at 2.9 Gy = 57.0 Gy_3
Overall BED = 109.7 Gy_3 (+4.1%)

Discussion:
Option 2 or option 4 would both be reasonable alternative approaches to take, producing similar reductions in tumour BED and an increase in normal tissue BED. These options would be discussed with the Clinical Oncologist with the understanding that the evidence for radiobiological parameters is sparse and corrections are poorly understood, so these calculations have a large degree of uncertainty. Ultimately, the clinical oncologist would be responsible for prescribing changes in the treatment regime based on these calculations.

6.5 LINEAR ACCELERATORS

In order to understand some of the concepts introduced later relating to planning a patient's treatment, it is useful to briefly review some of the technology behind modern treatment machines. Linear accelerators are the "work horses" of external beam radiotherapy (EBRT). Modern linear accelerators are technically advanced machines capable of delivering high energy X-rays (typically 6 and 10 MV in energy) and electron beams with great precision. Beams can be delivered to pinpoint (mm) accuracy with dosimetric parameters reproducible to within 1% of baseline measurements in addition to treatment beam delivery. Linear accelerators also incorporate imaging capabilities to aid accurate patient setup. More detail can be found in the Handbook of Radiotherapy Physics (Mayles et al. 2007).

The role of the Clinical Scientist is to ensure that Linear accelerators are commissioned correctly after installation to establish baseline performance, and to calibrate the machine in line with national guidelines and standards. It is essential that the dose delivered to a patient is characterised and understood with accuracy and precision and that the dose from one machine is matched to that of another not only within one hospital but also between hospitals. Following installation, it is important to oversee ongoing QC procedures to ensure that the machine continues to perform as expected. Calibration and QC requirements are discussed in more detail in Section 6.10.3.

The Clinical Scientist also has responsibility for ensuring that the radiotherapy service is run safely and complies with regulatory requirements (e.g. Ionising Radiation (Medical Exposures) Regulations [IRMER] and Ionising Radiation Regulations [IRR] as described in Chapter 7). An understanding of the limitations of treatment technique, including the physics of beam delivery and radiation interactions within the patient is also important, so that appropriate advice can be given to clinical colleagues as to the optimum way to use EBRT for the benefit of cancer patients.

Clinical Scientists also work with other staff groups, such as engineering technicians responsible for repair and maintenance so that any performance deviations can be quickly identified and corrected if necessary. This information, along with a detailed understanding of the likely consequence of deviations in equipment performance on patient treatment is then used to devise an appropriate course of action. It is not always practical to initiate adjustment and repair processes as soon as deviations in performance are identified but assessing the consequences of any deviations can be used to inform an appropriate course of action. This could be to justify taking the machine out of clinical use, to plan maintenance and QC, or to limit a machine to less demanding use cases. The Clinical Scientist's knowledge and experience of physics and engineering, coupled with the radiobiological and clinical knowledge of the beam interactions with the patient, underpins this decision-making process.

6.5.1 Major Linear Accelerator Components

A linear accelerator comprises a source of electrons, which are accelerated to the required energy using high powered microwaves. To produce a photon beam, the accelerated electrons are made to collide with a heavy metal target, typically tungsten. The resulting photon beam is shaped and directed towards the tumour in the patient. Figure 6.3 shows a schematic of a typical medical linear accelerator. The accelerator is designed so that the couch, collimator and gantry can all rotate around a single point in space, termed the isocentre. This allows the beam to be directed towards the patient from any angle. The distance from the photon source to the surface of the patient is called the Focus to Surface Distance (FSD). A light field is used to show the position of the beam on the surface and a projection of a crosswire is used to indicate the position of the beam central axis.

Major components of a linear accelerator include the following:

6.5.1.1 The Electron Gun

The electron gun produces electrons by the process of thermionic emission. There are two types of guns used in current accelerators. One design uses voltage pulses to the cathode, and the second type uses voltage pulses applied to a grid in order to control the electrons. It is important that electron production is synchronised with the microwave generator. It is the action of high-power microwaves on the electrons in a waveguide that results in the acceleration of the electrons.

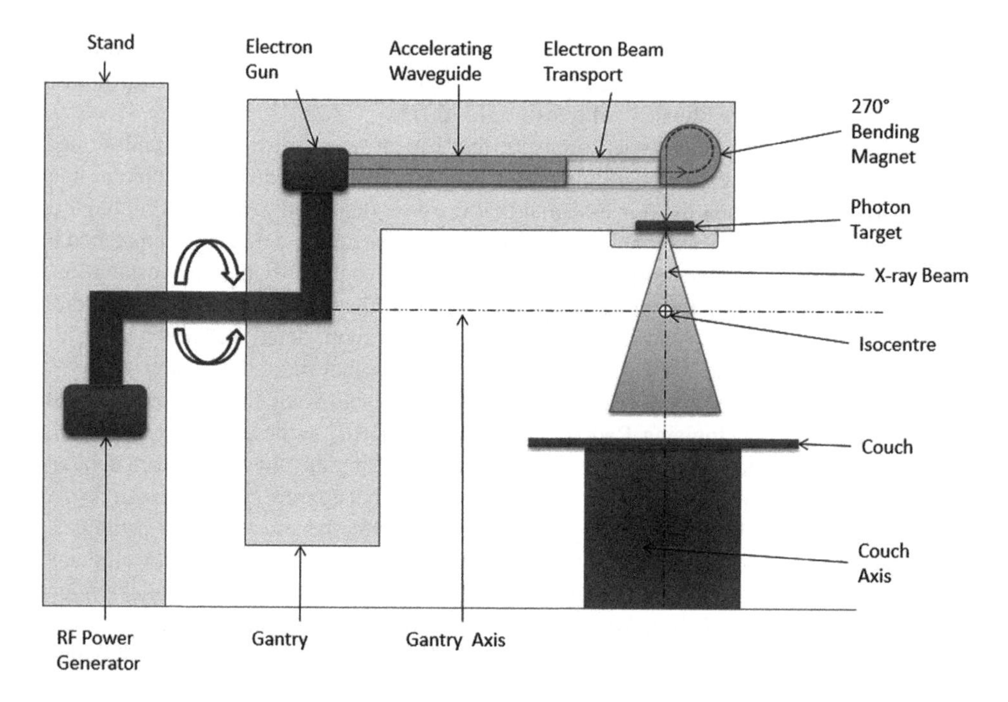

FIGURE 6.3 Linear accelerator schematic in photon delivery mode.

6.5.1.2 Microwave Generator

High powered microwaves are used to accelerate the electrons. These can be produced either by a magnetron, or a klystron. The magnetron is effectively a high-power microwave oscillator, whereas the klystron is used to amplify low power microwaves. In either case, the high-power microwaves are injected into the waveguide, synchronised with the electrons.

6.5.1.3 Modulator

The modulator is designed to provide high voltage pulses that allow the magnetron or klystron to produce high power microwaves. Line modulators used in modern linear accelerators are designed using an artificial transmission line (a series of capacitors and inductors) to store the required energy. The high voltage pulse is discharged through a pulse transformer via a thyratron. It is important that the resulting high voltage pulses are regulated since changes in the pulse parameters could ultimately result in a change in the output of the linear accelerator.

6.5.1.4 Accelerating Waveguide

The accelerating waveguide is a metal cylinder under vacuum. The cylinder is separated by washer-shaped dividing sections, which allow the electron beam to pass through the centre of the waveguide (Figure 6.4). The injected high-power microwaves set up oscillating electric fields within the waveguide and the electrons gain energy from this electric field. The dimensions of the waveguide sections are important. Initially, for a travelling waveguide, the shorter sections act to bunch the electrons and increase their energy by increasing their velocity. As the electrons approach the speed of light, the cavities become longer and more uniform in length, as the energy imparted to the electrons comes from increased relativistic mass as opposed to increased velocity.

6.5.1.5 Electron Beam Transport and Photon Production

A series of evacuated drift tubes, bending magnets, focusing coils and steering coils are used to direct the accelerated electron beam from the waveguide where it is directed towards the X-ray target. For efficient X-ray production, a high atomic number material is used, typically tungsten. At the energies typically used (6 or 10 MV), photons are predominately produced in a forward direction, mainly through bremsstrahlung interactions. It is important to control the position of the beam on the target in order to produce a consistent beam profile. Modern accelerators can switch between producing photons, as described previously, or producing electrons, by replacing the tungsten target with a scattering foil.

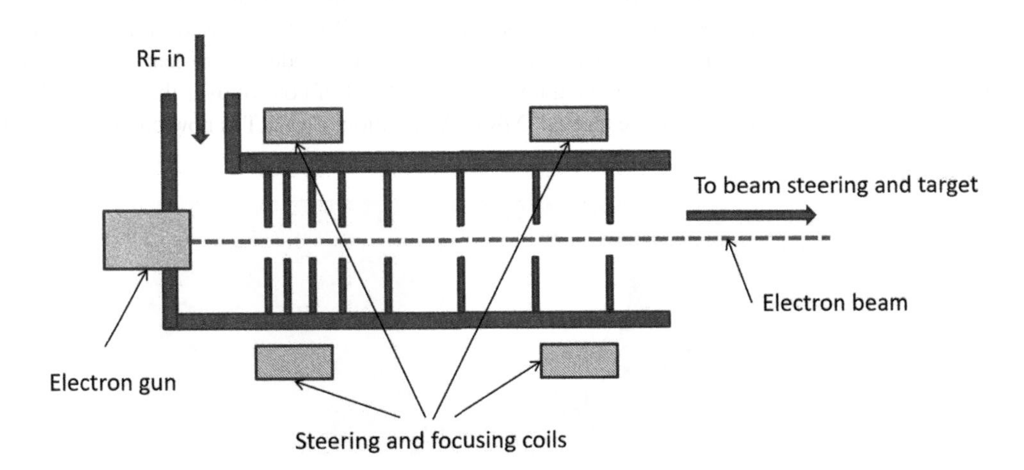

FIGURE 6.4 Schematic of an electron beam waveguide.

6.5.1.6 Primary Collimator

The primary collimator defines the largest field available. Conventionally, the beam intensity is made more uniform across the whole field using a flattening filter, although it is possible to run some machines in a "flattening filter free" mode, potentially resulting in a much higher dose rate. The flattening filter would also be removed from the beam path if the machine was producing an electron beam.

6.5.1.7 Dual Ionisation Chamber

The ionisation chamber is an integral part of the dose monitoring and control system. It is designed to provide signals in four quadrants, which are used in feedback circuits connected to the beam steering system. If the beam transport system is not set up correctly or has a fault, baseline properties of the beam such as its symmetry and flatness can go out of tolerance making it clinically unacceptable. Such fault conditions can be monitored using this ionisation chamber and feedback circuits. The ionisation chamber is also designed to define and control the amount of radiation delivered by the accelerator. The dual design is a safety feature to provide resilience, so a fault occurring in one ionisation chamber won't cause excess dose delivery to the patient.

6.5.1.8 Secondary Collimators

The secondary collimators consist of four blocks or jaws, typically manufactured from tungsten. The relative positions of the blocks are adjustable and used to define a rectangular or square field at the linear accelerator isocentre.

6.5.1.9 Wedges

In many cases, a flat beam does not give the optimum dose and the beam profile needs to be shaped. A physical wedge can be used to alter the dose distribution and attenuate the beam progressively across the entire field. On some machines, similar functionality is provided by using dynamic wedges, where one of the secondary collimator jaws is automatically driven across the field while the beam is on, producing a similar change in the resulting dose distribution compared to the physical wedge.

6.5.1.10 Multi-Leaf Collimator (MLC)

A Multi-Leaf Collimator (MLC), such as that shown in Figure 6.5, provides yet more control over beam shaping. The MLC allows the treatment fields to conform closely to the target volume, compared to simply using square fields defined by the secondary collimators.

There are different designs with different numbers of leaves, and leaf widths, but for all systems, the position of each leaf can be independently controlled. The MLC can be used in a static mode to define a complex shape. They can also be used in a dynamic mode, where leaf positions are adjusted continuously while the beam is on, allowing a high level of control over the dose-gradient and conformality – a delivery technique termed IMRT. A variation of IMRT is now commonly used

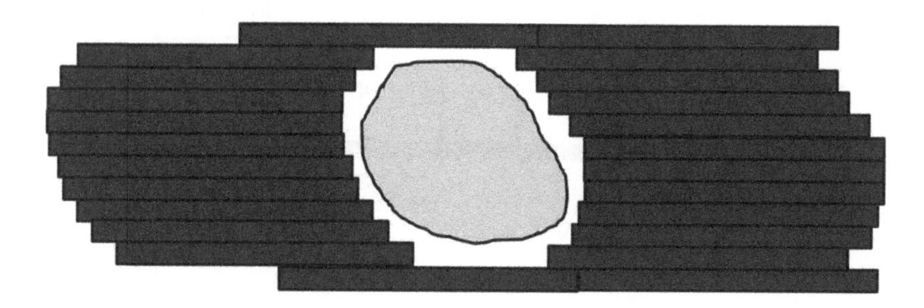

FIGURE 6.5 Schematic of some Multi-Leaf Collimator segments conforming to the shape of the target.

where the gantry rotates around the patient while modulating the beam using MLCs, varying the dose rate and the gantry speed – a delivery technique termed Volumetric Arc Therapy (VMAT).

6.6 TREATMENT PLANNING

The aim of treatment planning is to treat the tumour target volume (sometimes termed the Planning Target Volume [PTV]) with a homogeneous dose while minimising the dose to Organs at risk (OAR) and healthy tissues as described in detail in Clinical Treatment Planning by Parker and Patrocinio (2005). This can be achieved by directing the beam at various angles around the patient; this tends to spread a low dose throughout the tumour's surroundings but prevents localised high doses from accumulating at the entry and exit points of the beam.

When working in treatment planning, Clinical Scientists are involved in several aspects of the treatment planning process, including:

- Devising local protocols (Clinical Scientists can be involved in clinical trials, devising protocols and assessing treatment plans)
- Implementing new treatment techniques, for example, breath-hold treatment methods, IMRT planning, or using Flattening Filter Free beam data for treatment planning
- Performing independent dose checks on treatment plans
- Checking planning system integrity
- Auditing setup errors, to inform PTV margin recipes
- Advising on dose calculations: e.g. dose to a pacemaker, BED/EQD2 calculations, dose assessments for previous treatment/overlap considerations, dose assessment for unintended treatment exposure. The dose calculations are used to inform the treating clinician of likely dose limitations
- Advising on techniques or modalities to use for a treatment case
- Performing QC when software upgrades are performed, or new software is implemented

This list is not exhaustive – there are many aspects to the treatment planning processes that require the attention of a Clinical Scientist to enable the safe delivery of radiotherapy.

6.6.1 Principles Underlying Treatment Planning

Most radiotherapy treatments are delivered using external beam MV X-rays, although electron beams and kV energy X-rays are also used for treating superficial tumours. The choice of treatment modality depends on the position of the target in the patient, guided by considering how the energy of the radiation beam needs to be deposited. Typically, 6–10 MV photons are used for deep-seated tumours as these will give the required penetration depth and acceptable surface dose. A fundamental property of MV beams is that the position of maximum dose occurs a few centimetres below the surface, with the actual depth dependent on the nominal beam energy. In terms of treatment, this is termed the skin sparing effect. The skin sparing effect is beneficial as it can reduce the effect of radiation beam entry in skin tissue. If skin dose is not controlled, reactions of the skin to radiation can be so severe that the patient's treatment must be paused for the skin to recover. Measurement of Percentage Depth Dose (PDD) curves provides a means of quantifying the position of the maximum dose in water. As a Clinical Scientist, you may be involved with explaining PDD measurements to other staff groups to help with treatment modality and energy choice for a given clinical situation.

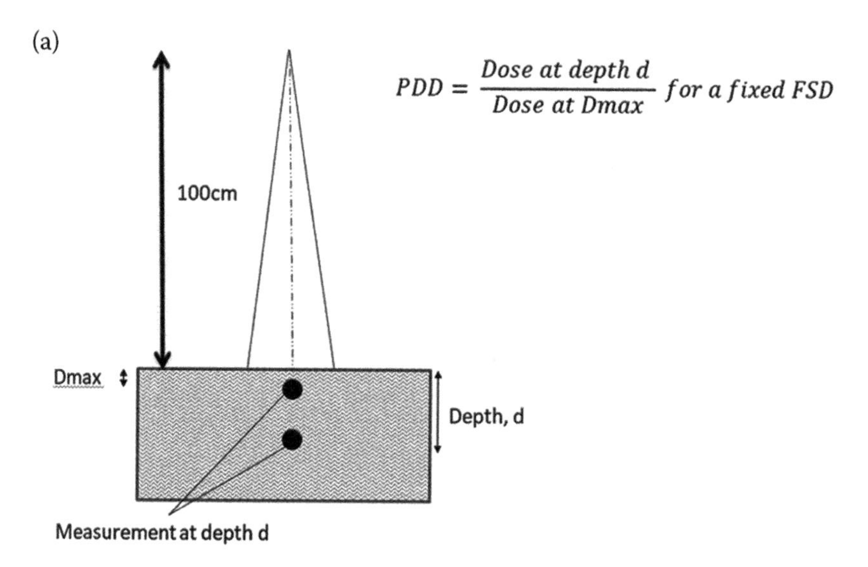

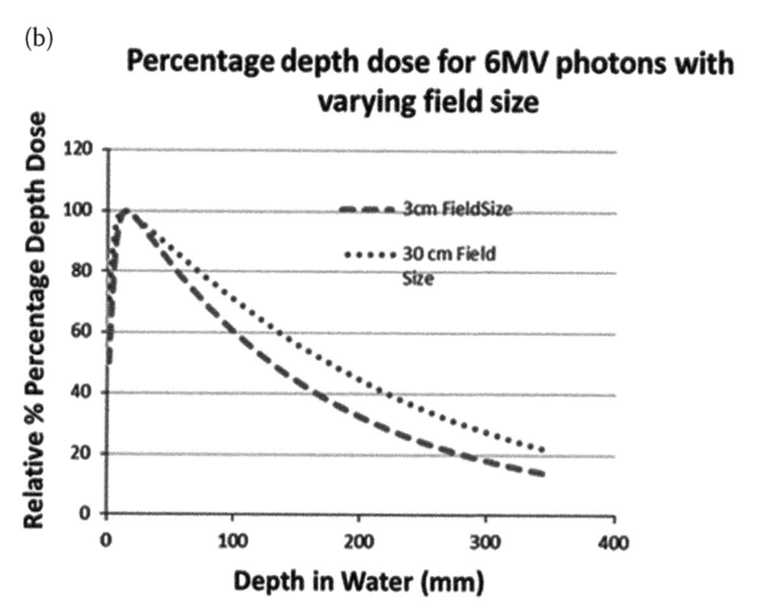

FIGURE 6.6 (a) Percentage Depth Dose measurement setup; (b) typical 6 MV PDD curve.

6.6.1.1 The PDD

The PDD describes the change in dose with depth, usually on the central axis, normalised to 100% at the maximum dose, Dmax. The PDD is measured in a water tank by scanning with a suitable detector in the depth direction for a fixed focus to skin distance (FSD) [Figure 6.6(a)]. A typical PDD for a 6 MV beam is shown in Figure 6.6(b).

Compton scattering is the main interaction of MV photons with tissue. The resulting electrons deposit dose as they travel away from the point of interaction, so the maximum dose is at depth. As more electrons scatter towards the central axis than away from it, there is initially an increase in dose with distance from the surface. The position of maximum dose occurs when electronic scatter has reached its equilibrium, with the same number of electrons scattered towards and away from the central axis. Surface dose is due to electronic contamination of the photon beam due to interactions in the linear accelerator head.

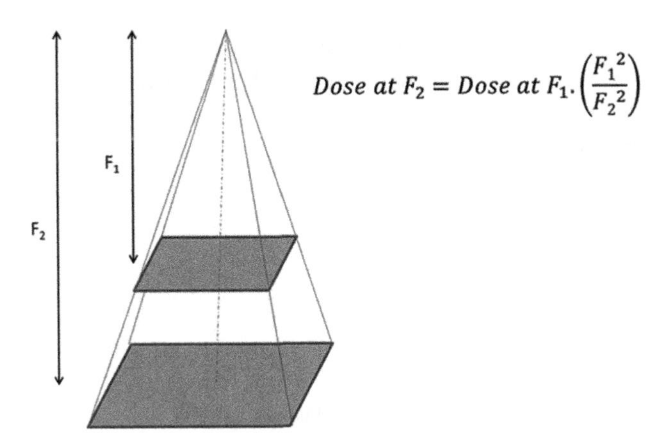

FIGURE 6.7 The inverse square law.

The dose changes with depth for two reasons: firstly, photons are absorbed and scattered out of the beam, and secondly due to divergence of the beam as the distance from the target increases. The primary photon source is the linear accelerator target and is 1–2 mm in diameter. X-rays emerging from the source diverge as the distance from the source increases. This leads to an increase in the field size with increasing distance, and correspondingly a decrease in the photon fluence. If the X-ray source is approximated to be a point source, the change in field size is proportional to the square of the distance from the source. The decrease in fluence with distance, and hence the dose, is therefore given by the square of the ratio of the distance from the source and is termed the inverse square law (Figure 6.7).

Both scattering and the inverse square law contribute to a decrease in the deposited dose as depth increases. The dose at depth is also affected by photons scattered into the central axis from elsewhere in the phantom, and for this reason the PDD should be measured in a phantom larger than the size of the field. This latter effect means that the gradient of the PDD decreases with increasing field size, as shown in Figure 6.6(b).

6.6.1.2 PDD for Electron Beams

Electrons are the treatment modality of choice when the tumour is closer to the skin surface and up to 6 cm deep or the tissue beyond the tumour is of concern. Typically, electrons (4–20 MeV) are used for their PDD characteristics [Figure 6.8(a)]. Electron beams allow for the treatment of superficial tumours with the benefit of sparing deeper underlying healthy tissues. This is due to the moderately flat plateau in the first few centimetres of the electron PDD, followed by a rapid drop in dose. This fall-off is less steep for higher energies (>15 MeV) and so the healthy tissue beyond the tumour may receive a significant dose when treating with higher electron beam energies. The most useful energies are 4–20 MeV, which can be used to treat tumours up to a depth of ~6 cm. Electrons are ideal for treating skin cancers especially lip, nose and ear, as well as operation scars and residual tumours.

6.6.1.3 PDD Characteristics for Superficial X-rays

Superficial X-rays have limited penetration so can only be used to treat lesions near, or on, the skin surface, typically at depths in tissue of 2–10 mm [Figure 6.8(b)]. Typically, 80–220 keV X-rays can be used. In the UK, most superficial treatments are for basal cell carcinomas.

(a)

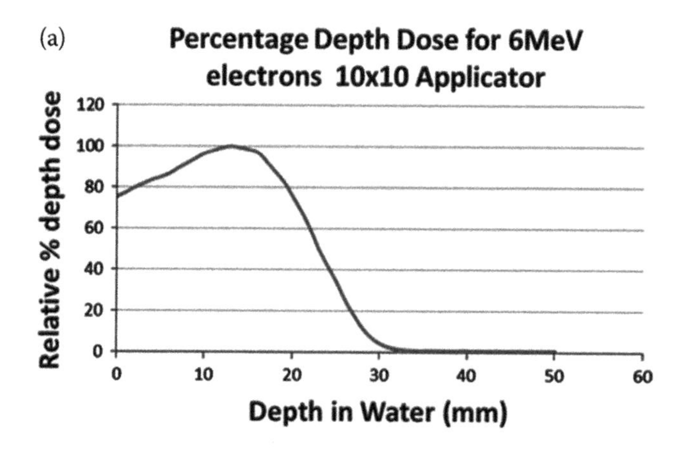

(b)

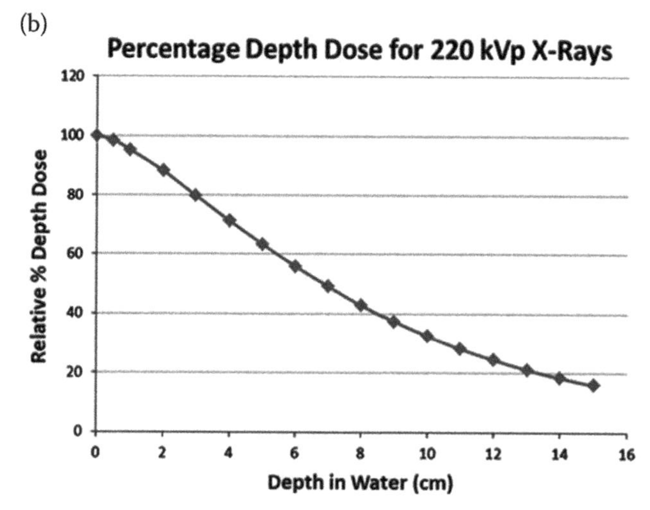

FIGURE 6.8 (a) PDD for a 6 MeV electron beam; (b) PDD for superficial X-rays.

6.7 PREPARATION FOR TREATMENT

Figure 6.9 and Table 6.2 show typical steps that occur during treatment preparation. It should be noted that the pathway summarised here can vary from centre to centre.

6.7.1 Patient Immobilisation

For radiotherapy treatment to be accurately delivered, the patient must be positioned in a comfortable and stable position that gives good access to the treatment area and can be comfortably maintained during a 10–20 minutes treatment session. Several strategies may be employed, depending on the treatment site, the most common of which are described later.

Vacuum bags are bean bags from which the air can be expelled and then sealed, such that the resulting shape of the bag conforms to the patient's contours. These are particularly useful for treating tumours in chest and abdominal regions [Figure 6.10(a)]. Breast and chest boards may be employed with adjustable supports to hold the patient's arms out of the way for unobstructed access to the target [Figure 6.10(b)].

In some cases, particularly for the head and neck that is difficult to keep still, it is necessary to implement more robust physical immobilisation. Thermoplastic is commonly employed to create a shell that attaches to the treatment couch while also being designed to be easily removed if the patient

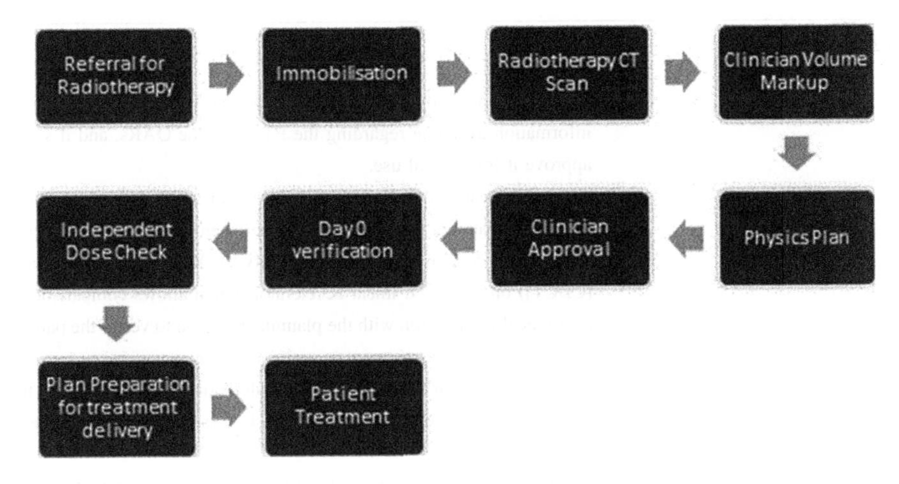

FIGURE 6.9 Preparation for treatment.

TABLE 6.2
Patient Treatment Pathway for External Beam Radiotherapy

Referral for Radiotherapy	The clinician refers the patient for treatment. Under IRMER they are acting as the Referrer as well as the Practitioner, so will need to justify and balance the need for radiation exposure against potential clinical benefits.
Immobilisation	If required, the patient may have to attend an appointment where their immobilisation needs are assessed. The team responsible for creating patient-specific immobilisation aids can vary from centre to centre. Dosimetrists or Therapeutic Radiographers are often involved in this work.
Radiotherapy CT scan (also known as "the planning scan")	In order to produce a treatment plan, and simulate treatment, a detailed CT scan is needed. This should be obtained with the patient positioned using the immobilisation equipment that will be used during treatment. This is usually acquired on a wide-bore CT scanner within the radiotherapy department which has a flat couch top to mimic that of the linear accelerator. The Hounsfield units of the CT scan will have been measured for specific materials when the CT scanner was first commissioned which allows a Hounsfield Unit vs electron density calibration curve to be produced. This information is essential for the dose calculation engine of the treatment planning system to model radiation interactions in human tissue. Clinical Scientists are responsible for the commissioning of the CT scanner and implementation of the calibration curve within the treatment planning system.
Clinician Volume Mark-up	The Clinical Oncologist will use the planning CT and any further relevant diagnostic images in the treatment planning software to mark the Planning Target Volume (PTV) that needs to be treated. Organs at risk (OARs) are also marked on the CT scan, often by dosimetrists.
Physics Plan	Dosimetrists, Clinical Scientists or Therapeutic Radiographers use a treatment planning computer to simulate different treatment options to produce a patient-specific, optimised plan that balances the aim to treat the PTV while sparing the OARs as much as possible. Once a plan has been produced, it is important that a separate person performs independent checks on the plan to check its integrity. This task is often performed by a Clinical Scientist.

(Continued)

TABLE 6.2 (Continued)

Clinician Approval	When the plan is ready, the Clinical Oncologist will review the plan and dose information available regarding the PTV and the OARs, and if satisfied, will approve it for clinical use.
Day 0 Verification	For complex treatments, the patient setup is verified before the patient attends their first treatment to ensure that the treatment is directed to the correct area. Often this is done using diagnostic quality kV X-rays (planar or cone-beam CT [CBCT]) on board the linear accelerator which allows comparison of anatomical information with the planning CT scan to verify the patient is in the correct position for treatment.
Independent Dose Check	The linear accelerator settings and dose calculation produced by the treatment planning system are independently checked as a safety critical check of the plan.
Plan Preparation for Treatment Delivery	A "record and verify" system is used to transfer setup parameters to the linear accelerator. These parameters and setup information are checked again by Therapeutic Radiographers as part of pre-treatment checks.
Treatment	When all checks have been performed and the plan is considered safe for use, the patient attends for treatment. which will often include some form of image guidance to confirm patient setup. If imaging shows the patient's anatomy has significantly altered (e.g. due to a change in tumour size, weight loss, or movement of the tumour), other strategies could be employed which may include re-planning the patient, using a pre-determined alternative plan for that patient (so called 'plan of the day') or real-time on-line adaptive radiotherapy, although the latter technique is currently not widely available.

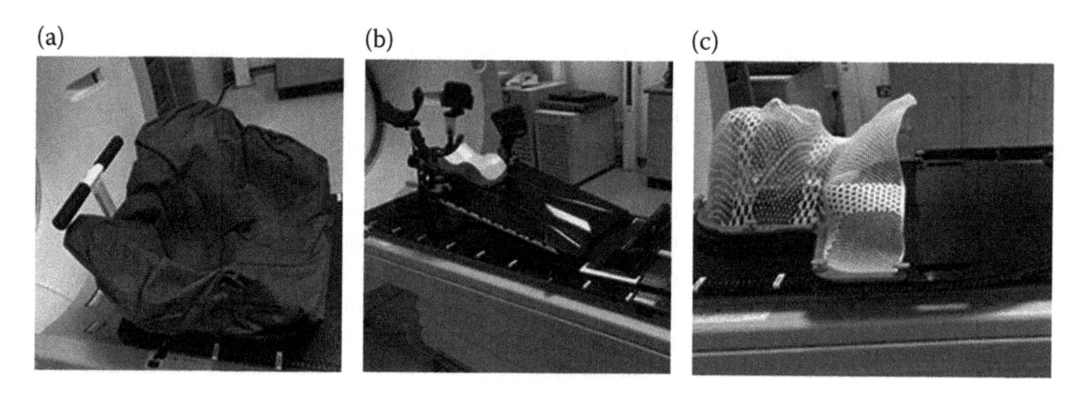

FIGURE 6.10 (a) A vacuum bag used in treatment of a lung tumour, (b) a breast board used in treatment of breast tumours, (c) a shell used for immobilising the head and neck region.

experiences difficulties. To produce an appropriate shell a commercially manufactured sheet of thermoplastic, mounted in a frame compatible with the treatment couch, is heated in a water bath or hot air oven until it becomes pliable. This is then draped over the head and face of the patient and pressure applied so it follows the contours of the patient. When cool, this forms a rigid mask the same shape as the patient which will immobilise the patient in a reproducible position for the duration of the treatment [Figure 6.10(c)].

6.7.1.2 Use of Imaging Data for the Treatment Planning Process

Clinical Scientists are responsible for checking that equipment being used in the treatment planning process is safe and provides the correct information. Treatment planning is routinely performed using a planning CT scan. CT scanners produce anatomical images of X-ray attenuation relative to water in Hounsfield Units. It is vital that the data transfer of CT information and conversion of Hounsfield Units to electron density is correct to ensure accurate dose calculations. To check this, data are exported from the CT scanner (including orientation and basic patient information) in a Digital Imaging and Communications in Medicine (DICOM) format. The CT images are transferred to an import folder and subsequently imported into the Virtual Simulation software or the treatment planning and dose calculation software via DICOM transfer. Specific rounding errors during transfer can cause potential changes to pixel size, the number of pixels, slice thickness, orientation and CT number scale which can lead to uncertainties in the relationship of the treatment beams to patient anatomy, and why data transfer checks between systems form part of ongoing QC processes.

Clinical Scientists are involved in the setup of imaging protocols used on the CT scanner to ensure doses are optimised for the appropriate image quality, as required under IR(ME)R (2017), as well as ongoing routine QC to ensure consistent performance. Dose optimisation could involve support from Clinical Scientist colleagues in Radiation Safety.

6.7.2 Data Pathway

Data are transferred between systems in DICOM format as described in Chapter 1. Clinical Scientists are required to understand networking and systems requirements to ensure data integrity. Figure 6.11 summarises the data pathway corresponding to a single radiotherapy treatment. Transfer of data throughout the whole process among the differing systems must be accurate to ensure optimal patient outcome.

Data transfer QC should be undertaken before and after upgrades to software/hardware and this is managed by a Clinical Scientist. This can be done by comparing DICOM files before and after transfer (for example, by comparing v.1, v.2 and v.3 of DICOM files in Figure 6.11). This would identify any issues relating to rounding errors, or translation and naming conventions, as well as deletion of data. Rounding or incorrect transfer of parameters such as pixel size, slice thickness and slice spacing throughout the process could introduce systematic errors, meaning that a treatment plan could be incorrect. Incorrect naming could cause misinterpretation of the data later in the treatment process.

6.7.3 Clinical Protocols and Dose Prescription

Local protocols are constructed by a multidisciplinary team and it is important to include all members of the radiotherapy team when devising clinical protocols. The team will work together to review any updated literature, for example, RCR guidelines (such as the RCR Fractionation Guidelines (RCR 2019a)), clinical trial results, the International Commission on Radiation Units and Measurements reports 50, 62 and 83 (ICRU, 1993, 1999, 2010), Radiation Therapy Oncology Group (RTOG 2020) contouring atlases (RTOG 2020), "Quantec" papers (Bentzen et al. 2010) describing the quantitative analyses of normal tissue effects, and any other published working group information. The techniques used at a local centre will depend on the equipment available at that centre. Treatment setup audits may be performed to inform the PTV margin used for a specific site and technique. If a dose prescription deviates from the protocol, the plan will need to be reviewed and the reason for the deviation documented.

6.7.3.1 Capabilities and Limitations of Treatment Machines and Associated Equipment

Establishing documented treatment restrictions is helpful, as in some setup situations, the linear accelerator will not be physically able to achieve the required setup. Typical examples include:

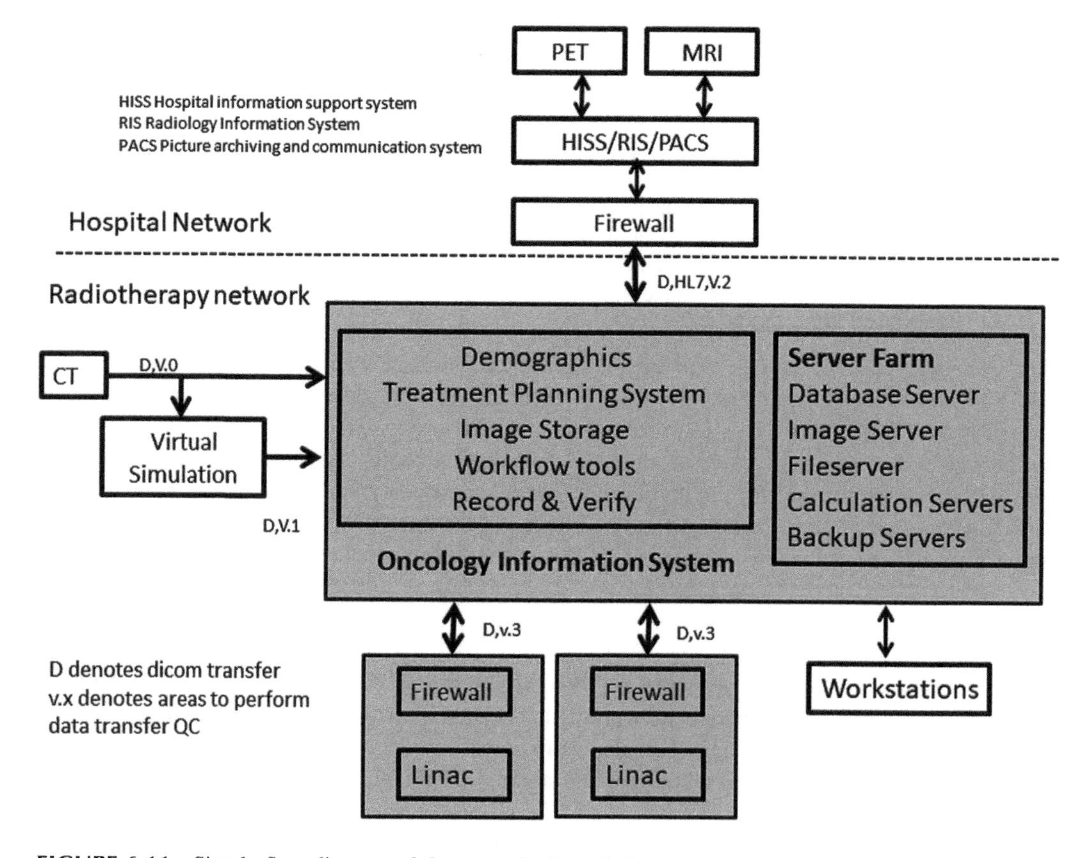

FIGURE 6.11 Simple flow diagram of data transfer in radiotherapy.

- Field size limits for wedged beams
- Field size limits for the range of motion of MLCs in the field
- Position of isocentre to allow rotation of linear accelerator gantry around the patient
- Minimum number of MU for a dynamic wedged beam

Sometimes these restrictions are hard coded into the treatment planning system and so plans cannot be produced with these treatment restrictions. However, not all conditions are hard coded in the planning system, so knowing these limitations at the planning stage of the patient's treatment can ensure that the plans that are produced are deliverable on the linear accelerator.

6.7.3.2 Steps Taken to Produce a Treatment Plan

Several steps are involved with generating a treatment plan:

6.7.4 Outlining

The treatment planning process involves finding the optimal beam arrangement and beam treatment technique to treat the tumour while sparing the OAR. It is important that the target treatment site and OAR have been outlined accurately. In order to outline the target tumour, Clinicians need all available diagnostic information. The Radiotherapy CT scan is obtained with the patient in their treatment position, and, as mentioned earlier, provides a snapshot of the patient's anatomy at the time of the scan. Image registration

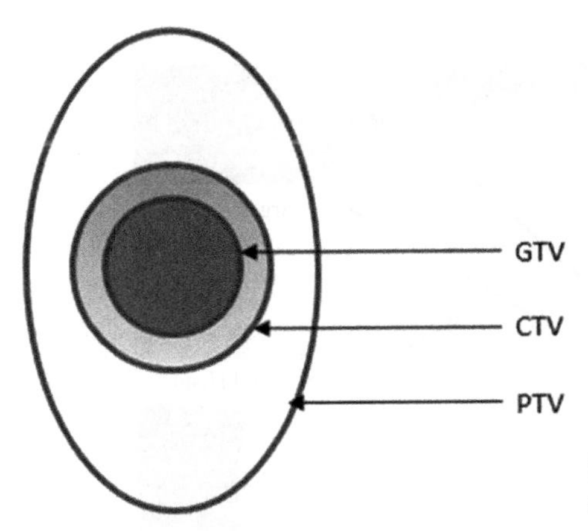

FIGURE 6.12 A schematic illustration of the GTV, CTV and PTV used in treatment planning.

is a useful tool that allows diagnostic scans, for example, PET-CT images or MR images to be matched to the Radiotherapy CT, which can aid the target delineation (and OAR delineation) process.

When marking the treatment volume the Clinician needs to identify the Gross Tumour Volume (GTV) and the Clinical Target Volume (CTV). The GTV is the disease that can be seen, palpated, or seen on images. The CTV includes the GTV as well as an additional margin to allow for the spread of subclinical disease, where there is a risk that the tumour has spread.

The third volume is the PTV. The PTV is the volume that the dose distribution will be planned to cover. The PTV includes a further margin that allows for uncertainties in planning, patient setup and treatment delivery (Figure 6.12).

6.7.5 Image Registration

Image registration can be used to fuse multiple modality images to provide additional information that is not available from the CT scan alone. Although CT has excellent geometrical accuracy and provides electron density information, soft tissue contrast is poor. MR images are typically used for brain and pelvic regions, and sometimes for the head and neck, to take advantage of the better soft-tissue contrast available from MR scans. This aids delineation of the treatment target and OAR. PET scans are also sometimes used, mainly for imaging the lung, head and neck and pelvic regions in order to provide functional information. Further information is available in AAPM (2017), report 132, on the use of image registration and fusion algorithms and techniques.

CASE STUDY: PROSTATE RADIOTHERAPY PLANNING

The following diagram shows the typical anatomy for a prostate patient. The prostate is the organ requiring treatment and in this example it is marked by the clinical oncologist as a CTV. The CTV is grown to form a PTV. The OAR in this example are the bladder, rectum and the left and right femoral heads (Figure 6.13a).

The planned dose to the patient is built up by adding multiple beams together. We use planning software to model the beam combinations and display the predicted dose to the target volume and the OAR.

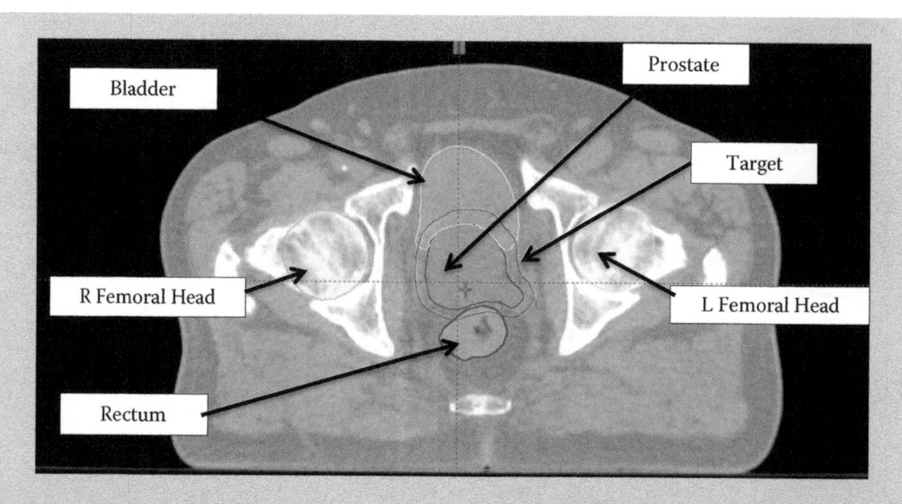

FIGURE 6.13A Prostate image used for radiotherapy planning.

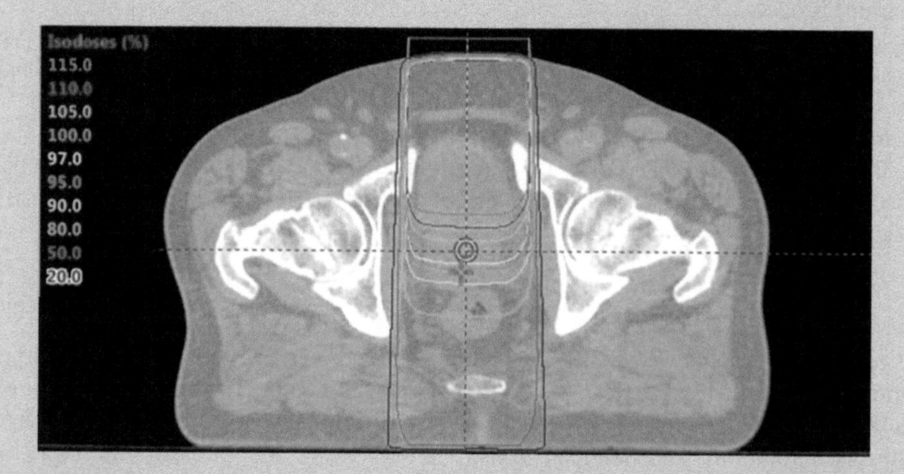

FIGURE 6.13B Dose distribution for a single 6 MV beam arrangement.

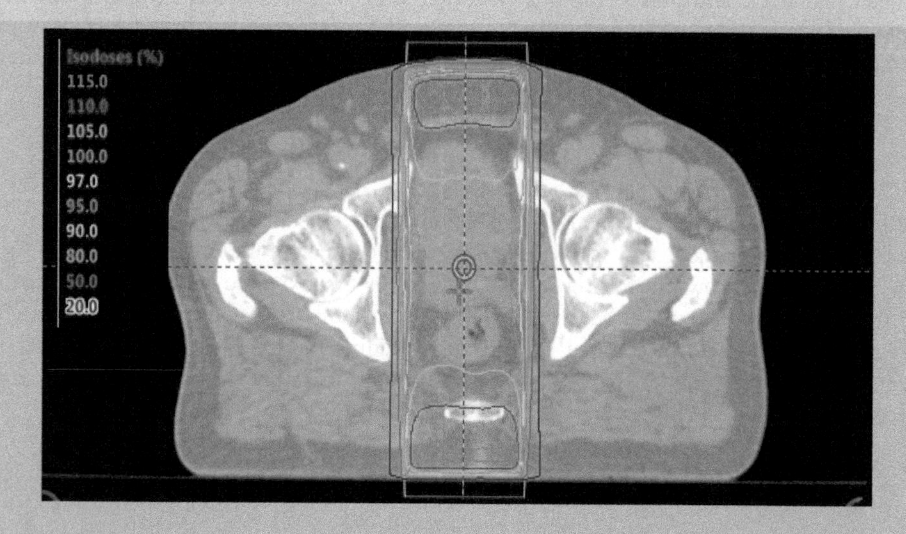

FIGURE 6.13C A two-field parallel opposed beam arrangement.

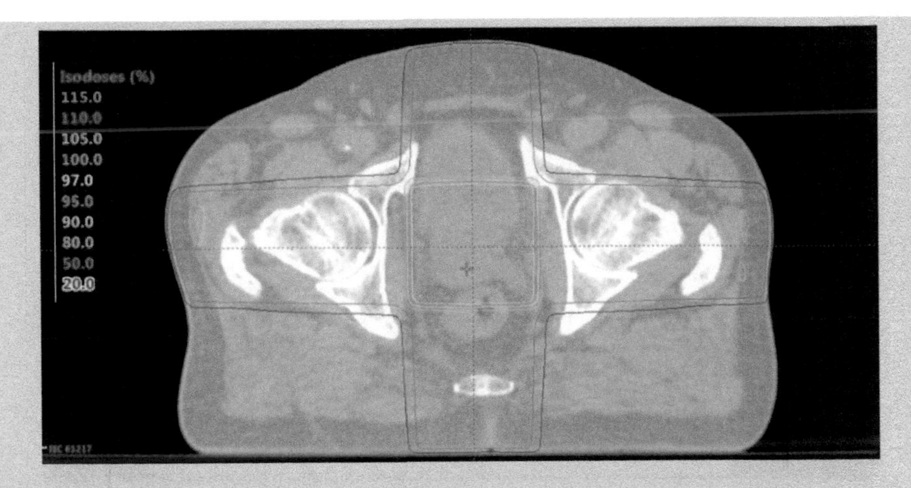

FIGURE 6.13D A four-field box arrangement.

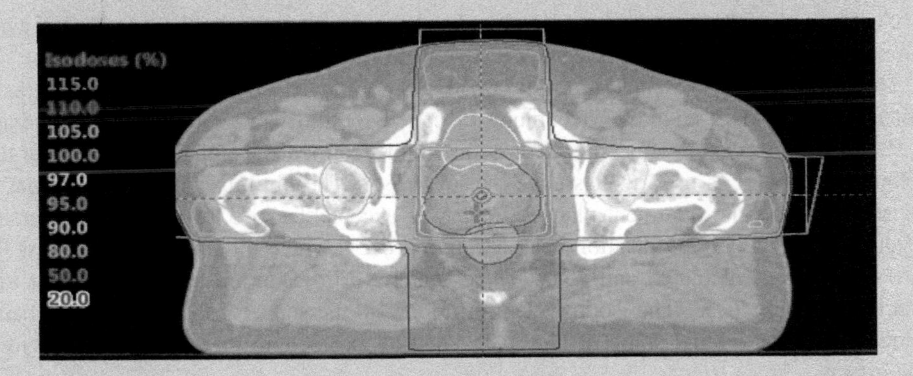

FIGURE 6.13E Isodose distribution for a three-field arrangement with wedges.

Figures 6.13B–6.13E shows a typical dose distribution for a single beam 6 MV photon field directed through the anterior of the patient. The lines represent different isodose levels representing equal dose levels, so in this example, any point on the green 95% isodose line represents 95% of the treatment dose.

Figure 6.13C shows a typical two field parallel opposed beam arrangement. One beam treating through the anterior of the patient and the opposing beam treating through the posterior of the patient. Here you see typical hot regions of dose (110–115%) received by the anterior and posterior of the patient. This occurs because the entrance dose from the anterior beam plus the exit dose from the posterior beam add together and are greater than the dose in the centre of the patient.

The next beam arrangement (Figure 6.13D) shows a four-field box arrangement. By adding more beams around the patient, the dose level in the centre of the patient becomes higher than the dose levels at the beam entrance and exit positions.

A typical, traditionally planned, prostate plan would use a three-field technique. An example is given in Figure 6.13E. The three-field technique uses wedged beams to compensate for the missing posterior field. One of the benefits of the three-field technique is that the dose to the rectum OAR is reduced when compared to a four-field configuration.

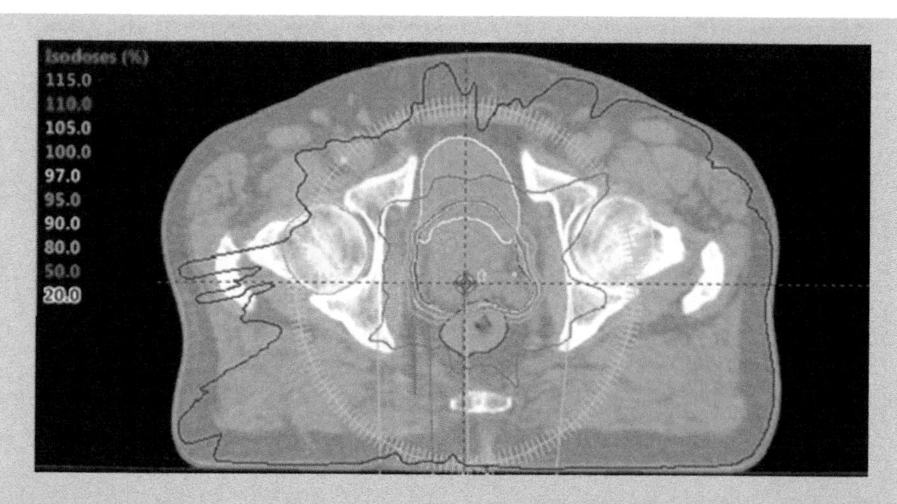

FIGURE 6.13F Prostate plan generated using the VMAT technique.

Note how the 50% light blue isodose line has been drawn up in the three-field plan when compared to the four-field plan. This will result in a lower dose to the rectum.

The final example presented in Figure 6.13F shows a more advanced technique to treat a prostate patient. The advanced technique is called Volumetric Modulated Arc Therapy (VMAT) and this technique is now used routinely for treatments. To plan the delivery of this technique, software is used that is called an optimiser. The optimiser helps the planner to devise an appropriate treatment plan by indicating acceptable dose levels for the treatment target and OAR. Plan optimisation is an iterative process where the planner will review the dose distribution calculated and adjust values entered to the optimiser until a satisfactory plan has been produced. The isodoses conform more to the shape of the target that is being treated and shape around some of the OAR, such as the rectum, thereby reducing adverse side-effects.

6.7.6 Criteria for an Acceptable Plan

Acceptable plans must meet locally agreed criteria. Sometimes the criteria will be influenced by a Clinicians experience of outcomes for their patients, backed up by evidence-based references.

A good treatment plan is one that includes the following criteria:

- Delivers uniform dose to the PTV according to the ICRU guideline in Report 50 (ICRU 1993)
- The 95% isodose should conform closely to the size and shape of the PTV
- PTV dose should be kept within +7% and –5% of the prescribed dose
- Any significant dose to OAR should be avoided, and critical doses should not be exceeded
- The setup accuracy, reproducibility, and time it takes to treat must also be acceptable.

All conventional external beam plans have a dose reference point in accordance with ICRU 50 recommendations (ICRU 1993), which state that the reference point should be selected according to the following criteria:

- The dose at the point should be clinically relevant and representative of the dose throughout the PTV
- The point should be easy to define in a clear and unambiguous way

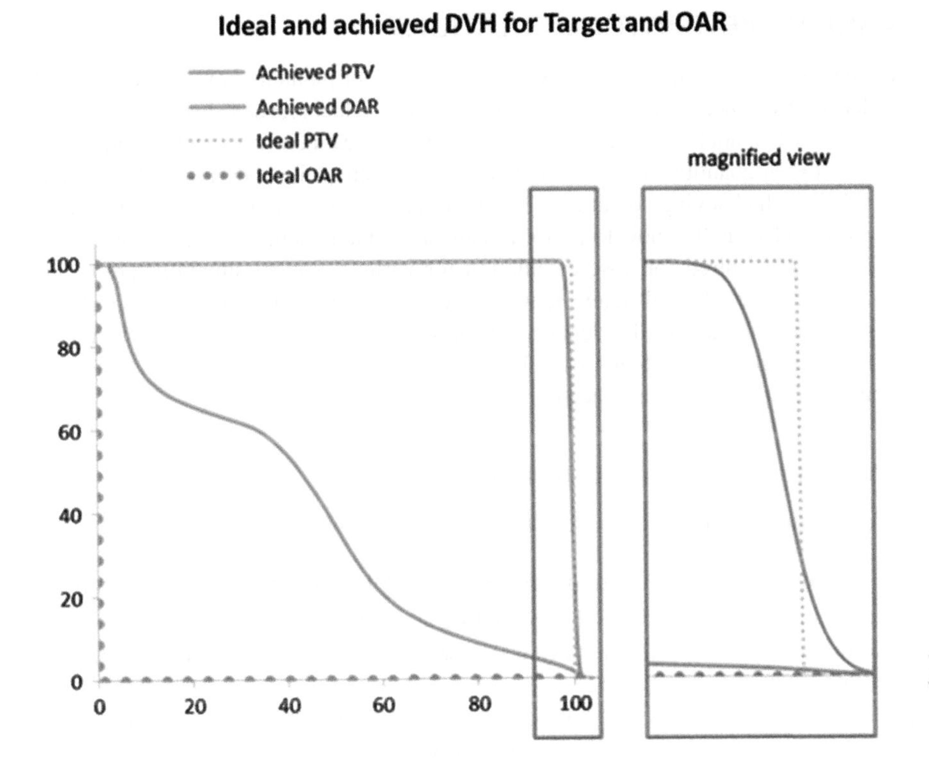

FIGURE 6.14 Ideal and achieved DVH for target and OAR.

- The point should be selected where dose can be accurately determined (physical accuracy)
- The point should be selected in a region where there is no steep dose gradient
- The recommendations are that it is placed centrally within the PTV if possible. This point is normalised to 100%.

6.7.7 Appraisal of Treatment Plans

Appraising a treatment plan involves reviewing the isodose distribution on the CT scan slices and may utilise a DVH. PTVs, OAR volumes and a dose calculation are required to generate a DVH in the treatment planning software. Using a DVH allows the evaluation of a treatment plan quantitatively. DVHs summarise information contained in the 3D dose distribution and can be used to check for coverage and uniformity, and to compare treatment plans. However, there is no spatial information contained in the DVH, so a combination of review of the isodoses and review of the DVH is useful when appraising a treatment plan. A cumulative DVH is the most used, as it is easy to see how much of the target volume is covered by 95% of the dose.

Ideally, the PTV coverage would be 100% of the dose in 100% of the volume, resembling a step function, and the critical structure (OAR) would receive no dose at all, but this is not achievable in practice. Figure 6.14 shows ideal and typical DVHs, Dose (in Gy) is shown on the x-axis and percentage of total volume receiving greater than this dose on the y-axis. Planning is often a balancing act between PTV coverage and sparing of OAR. Acceptable dose to the OAR depends on the type of normal tissue. For a parallel organ such as the lung, high dose to a small volume has no clinical toxicity; whereas for serial organs such as the spinal cord, a high dose to a small fraction could lead to organ failure, so it is preferable to administer a smaller dose to a larger volume.

6.8 MANUAL TREATMENT CALCULATIONS

Modern planning systems and the ubiquity of CT imaging for planning have reduced the reliance on manual calculations. However, there is still a role for simple calculations in checking the integrity of the planning system calculation and may be used for palliative treatments. An understanding of the basic relationship between quantities is also essential for appreciating the underlying physics that governs changes in dose with changing conditions. For these reasons, it is essential for a Clinical Scientist in radiotherapy to understanding how to perform manual calculations in a variety of situations.

The parameters required for dose calculation for various clinical situations are defined in the following section. The Clinical Scientist will have responsibility for producing charts and tables containing this data for use in the department, based on measurements taken when a new linear accelerator is commissioned. Scientists are also responsible for ensuring that the methods of calculation are clearly defined and accurately applied.

6.8.1 MV Photon Treatment Calculations

MV photon treatments require an understanding of the following factors:

6.8.1.1 Output or Field Size Factor (Incorporating Head Scatter and Phantom Scatter)

The output factor (OF) describes the change in dose at depth on the central axis when the field size, as defined by the secondary collimators (jaws) and/or MLCs, is changed. The definition of the OF is shown later.

Typical photon dose on the central axis is attributed to two sources: primary radiation from the linear accelerator target and scattered radiation. Scattered radiation is generated within the head of the linear accelerator (head scatter) and produced within the phantom (or patient), referred to as phantom or patient scatter. Primary radiation is unchanged by changing field size, but the latter types of scatter both increase as the field size increases. The flattening filter is the main source of scatter in the linear accelerator head, and the head scatter is largely determined by how much of the flattening filter is exposed by the jaws. Phantom scatter is dependent on the size and density of the phantom or irradiated body; if the phantom is smaller than the size of the radiation field, then the amount of phantom scatter will be determined by the size of the phantom and not the jaw settings.

The variation in OF with field size is illustrated in Figure 6.15. As the field size increases the dose at the field's centre is impacted to lesser extent, but at smaller field sizes a slight change can have a considerable effect on the dose. This is particularly important for treatments that use small fields such as IMRT and VMAT.

6.8.1.2 Equivalent Square

The OF will be measured by the Clinical Scientist when a new linear accelerator is commissioned. For practical reasons only square fields are measured in the range from the smallest clinically used field to the maximum (usually 40 × 40 cm at the isocentre). A clinical field may be of any rectangular size achievable by the linear accelerator jaws and could be further modified by the MLCs in order to avoid an OAR. The equivalent square is the square field size that delivers the same scattered dose to the central axis as the clinical field being considered. Note that this does not mean that the areas are equivalent, as scatter generated a long way from the central axis will have less effect than that close to it. A table of equivalent square values for a range of rectangular fields is available in BJR Supplement 25 (Day and Aird 1996). Alternatively, the following expression may be used:

$$\text{Equ Sq} = 2AB/(A + B),$$

where A and B are the length and width of the field. Where there is significant shielding with MLCs, the equivalent square field size should be reduced accordingly.

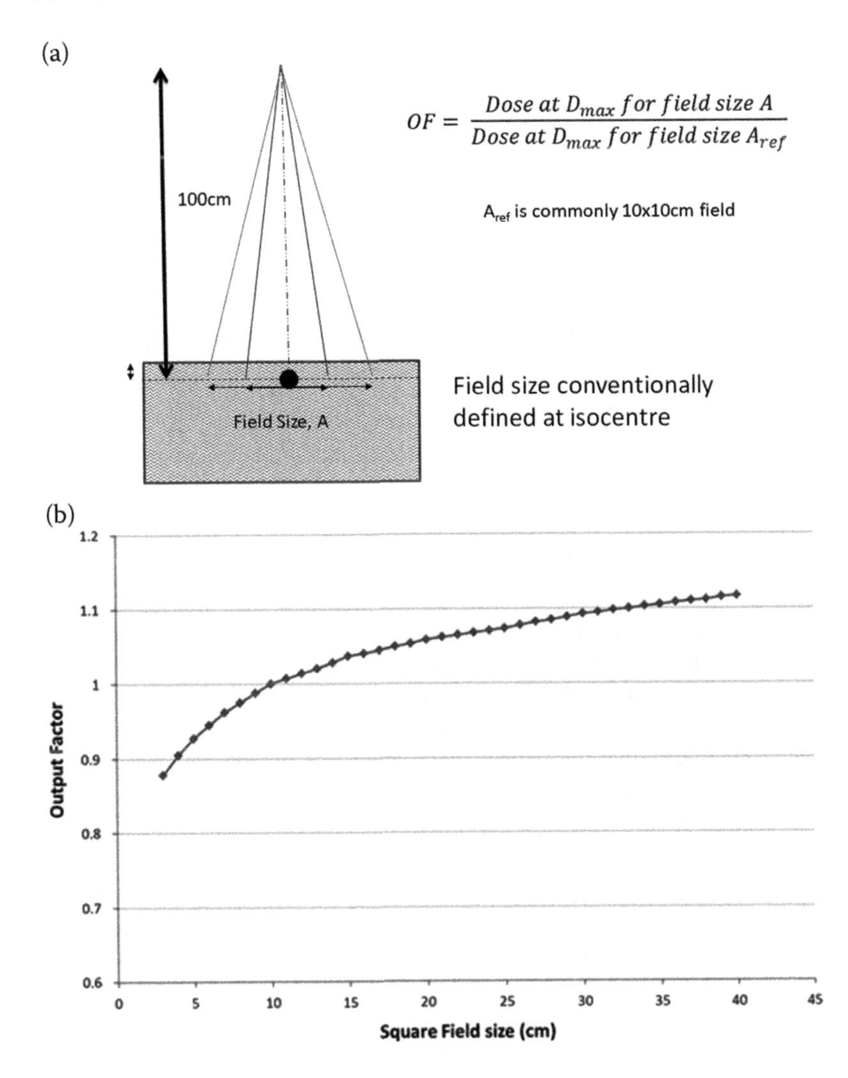

FIGURE 6.15 (a) Setup to measure an output factor; (b) output factor vs field size graph.

6.8.1.3 Tissue Phantom Ratio and Tissue Maximum Ratio

The Tissue Phantom Ratio (TPR) is the ratio of the dose at the isocentre at a given depth "*d*" to the dose at a reference depth. It is distinct from the PDD in that because it is related to a point at a fixed distance from the source, neither beam divergence nor the inverse square law contribute to the change in dose. Tissue Maximum Ratio (TMR) is a special case of the TPR, where the reference depth is the depth of the maximum dose (Figure 6.16).

6.8.1.4 Manual Calculation of Monitor Units

The dose delivered for a given treatment is controlled using the charge recorded by the ionisation chamber in the linear accelerator head. The charge is related to a generic quantity known as Monitor Units (MU). The linear accelerator is calibrated so that 100 MU delivers 1 Gy to a fixed point under reference conditions. This reference value is then used as a calibration point for calculating the number of MU needed to deliver the required dose in the conditions (field size, depth, etc.) to be used for treatment. Additional description is given in the ESTRO report by Mijnheer et al. (2001) on MU calculations. The method of calculating MU depends on whether the treatment setup is isocentric or at fixed FSD:

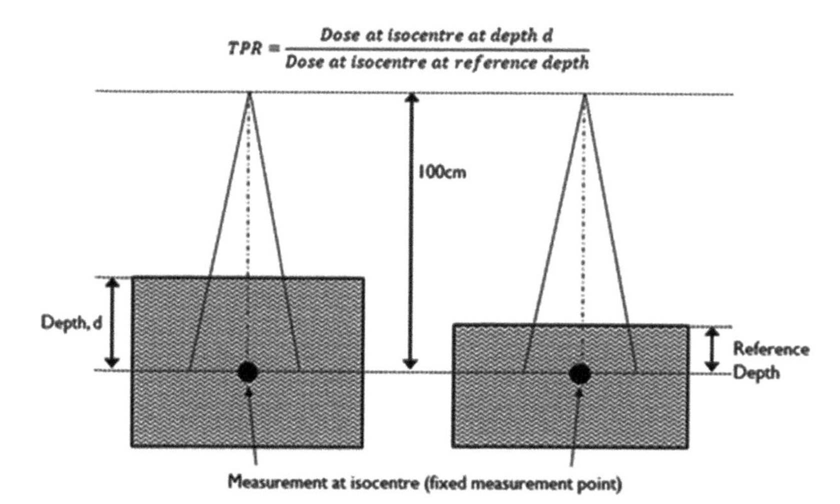

FIGURE 6.16 Illustration of Tissue Phantom Ratio (TPR).

- Isocentric treatments involve positioning the target volume or region to be treated at the linear accelerator isocentre. This has the considerable advantage that the patient only needs to be positioned once. Calculating MU for isocentric treatments requires the use of TPR or TMR.
- A fixed FSD treatment is delivered with the surface of the patient the same FSD for all beam directions. This means that the patient may have to be moved between beams to maintain same FSD, and hence fixed FSD treatments are now generally only used for treatments involving a single field. Fixed FSD treatment MU are calculated using PDD.

A simple calculation of MU for a given dose for a single field at fixed FSD will have the form:

MU = required dose at prescription point (Gy)/(PDD × OF × Reference dose (Gy/MU))

For an isocentric treatment this will have the form:

MU = required dose at prescription point (Gy)/(TPR × OF × Reference dose (Gy/MU))

CASE STUDY: MANUAL CALCULATION OF MU FOR AN ISOCENTRIC TREATMENT

A patient is to be treated isocentrically with two parallel opposed beams. The field size is 25 cm x 16 cm and the patient separation (overall thickness) is 26 cm. The required dose is 20 Gy in five fractions. How many MU are required from each beam?

The equivalent square for a 25 cm x 16 cm field = (2 × 25 x 16)/(25 + 16) = 19.5 cm x 19.5 cm square field. The OF of this field is 1.055.

The patient separation is 26 cm, so the isocentre is at 13 cm deep for each beam. Table 6.3 gives a TPR of 0.827.

The dose per fraction is 4 Gy, so each beam delivers 2 Gy to the isocentre.

Therefore, the required MU are 2 Gy/(0.827 * 1.055 * 0.01 Gy/MU) = 229 MU

TABLE 6.3
Values for Field-to-Surface Distance (FSD), Depth and TPR

FSD	Depth (cm)	TPR
89.5	10.5	0.883
89	11.0	0.871
88.5	11.5	0.860
88.0	12.0	0.850
87.5	12.5	0.839
87.0	13.0	0.827
86.5	13.5	0.816
86.0	14.0	0.805
85.5	14.5	0.793
85.0	15.0	0.782

Further corrections are needed if a wedge is present, or to account for inhomogeneities. More examples can be found in the ESTRO report describing practical examples by Mijnheer et al. (2001).

6.8.1.5 Radiological Depth

Data tables used for manual calculations are based on measurements in water and assume a tissue density of 1 g/cm^3. This is acceptable for muscle, but if there is considerable inhomogeneity, in the form of bone, lung or air cavities, then this may need to be considered. This involves calculating the effective, or radiological depth; the equivalent depth of unit density material which results in the same attenuation. The radiological depth can be calculated by summing the product of the depth of an inhomogeneity by this density:

$$\text{Radiological depth} = \sum (\text{depth of inhomogeneity}) \times \text{density}$$

6.8.1.6 Wedge Factor

The introduction of a wedge to the field will lead to a reduction in dose. This reduction is quantified by the wedge factor:

$$\text{Wedge factor} = \text{dose at a point in a wedged field relative to the dose in an open field for a given field size.}$$

As described earlier, a wedged beam can be produced by two different methods, depending on the linear accelerator manufacturer. Some linear accelerators automatically insert a physical wedge-shaped piece of material, usually tungsten, into the beam. Alternatively, a dynamic wedge will produce a gradient across the beam profile by gradually moving one jaw across the beam as the treatment is delivered. The wedge factor for a physical 60° wedge will be of the order of 0.25 to 0.33due to the definiton of the wedge factor being dose in wedged field / dose in the open field. The wedge factor for a dynamic wedge will be larger.

6.8.1.7 Off-Axis Factors

The calculations given previously can be used to calculate dose on the central axis. The dose at a point away from the central axis will be different due to different scatter conditions, and if applicable, because of a wedge. In this case, other corrections will be needed, such as an off-axis wedge factor, or methods for determining scatter at the point in question.

6.9 CHECKS ON COMPLEX PLANS

Most radical treatments are now delivered using IMRT or VMAT techniques. These can involve many small treatment fields to deliver the prescribed dose, which would be impractical to calculate manually. However, recommendations currently advise that every plan should have a check that is independent of the primary treatment planning system. Therefore, an alternative method is required, either by a secondary calculation or by measurement. There are a range of systems and approaches for performing this independent check, with varying degrees of complexity. A Clinical Scientist would be involved in evaluating the various methods and determining which would be most appropriate for their department's requirements and individual situation. This may be different for different clinical situations and according to how much experience the department has with each technique. They would also be required to advise on appropriate tolerances and criteria for a plan to pass this check, and to discuss with the clinical oncologist the action to be taken in the event of an unacceptable result.

6.9.1 Electron Treatment Calculations and Measurements

Electrons lose energy continuously through multiple interactions until they are completely absorbed. Unlike photons, they have a finite range in tissue, and are therefore useful for treating superficial lesions such as skin tumours while sparing the underlying tissue. The range of an electron beam is determined by its energy, and so several beam energies will be available for treatment of lesions of varying depths.

With electrons, the size of the treatment field is additionally defined by an applicator which is mounted into the linear accelerator head, containing an end-frame or cut out which is positioned at or close to the patient surface (MLCs are not used). Several different size applicators will be available, each having a range of end-frames that can be a standard size or custom made by clinical technologists for an individual patient treatment. An electron applicator and end-frame are shown in Figure 6.17. The Clinical Scientist will measure OFs and depth doses for standard applicators during commissioning and make this data available in tables for Clinical Oncologists to select the appropriate energy for treatment and for calculating treatment MU. Some treatment planning systems also have algorithms (usually Monte Carlo) for calculating electron dose distributions.

The change in dose at the central axis with field size for electron beams is governed by their finite range. Provided the field is larger than the maximum range of the electrons, the dose at the central axis will be unchanged and no field size correction is needed. Each treatment type has a "Code of Practice" for the dosimetry of the beams (McEwen et al. 2003).

Electron treatments are usually delivered at a fixed distance from the end of the applicator. However, in some situations, it is not possible to maintain a fixed distance across the whole field due to physical restrictions of the linear accelerator and patient anatomy or a change in the patient contour. In this case, the MU must be corrected to account for the change in distance. Electron beams are produced by replacing the photon target in the linear accelerator head with scattering foils and emerge from a distributed source rather than from a point. Therefore, they do not fundamentally obey the inverse square law. The reduction in dose with distance does approximate to a $1/x^2$ relationship up to about 10 cm beyond the end of the applicator. The quantity x in this relationship is defined as the effective source position. This is an abstract point that has no physical significance but can be used to correct MU calculations. Its value will be different for each beam energy and applicator and is determined by measurement by the Clinical Scientist.

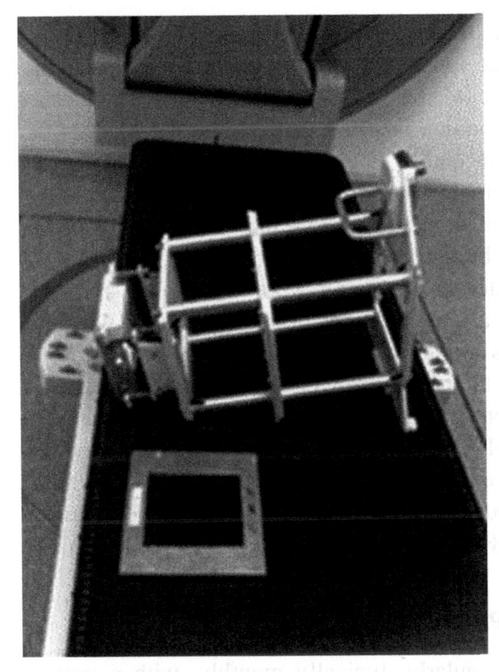

FIGURE 6.17 Electron applicator with the end frame removed.

6.9.2 kV PHOTON TREATMENT QC

Treatments using photons in the kilovoltage range are used for treating superficial tumours such as skin. An X-ray tube is used with different sized applicators to define the field size. Factors for calculating MU for different applicators will be measured by a Clinical Scientist. Treatments involve a single field delivered with the applicator against or close to the skin surface. The FSD of kV treatments is much shorter than for MV, typically around 20–30 cm. For this reason, differences in FSD have a greater effect on the dose than for MV photon treatments and corrections using the inverse square law can be quite significant. A good summary of the use of kV beams in radiotherapy is given by Palmer et al. (2016).

6.10 BEAM CHARACTERISTICS

6.10.1 Absolute Dosimetry

Absolute dosimetry is the process of making a measurement with a radiation detector and converting this measurement to absorbed dose. This process ensures that two patients prescribed the same treatment dose in two different centres are treated to the same dose. The chain of calibration begins with primary standards laboratories, such as the National Physical Laboratory (NPL, UK), National Institute of Standards and Technology (NIST, USA), the Laboratoire national de métrologie et d'essais (LNE, France) and many others. These laboratories work from first principles using primary standards to measure absorbed dose. Centres regularly compare and validate their standards to ensure consistency between laboratories.

A Clinical Scientist working in a radiotherapy department will generally be working with secondary and tertiary standards. A secondary standard is one that has been calibrated against a primary standard, and Clinical Scientists will use both an ionisation chamber and an electrometer secondary standard. In the UK, they will be sent to NPL for calibration, and the ion chamber will be calibrated using calorimetry.

The secondary standard chamber is placed in a graphite calorimeter and irradiated using a linear accelerator. The charge collected by the chamber and the temperature rise of the calorimeter are accurately measured. A current source is then used to heat the calorimeter to produce the same temperature rise the radiation produced, allowing the energy required to raise the temperature this amount to be calculated. The calibration factor of the ionisation chamber can be calculated, considering corrections including temperature and pressure (to correct for the mass of air in the chamber which is unsealed), and ion recombination. This calibration factor is valid only for a beam of the quality used to perform the calibration, and the beam quality is characterised by the ratio of the ionisations measured by an isocentric chamber at 20 cm deep to one at 10 cm deep (TPR 20/10). By performing the aforementioned procedure at different beam energies, a table is created allowing calibration factors for different qualities of beams to be interpolated.

The secondary standard electrometer is calibrated by applying a known charge and then comparing the measured reading with this, issuing a table of correction factors. NPL will then return the secondary standards with their certificates of calibration.

Once the secondary standards are received back in the radiotherapy centre, Clinical Scientists use these to calibrate other ionisation chamber and electrometer calibrations. The procedures to calibrate these tertiary standards (field instruments) for use in MV photon and electron beams are documented in Codes of Practice (Eaton et al. (2020) and McEwen et al. (2003), respectively). This is an important process as a systematic error could be introduced which would affect all patients subsequently treated and therefore it is essential that an MPE is involved in all aspects of the procedure.

6.10.2 Standard Output and Relation to TPS

The output of the linear accelerator is measured regularly, typically monthly, with a tertiary standard field chamber. This is performed by measuring the air temperature and pressure in the room, and then placing the calibrated ion chamber in a phantom (either water equivalent plastic or water) at the required position. After connecting the chamber to the electrometer and setting the appropriate voltage, the chamber is pre-irradiated and nulled. Then measurements are taken at a standard field size and set number of MU, and the machine output can be calculated.

Fundamental in this process is that the definition of a MU must be the same on the linear accelerator as in the treatment planning system. For example, when discussing MV photons, 1 MU can be defined as being the amount of radiation that gives 1 cGy to the isocentre when a 10 × 10 cm field is used to irradiate a chamber at either isocentre 5 cm deep, or at isocentre 10 cm deep. Although not recommended in the latest Code of Practice, some centres calibrate at the position of maximum dose with the surface of the phantom at 100 SSD.

CASE STUDY OUTPUT EXAMPLE

Measurements taken for an output measurement using 100 MU	
Temperature:	22.1°C
Pressure:	1007 mbar
Temperature and pressure correction factor =	1.013
Chamber readings uncorrected:	21.45 nC, 21.46 nC, 21.47 nC, 21.45 nC (fourth reading taken as first 3 showed a trend)
Mean chamber reading:	21.46 nC
Mean chamber reading corrected for temperature and pressure:	21.74 nC
Calibration factor for chamber:	4.662 cGy/nC
Dose recorded by chamber:	21.74 nC x 4.662 cGy/nC = 101.4 cGy
Linear accelerator output:	1.013 cGy/MU

6.10.3 Quality Control

After installation, the Clinical Scientist will work with the machine vendor and undertake acceptance testing, ensuring that the machine operates according to manufacturer specifications for a defined set of situations. Following acceptance, Clinical Scientists undertake a detailed commissioning process that is overseen by an MPE. The purpose of commissioning is to define baseline performance characteristics of the linear accelerator and to establish the required procedures, protocols and instructions for safe clinical operation. Establishing baseline performance parameters is of the utmost importance because dosimetric parameters characterising the beam are used within the treatment planning system, so the use of incorrect commissioning data could lead to systematic errors in plans generated for all subsequent patients treated.

A suitable radiation detector such as a photon diode is mounted onto the gantry of the plotting tank. Under computer control, the detector can scan precisely in three dimensions, allowing beam profiles in either vertical, horizontal or diagonal directions to be measured. A plotting tank is shown in Figure 6.18(a).

(a)

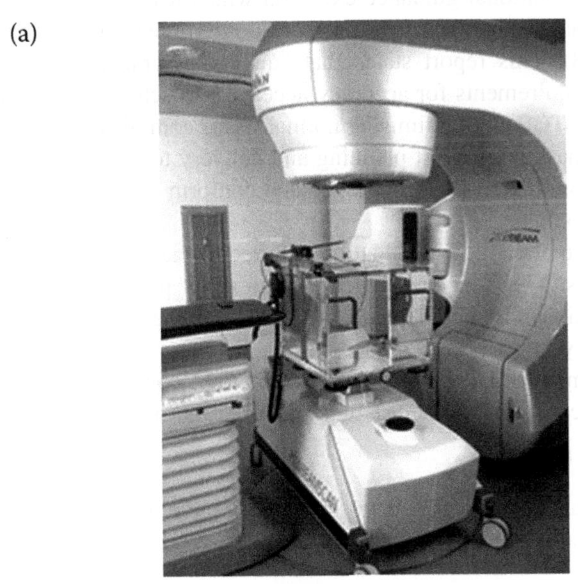

(b)

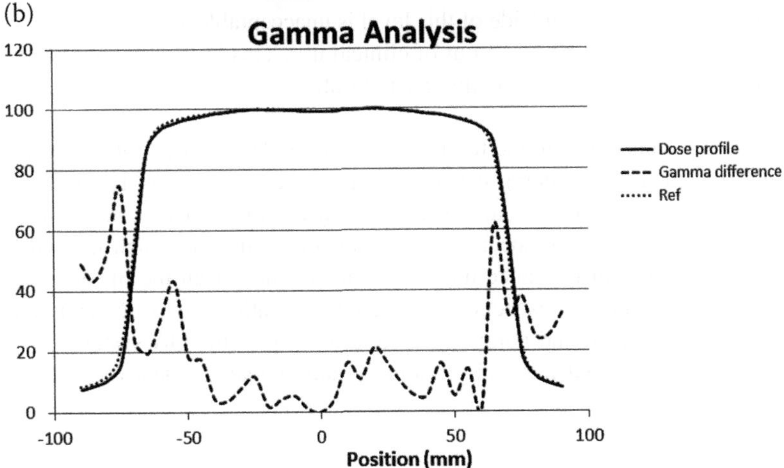

FIGURE 6.18 (a) Example of a plotting tank used in the commissioning process. (b) Beam profile and gamma analysis.

Figure 6.18(b) shows an example of a horizontal beam profile acquired using the plotting tank and the result of a comparison made with the profile calculated by the planning system using a common tool called Gamma analysis (Low and Dempsey 2003). Gamma analysis is a method of quantifying agreement between two dose distributions, taking into account two metrics – dose difference (important in uniform dose areas), and distance to agreement (of more significance in high gradient dose areas, where clinically insignificant small misalignments would have a large effect on dose difference results).

Once placed into clinical use, it is important to ensure that the linear accelerator performance characteristics do not change significantly from the baseline values measured during the commissioning process. Potential reasons for changes in performance characteristics could include mechanical wear, component ageing or component failure.

It is a legal requirement under IRMER (2017) to have an appropriate QA program in place to ensure linear accelerators are working correctly to deliver the right amount of radiation to the right location, so that patients have the best possible outcomes from their treatment.

The QC checks that need to be regularly performed on a linear accelerator take place at different frequencies and national and international guidance exists on which tests need to be undertaken, how often, and what the reasonable tolerances are. In the UK, the most important document is IPEM Report 81-2 (Patel 2018). This report states that "quality assurance programs must be structured to meet the clinical requirements for accuracy necessary to achieve optimum treatment outcome in terms of maximising TCP and minimising normal tissue complications". This requires uncertainties and errors in dosimetry, treatment planning and delivery to be reduced such that the accuracy and precision of the dose delivery are improved and conform to the ICRU guidelines of an accuracy of 5% (ICRU 1993).

IPEM (2018) Report 81-2 provides recommendations of frequencies and tolerance levels for alignment and dosimetry checks. They may not represent the exact procedures used in all departments and the recommendations can be adapted by an MPE for the different equipment and treatment techniques used locally.

QC checks on machine alignment and dosimetry need criteria to be compared to decide whether deviations from baseline are acceptable or whether any action needs to be taken. There are two levels:

- **Tolerance level:** performance within this level gives acceptable accuracy and clinical use can be continued. The manufacturer's specification must be considered when tolerances are set.
- **Action level:** performance outside of this level is unacceptable and immediate action must be taken. The machine must be taken out of clinical use. This level must be locally determined based on the types of treatment being carried out.

For QC tests giving results between the tolerance and action levels, it is important for the Clinical Scientist to understand the tests being carried out and to exercise good judgment as to the appropriate action to be taken. For example, a linear accelerator that is being used for SABR treatments might need to have an isocentre that is smaller than one which is used for palliative treatments. As such, an out of tolerance measurement may not mean the linear accelerator being taken out of clinical use, but any restrictions on the type of patient plans that can be delivered are carefully justified, documented and communicated according to the QA system.

Checks are typically grouped into daily, monthly and annual frequency and can be split into four main groups:

- Safe operation
- Optical and mechanical performance

- Imaging performance
- Beam setup/dosimetry

Clinical Scientists may be responsible for carrying out the QC tests, but at some centres Radiographers, Technicians or Dosimetrists may perform some or all of them. However, as with all ionising radiation medical exposures, an MPE is required to maintain overall responsibility. A full description of how these tests are performed and their rationale is beyond the scope of this book and the reader is referred to guidelines in relevant documentation such as IPEM (2018).

After routine linear accelerator servicing or repair after breakdown, the Clinical Scientist has the role of either performing or overseeing QC checks to ensure the linear accelerator is performing within the prescribed tolerances before being returned to clinical use. It is important that effective communication occurs between engineering staff or manufacturers' service engineers responsible for rectification work and the Clinical Scientist involved so that all appropriate QC checks required can be identified. For example, it would not be practical or necessary to perform all daily, monthly and annual QC checks after replacement of a field lamp bulb, but checking crosswire and field edge position changes while rotating the collimator at two different couch heights and checking optical field sizes would be an appropriate response to ensure that the bulb had been located correctly.

When interpreting a set of results, it is also important to consider any interdependency. For example, the value of performing an image quality test by setting a phantom up to align with the lasers will rely on the laser accuracy being in tolerance. Routine QC checks should therefore be designed to follow a logical order. Following breakdown and repair, the situation may be more complex. For example, measuring field sizes based on the light field and projected crosswire position will depend on the jaw position calibration, but the measurement technique would also depend on setting an appropriate couch height (using the optical distance indicator) and would also rely on the crosswire correctly indicating the centre of the radiation field. Investigating an out of tolerance field size measurement in such circumstances would require deeper investigation before calibration adjustments are made.

In summary, QC testing of linear accelerators is a legal requirement and an essential component of a radiotherapy service. They serve to maintain geometric and dosimetric accuracy within defined tolerances.

6.11 SYSTEM UPGRADES

The pre-treatment planning process involves several steps, as described in Figure 6.9. A change or routine upgrade to any component needs to be evaluated and checked thoroughly to ensure the integrity and accuracy of the whole process. This task is the responsibility of the Clinical Scientist.

The actual process will depend on the component being upgraded and how it is used in the department, but should consider:

- Reviewing release notes and publications from the manufacturer to identify expected changes and limitations
- Thorough testing of the new system on a stand-alone test system, which is separate from the clinically used system
- A visit to observe the new system at another centre
- Training directly from the manufacturer and cascaded to all relevant staff
- Communication to all relevant parties as to the impact of the change (e.g. a change to the planning system dose calculation algorithm may impact on the dose distribution and needs to be communicated to the Clinical Oncologists)
- A risk assessment showing that possible risks have been evaluated and adequately managed
- An external audit from another centre

- An end to end test of the whole process to check how the new component interacts with other components
- Updating any relevant documentation in the Quality System
- Additional checks on the first few plans once the new system is in clinical use

The Clinical Scientist will be required to assess which of these steps is necessary for the safe and accurate roll out of a new system. Problem-solving skills will be required to resolve any issues that arise during this process and good communication with other staff groups is essential.

6.12 INCIDENTS IN RADIOTHERAPY

The processes in radiotherapy are very involved requiring great attention to detail and care from everyone in the multidisciplinary team. Despite this, systems and processes can break down and human error can have an impact. Described more fully in Chapter 7, incidents are taken extremely seriously in radiotherapy and will involve Clinical Scientists and MPEs as part of any investigation – and are sometimes led by senior Scientists. There is a national system to log incidents and near misses in radiotherapy enabling centres to learn from each other and review their processes before incidents happen. The guidance document "Towards Safer Radiotherapy", by the British Institute of Radiology (BIR), Institute of Physics and Engineering in Medicine (IPEM), National Patient Safety Agency (NPSA), Society and College of Radiographers (SCoR) and the RCR, is an important review document that describes how incidents are managed (RCR 2008a).

6.13 BRACHYTHERAPY AND SEALED SOURCES

The term "brachytherapy" comes from the Greek word for close and is used to describe a type of treatment where radiation dose is delivered to tumours by placing a source of radiation in close proximity. In most cases, this would be achieved using radioactive sources emitting gamma rays but more recently miniaturised X-ray tubes have begun to be used. In both cases, the close proximity allows treatment to take advantage of the inverse square law so that large doses are delivered directly to the tumour but dose drops off quickly with distance from the source of the radiation. This means that the dose to healthy tissue surrounding the tumour is far lower than with EBRT described thus far. Brachytherapy may form the sole treatment or be used as a boost for a particular region with delivery being concurrent with or sequential to EBRT.

Brachytherapy is best suited to treatment sites that are easily accessible, such as within body cavities or on the surface of the skin. Common types of brachytherapy include intracavitary (e.g. cervix), intraluminal (e.g. oesophagus), surface (e.g. skin) and interstitial (e.g. prostate), and is most often used to treat gynaecological cancers. Depending on the technique being used, a general anaesthetic may be required and this may affect patient suitability.

6.13.1 Radionuclides Used

The most common brachytherapy treatment systems currently in use are the HDR afterloader systems that utilise a single, HDR-sealed source of Iridium-192 or less commonly, Cobalt-60. More information is given in chapter "brachytherapy and unsealed radionuclide therapy" by Horton et al. in Practical Radiation protection by Martin and Sutton (2015). The source is welded to the end of a cable attached to a stepper motor such that the source can be positioned with great accuracy within a patient. The source is typically stepped to different positions where it dwells for the required time gradually delivering the prescribed dose and required dose distribution. A typical Iridium-192 source-based HDR treatment unit is shown in Figure 6.19.

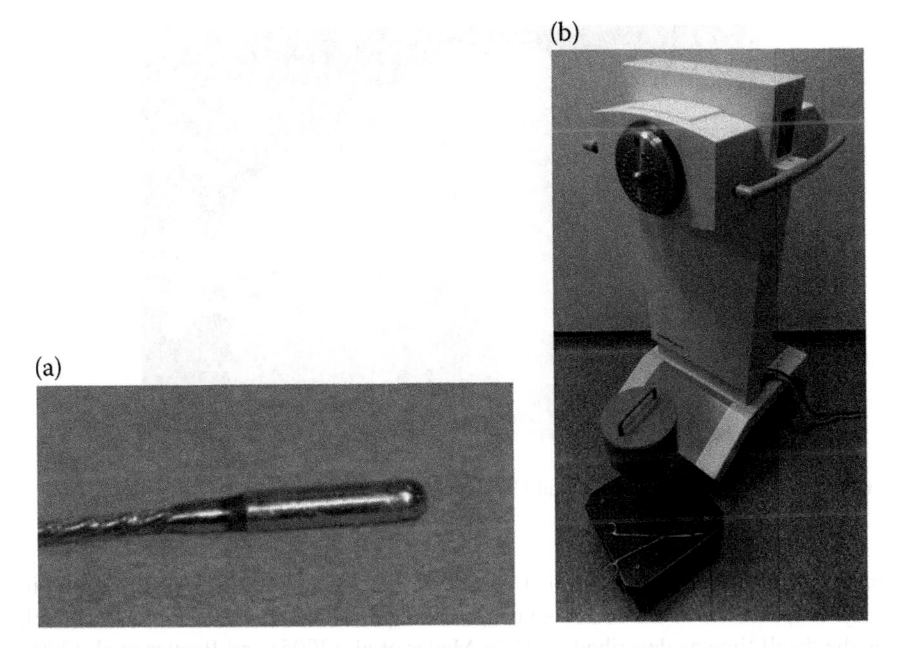

FIGURE 6.19 (a) Iridium-192 brachytherapy source. (b) Example of a commonly used HDR unit (Gammamedplus IX, Varian).

As an alternative to HDR, a similar technique is called medium dose rate (MDR), which as the name suggests uses sources with a lower dose rate. Pulsed dose rate (PDR) is a technique used to bridge the gap between HDR and MDR by giving many fractions but with breaks in between using an HDR source to mimic the radiobiology of MDR source. Both MDR and PDR are rarely used in the UK today. LDR uses low dose rate seeds such as Iodine-125. These seeds may be placed in the tumour and left *in situ* for the rest of a patient's life. This is a common treatment for prostate cancer (alternatively interstitial needles with HDR brachytherapy can also be used when treating the prostate).

6.13.2 Clinical Applications

In the case of vaginal treatments, applicators can be inserted without the need to attend an operating theatre. Planning may be based only on a standard template for these simple treatments, selection being dependent on the length and diameter of vaginal applicator used. For skin treatments, applicators can simply be placed over the lesion directly before treatment. For a small skin lesion planning may consist only of a manual dose calculation based on very few parameters.

For complex treatment plans, the first step is to position the applicators and/or hollow needles in or around the tumour under anaesthetic. After the patient has recovered, a scan is taken to allow visualisation of the treatment area including the applicator placement in order to plan the treatment.

This part of the treatment process is very similar to treatment planning for EBRT. A CT scan is used for geometric accuracy and additionally, MR imaging may be used for better visualisation of the tumour with images being registered to allow the advantages of the two modalities to be combined as described previously (Figure 6.20). The CTV and OARs are outlined to allow the doses to these structures to be assessed. As the tumour will not move in relation to the source, the CTV represents the PTV. In addition to these standard structures, the applicator needs to be identified to allow the positions of the source within it to be modelled.

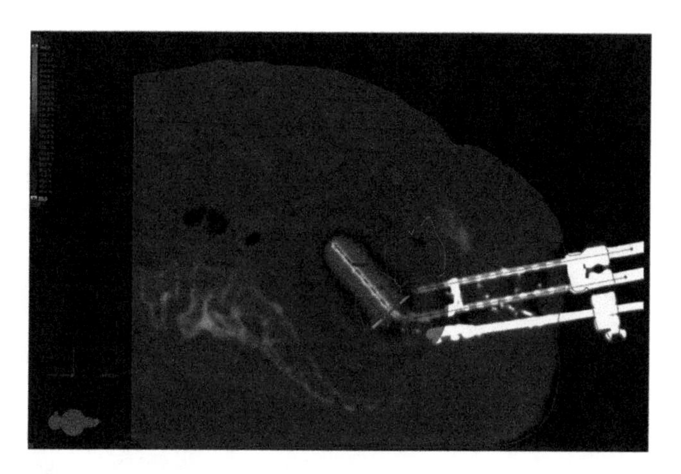

FIGURE 6.20 CT of patient with PTV and OAR outlined.

Treatment planning is often performed by a Dosimetrist or Clinical Scientist. Resulting plans will specify the positions of the sealed source and the period of time that the source dwells there, known as the dwell time as described by Haie-Meder et al. (2005) and Poetter et al. (2006). Plans usually start from templates known to provide uniform dose distributions over the intended treatment area based on the geometry of applicators and or needles selected. Optimisation is then achieved by adjusting the source positions and dwell times so that the dose distribution is adapted to the individual anatomy of the specific patient. In common with standard external beam techniques, independent checks are performed on all plans before they are used to treat patients and this task is usually performed by a Clinical Scientist.

The treatment unit is connected to the needles and applicators using guide tubes. The source is driven under computer control through the guide tubes and into the needles and applicators to the planned positions. The term "afterloader" refers to the fact that applicators are placed first and then the source is loaded afterwards. Previous methods included the manual insertion of radioactive sources or wires in an operating theatre and therefore the introduction of afterloaders significantly reduced the dose to staff placing the sources or wires.

CASE STUDY: BRACHYTHERAPY BOOST FOR CERVICAL CANCER

A patient was diagnosed with cervical cancer and has already undergone EBRT with a dose of 45 Gy in 25 fractions. She is now having a brachytherapy boost of 21Gy in three fractions (#). While planning the second fraction, a Clinical Scientist calculated that the dose to the portion of the bowel in close proximity to the tumour would be outside the recommended tolerance (75 Gy to no more than 2cc volume) if the bowel received the same dose again in the final fraction. The dose delivered could simply be reduced to prevent this but this would also compromise the dose to the tumour. Instead, before the next scan and fraction, saline was pumped into the bladder which once full pushed the bowel out of the way and allowed the full dose to be delivered without compromising the bowel. For more information on gynaecological cancer planning, see the ESGO (2020).

6.13.3 Sealed Sources – Checks and Calculations Required

A sealed source (such as that used in an HDR treatment unit) is a radioactive source that is contained in a way that prevents contamination. For this reason, where sources are used it is required by law that they are sealed whenever possible. Usually, sealed sources are made from an isotope in a solid form encapsulated in metal. Even when sealed, radioactive materials pose a greater risk to individuals and the environment than other X-ray equipment, as the radiation cannot be switched off. It is important to understand the additional legislative and security requirements if radioactive sources are to be used. The legislation requires acquiring Environmental Permits and ARSAC licenses, timely reporting, detailed record-keeping, strict access controls and contingency planning which is independently audited by the national competent authority. In the UK, the competent authority is the Environment Agency (EA).

It is the role of the Clinical Scientist to have a detailed knowledge of the risks posed by radionuclides used for treatment and the safety measures required. They also ensure compliance related to the legislation surrounding holding, using and disposing of sources. Dependent on the activity and concentration of the sources, it may be necessary to gain a permit from the EA, under the Environmental Permitting Regulations (EPR 2016) and notify, register with or obtain consent from the Health and Safety Executive (HSE) under IRMER (2017) before they can be held and used. A permit specifies the operator, the source (including the isotope, the number of sources and their activity), what the source may be used for, how sources may be stored and disposed of and a map showing the site where the source may be utilised.

In the UK, hospitals where brachytherapy is performed are required to demonstrate that they have adequate facilities and expertise to acquire a site-specific ARSAC license, which specifies the radionuclides and the treatments that can be performed. Clinical Oncologists performing brachytherapy must also hold their own ARSAC license which ensures they have the training and experience to act as practitioners when delivering these treatments.

The operator is legally responsible for choosing a source of an appropriate design and construction for the task. There is then further legislation controlling how the source is brought to the facility. Carriage of Dangerous Good Regulations (2019) defines how packages containing radioactive materials should be packaged, labelled and transported. They also ensure that the couriers are trained and provided with information and procedures to allow them to ensure that the public and the environment are protected in the event of an incident during transport.

Once the source arrives, it must be checked to ensure that it meets the conditions of the permit and is not leaking. Leak tests are required by IRMER (2017). Sealed sources can be damaged allowing the radioactive materials to be released and contamination to occur. Leak tests involve using a swab to sample the surface of the source and then monitoring the swab for signs of contamination. Leak tests must be performed regularly to ensure the integrity of the source.

Sealed sources need to be handled safely to minimise the dose received by the operator following three basic principles:

- Distance: Maximise the distance from the source; doubling the distance to the source gives a fourfold reduction in dose according to the inverse square law.
- Shielding: Use recommended shielding, which is a legal requirement if specified in the Local Rules described in Chapter 7.
- Time: Minimise the time interacting with the source.

If the operator of a facility breaches any limit of the permit, or the source is released, lost or stolen, they are required to inform the EA, minimise the effect of the incident, return to compliance as soon as possible, and act to ensure that the incident is not repeated in the future. Any incident must be thoroughly investigated, and records of the investigation maintained.

Access always need to be strictly controlled, and the location of the sources must be regularly checked, the sources accounted for and the check recorded. When not in use, they must be stored and transferred securely. Finally, when sources are no longer required, they must be disposed of responsibly through a licensed radioactive waste management service.

HDR sources are usually classified as High Activity Sealed Sources (HASS). The activity at which designation as a HASS occurs is isotope dependent but can be found tabulated in the legislation. For Ir-192 source, the threshold is 80 GBq and a typical HDR source has an activity between 400 and 500 GBq. As HASS sources pose a higher risk, they are controlled by stricter regulations. The employer has a legal responsibility to provide regular training to staff involved in the use of HASS sources in how to manage, control and use them safely as well as the potential risks if control of the HASS source is lost. Regulators must inspect the premises to ensure adequate security is in place to protect sources, and operators must facilitate this. The regulators must do this under the support of other security experts where necessary and are legally obliged to follow their advice. The security measures considered include physical barriers, alarms and early detection systems. They will also need to include procedures for controlling access, approving individual access to HASS, preventing and detecting loss or theft and to tighten security in response to an increased threat. Financial provisions also need to be in place so that the source will always be disposed of appropriately.

Clinical Scientists in radiotherapy may be involved in ensuring compliance with legislation and environmental permits by providing training on safe use and the consequences of misuse of HASS, performing risk assessments, dose investigations and environmental surveys, generating contingency plans, maintaining records, meeting with the RPA and representatives from the EA and CTSA during audits, and generating action plans to improve compliance.

6.14 STRONTIUM-90 CHECKS

One use for sealed sources in radiotherapy is for performing constancy checks (QC) on radiation monitoring equipment (such as ionisation chambers). When performing QC on a linear accelerator, it is essential to know that the variation measured is a result of fluctuation in the output of the linear accelerator and not due to changes in the sensitivity of the chamber. As the activity of the Strontium-90 source decays at a predictable rate (with a half-life of 29.1 years), it can be used to ensure a consistent response from the measurement equipment over many years. Typical strontium sources used for radiation monitoring QC have activities in the range of 10–900 MBq, and so to use sources in this way the HSE must be notified as they are more radioactive than the notification quantity of 10 kBq.

CASE STUDY: STRONTIUM CHECK

During the strontium check of a farmer chamber, the mean charge recorded is: 217.4 nC (Figure 6.21).

These values are corrected for atmospheric conditions using the following equation:

$$f_{TP} = \left(\frac{T + 273.2}{293.2}\right)\left(\frac{1013}{P}\right)$$

If the temperature was 21.5°C and the pressure was 1009.5 mbar

$$f_{TP} = 1.009$$

and so the charge is now corrected to 219.3 nC.

FIGURE 6.21 Strontium check sources.

We now correct for radioactive decay. If A is the activity now and A_0 is the original activity, the relationship between those values is:

$$A = A_0 e^{-\lambda t}$$

where λ is the decay constant and t is the time that has passed between when A_0 and A were measured. After one half-life $T_{1/2}$, it must be true that $A/A_0 = 1/2$. Thus:

$$1/2 = e^{-\lambda T_{1/2}} \rightarrow \lambda = \frac{\ln 2}{T_{1/2}}$$

putting this back into the original equation:

$$A = A_0 e^{-\frac{\ln 2}{T_{1/2}} t} \rightarrow A_0 = A e^{\frac{\ln 2}{T_{1/2}} t}$$

Thus, the decay correction factor is $e^{\frac{\ln 2}{T_{1/2}} t}$.

The original reading when the chamber was first commissioned (10 years, 3 month and 17 days go) was 281.1 nC.

Thus, the decay correction is:

$$e^{\frac{\ln 2}{T_{1/2}} t} \rightarrow e^{\frac{\ln 2}{29.1} 10.297} \rightarrow 1.278$$

and the corrected reading is 280.2 nC.

The difference between the two readings is thus 0.3%. As the recommended tolerance is 1%, this chamber is functioning correctly.

Scientists working within radiotherapy will also be involved in radiation safety work. Radiation safety in radiotherapy is described in Chapter 7. In some centres, this work is carried out by radiation safety experts with some knowledge of radiotherapy, in others by radiotherapy experts with some knowledge of radiation safety. Scientists may be involved in design of facilities, licensing or permitting of those facilities, risk assessments, development of contingency plans, commissioning work, operational radiation protection, incidents, personal monitoring, environmental monitoring, audit, teaching and training, and involvement in decommissioning facilities or end of life of sources. In almost all instances, multidisciplinary and cross specialism working is required.

6.15 THE FUTURE OF RADIOTHERAPY

Advances in computing power and the development of sophisticated optimisation algorithms, together with improvements in linear accelerator engineering, have led to the ability to modulate the delivered dose using MLCs (Section 6.2) and tightly conform the dose distribution to a given target. The ability to more tightly control the dose distribution has improved treatment outcomes by minimising the doses delivered to healthy tissue, while also allowing higher doses to be delivered to the target volume. The use of highly conformal radiation dose distributions, however, makes treatment delivery less forgiving in terms of uncertainties relating to target localisation and control of organ movement, leading to the more widespread implementation of more complex image guided and motion management techniques. Although IMRT is now considered the standard of care for many tumour sites, rather than the future of radiotherapy, advances in technology coupled with the use of imaging in the treatment process has opened the door for more personalised treatments (e.g. on-line adaptive radiotherapy, where the original treatment plan can be modified to account for observed changes in the patient anatomy during the treatment session).

Currently, daily adaptive radiotherapy is constrained by workflow and technology limitations. Variations in organ shapes mean that in an ideal situation, the optimal treatment plan may need daily adjustments. In practice, the workload associated with re-planning, checking and re-approval of plans makes the decision to re-plan for most treatment sites to be made only for the most extreme cases. For daily adaptive radiotherapy to become mainstream, the sophistication and QA procedures surrounding image matching and resultant plan adaption need to develop to allow confidence in such techniques to be implemented.

The radiotherapy process typically involves the patient attending for a planning CT scan, which performed in the same position as when the patient attends for treatment. The CT scan helps the Clinical Oncologist and dosimetrists define the area to be treated including marking anatomical structures that should be avoided to minimise treatment toxicity. The CT data, along with the structures marked on the scan can then subsequently be used to plan the treatment and calculate the dose distribution. However, the target delineation process is subject to intra- and inter-operator variability. Other imaging modalities can be used to provide additional information to aid the delineation process. MR scans can provide improved soft-tissue contrast, while functional imaging techniques such as PET can provide additional information relating to cell metabolism and other biological processes occurring within the tumour cells. For example, using appropriate radiolabelled substances, it is possible to identify areas of hypoxia or areas of increased cell proliferation. Such information can be used to modify the areas to be treated compared to targets defined solely on morphological CT data.

We have seen how technical developments have led to the ability to deliver high precision radiotherapy. However, the ability to deliver a precise dose is only going to achieve the desired result if the dose is delivered to the correct place. The position of the target volume within the patient is subject to a degree of uncertainty. This uncertainty relates to small variations in the patient position on each treatment delivery (setup errors), changes in the position of the target volume because of other biological processes, for example, rectal filling, bladder filling and breathing motion. Anatomy changes can occur because of tumour response to the radiation, or because of weight loss/gain during the treatment process. Some of these uncertainties can be accounted for during the planning process, while further uncertainty reductions can be made by the use of pre-treatment imaging. A useful publication is the RCR's (2008b) publication *On Target: Ensuring Geometric Accuracy in Radiotherapy*. In such cases, images obtained immediately prior to treatment can be compared to those made during the planning process and the patient position can be adjusted accordingly. The last decade has seen a significant increase in the use of imaging immediately prior to treatment. Various systems have been used for imaging including kV planar imaging and cone-beam CT, which can provide superior soft-tissue contrast compared to using the treatment beam. Non-ionizing imaging modalities such as ultrasound and body surface imaging have also been used.

Although CT is likely to remain the dominant imaging modality to plan the treatment, MR is increasingly being used. Previously, MR was not used to plan the treatment because of limitations related to geometric integrity and MR not inherently giving an estimate of electron density which is important for accurate dose calculation. However, as techniques are developed to address these limitations, it is likely that the superior soft-tissue contrast offered by MR will become more widely used.

The advantages conferred by using MR for improving soft-tissue contrast has recently been extended to incorporate MR imaging with a linear accelerator. This means that high-quality images are available at the point of treatment delivery, potentially allowing real-time adaptation of the treatment plan to account for changes in anatomy. At the time of writing, there are only two such machines available in the UK, and more widespread adoption of such technologies depends on robust design of multi-centre trials to provide strong evidence of improved treatment outcomes to justify adoption given the spiralling health care costs and diminishing resources associated with the current economic climate.

Current radiotherapy paradigms typically personalise treatment plans based mainly on anatomical and morphological information and increasingly using functional imaging to define target volumes. The doses and fractionation schemes, however, are typically based on population data with little account made of individual responses to radiation. Every patient and every tumour is different. Some patients will be more sensitive to radiation, potentially increasing side effects and reducing quality of life, whereas some tumours may require higher radiation doses to effect control. Increasing the level of personalisation of treatment could involve the inclusion of phenotypic and genotypic differences that potentially exist between patients and their tumours. Kerns et al. (2014) described how normal-tissue adverse effects could not be completely accounted for by dosimetric, treatment or demographic factors. They argued that radiogenomics has two main goals:

- To develop assays for predicting which people may suffer increased toxicity.
- To identify molecular pathways for radiation-induced normal tissue toxicities.

Since the diagnostic power of genomic sequencing arises from the identification of variants, which in turn is derived from the comparison of sequences from affected and non-affected phenotypes, the availability of electronic health records linked to genomic information could lead to new associations of variations and disease risks. Current radiotherapy patient management systems store a wealth of information relating to the treatments that are delivered. Currently, linking such data to outcome information and external electronic patient records is limited, but increasing. As genetic screening and identification of candidate genes that influence radiosensitivity becomes more mainstream, integration of all these rich sources of data could potentially be used to further personalise radiotherapy treatments. Skripcak et al. (2014) described how creating a data exchange mechanism, with pre-clinical, clinical and outcome information integrated into large collaborative data sets, could lead to improved predictive models for decision support, discovery of prognostic features, and the potential to improve quality of life and increase survival times. There are potential benefits, therefore, to pooling data and increasing knowledge and understanding that could bring social and economic benefits including improving outcomes and reducing the number of ineffective treatments.

The increased use of imaging within the radiotherapy workflow potentially opens the door for the extraction of quantitative information from multi-modality images to incorporate into mathematical models for the prediction of treatment outcomes. Examples of suitable features could include location, texture and morphology of the delineated tumour volume. As an example, shape-based metrics based on the deviation from an ellipsoid structure was observed to have an association with survival in sarcoma patients. Further examples include the prediction of local control in lung cancer patients using PET/CT images (Oikonomou et al. 2018) and the prediction of metastatic disease in sarcoma patients using PET/MR (Vallières et al. 2015).

The scientist must typically engage with multiple staff groups, including clinicians, radiographers, dosimetrists and engineering staff to ensure robust and safe service delivery. As the scope of the radiotherapy process widens to include enhanced information, for example, increased use of alternative imaging modalities, the increased use of genomic information to personalise treatment, and improved use of biomarkers to monitor treatment progression, it is likely that more complex interactions with new staff groups such as bioinformatics will emerge. The increased complexity associated with modern radiotherapy delivery techniques means that the QA processes must continually change and adapt to the changing landscape. As new technologies and workflows are introduced, continued professional development and training are essential components to ensure that appropriate knowledge and understanding develop along with the new opportunities to improve treatment outcomes for cancer patients.

REFERENCES

AAPM (2017) *Use of Image Registration and Fusion Algorithms and Techniques in Radiotherapy*. Report of the AAPM Radiation Therapy Committee Task Group No. 132. Med. Phys., 44: e43–e76. Online Available from: https://doi.org/10.1002/mp.12256.

Bentzen SM, Constine LS, Deasy JO, et al. (2010) "Quantitative Analyses of Normal Tissue Effects in the Clinic (QUANTEC): An Introduction to the Scientific Issues". *Int J Radiat Oncol Biol Phys* **76**(3Suppl): S3–S9.

Carriage of Dangerous Good Regulations (2019) "SI 2009 No 1348". Online Available from: https://www.hse.gov.uk/cdg/regs.htm [accessed 9 November 2020].

CRUK (2020) "Cancer Statistics for the UK". Online Available from: https://www.cancerresearchuk.org/health-professional/cancer-statistics-for-the-uk [accessed 9 November 2020].

Dale RG and Jones B (2007) *Radiobiological Modelling in Radiation Oncology*. British Institute of Radiology, London.

Day MJ and Aird EGA (1996) "Central Axis Depth Dose Data for Use in Radiotherapy". *Br J Radiol Suppl* **25**: 84–151.

Eaton DJ, Bass GA, Booker P, et al. (2020) "IPEM Code of Practice for High-Energy Photon Therapy Dosimetry Based on the NPL Absorbed Dose Calibration Service". *Phys Med Biol* **65**(19): 195006.

EPR (2016). *The Environmental Permitting (England and Wales) Regulations*. SI 2010/67. The Stationery Office, London, England.

ESGO (2020) "Guidelines and Quality Indicators". Onine Available from: https://guidelines.esgo.org/ [accessed 9 November 2020].

Haie-Meder C, Poetter R, Van Limbergen E, et al. (2005) "Recommendations from Gynaecological (GYN) GEC-ESTRO Working Group (i): Concepts and Terms in 3D Image Based 3D Treatment Planning in

Cervix Cancer Brachytherapy with Emphasis on MR assessment of GTV and CTV". *Radiother Oncol* **74**: 235–245.

ICRU (1993) *Report 50 Prescribing, Recording and Reporting Photon Beam Therapy*. ICRU, Bethedsa, MD.

ICRU (1999) *Report 62, Prescribing, Recording and Reporting Photon Beam Therapy (Supplement to ICRU Report 50)*. ICRU, Bethedsa, MD.

ICRU (2010) *Report 83, Prescribing, Recording and Reporting Photon-Beam Intensity-Modulated Radiation Therapy (IMRT)*. Journal of the ICRU 10 No. 1. ICRU, Bethesda, MD.

IPEM (2018) *Physics Aspects of Quality Control in Radiotherapy*, IPEM Report 81, 2nd edition. IPEM, York.

IRMER (2017) *The Ionising Radiations (Medical Exposure) Regulations*. SI 2000/1059. The Stationery Office, London, England. Online Available from: http://www.legislation.gov.uk/uksi/2017/1075/contents/made [accessed 9 November 2020].

Joiner MC and van der Kogel AJ (2018) *Basic Clinical Radiobiology*. CRC Press, London.

Kerns SL, Ostrer H and Rosenstein BS (2014) "Radiogenomics: Using Genetics to Identify Cancer Patients at Risk for Development of Adverse Effects Following Radiotherapy". *Cancer Discov* **4**(2). Online Available from: http://cancerdiscovery.aacrjournals.org/content/4/2/155.short [accessed 9 November 2020].

Low DA and Dempsey JF (2003) "Evaluation of the Gamma Dose Distribution Comparison Method". *Med Phys* **30**(9): 2455–2464.

Martin CJ and Sutton DG (2015). *Practical Radiation Protection in Healthcare*, 2nd edition. Oxford University Press, Oxford, England.

Mayles P, Nahum A and Rosenwald JC (2007) *Handbook of Radiotherapy Physics Theory & Practice*. CRC Press, London.

McEwen MR, Nisbet A, Nahum AE and Pitchford WG (2003) "The IPEM Code of Practice for Electron Dosimetry for Radiotherapy Beams of Initial Energy from 4 to 25 MeV Based on an Absorbed Dose to Water Calibration". *Phys Med Biol* **48**: 2929–2970.

McKeown SR (2014) "Defining Normoxia, Physoxia and Hypoxia in Tumours - Implications for Treatment Response". *Br J Radiol* **87**(1035): 1–12.

Mijnheer B, Bridier A, Garibaldi C, Torzsok K and Venselaar J (2001) *Physics for Clinical Radiotherapy Booklet No. 6: "Monitor Unit Calculation for High Energy Photon Beams"*. The European Society for Therapeutic Radiology and Oncology (ESTRO).

NICE (2018) "Early and Locally Advanced Breast Cancer: Diagnosis and Management Guideline NG101". Online Available from: https://www.nice.org.uk/guidance/ng101 [accessed 9 November 2020].

Oikonomou A, Farzad K, Pascal N, et al. (2018) "Radiomics Analysis at PET/CT Contributes to Prognosis of Recurrence and Survival in Lung Cancer Treated with Stereotactic Body Radiotherapy". *Sci Rep* **8**(1): 1–11.

Palmer AL, Pearson M, Whittard P, McHugh KE and Eaton DJ (2016) "Current Status of Kilovoltage (kV) Radiotherapy in the UK: Installed Equipment, Clinical Workload, Physics Quality Control and Radiation Dosimetry". *Br J Radiol* **89**: 20160641.

Parker W and Patrocinio H (2005) "Clinical Treatment Planning in Externa; Photon Beam Radiotherapy". In EB Podgorsak ed. *Radiation Oncology Physics: A Handbook for Teachers and Students*, 260. International Atomic Energy Agency, Vienna.

Patel I (2018) *Physics Aspects of Quality Control in Radiotherapy - IPEM Report 81*, 2nd edition. IPEM, York.

Poetter R, Haie-Meder C, Van Limbergen E, et al. (2006) "Recommendations from Gynaecological (GYN) GEC-ESTRO Working Group (ii): Concepts and Terms in 3D Image Based Treatment Planning in Cervix Cancer Brachytherapy-3D Dose Volume Parameters and Aspects of 3D Image-Based Anatomy, Radiation Physics and Radiobiology". *Radiother Oncol* **78**: 67–77.

RCR (2008a) *Royal College of Radiologists, Society and College of Radiographers, Institute of Physics and Engineering in Medicine, and National Patient Safety Agency, British Institute of Radiology. "Towards Safer Radiotherapy"*. The Royal College of Radiologists, London.

RCR (2008b) *On Target: Ensuring Geometric Accuracy in Radiotherapy*. The Royal College of Radiologists, London.

RCR (2017) *Radiotherapy Target Volume Definition and Peer Review – RCR Guidance*. The Royal College of Radiologists, London.

RCR (2019a) *Radiology Dose Fractionation*, 3rd edition. The Royal College of Radiologists, London.

RCR (2019b) *The Timely Delivery of Radical Radiotherapy: Guidelines for the Management of Unscheduled Treatment Interruptions*, 4th edition. The Royal College of Radiologists, London.

RTOG (2020) "Contouring Atlases". Online Available from: https://www.nrgoncology.org/About-Us/Center-for-Innovation-in-Radiation-Oncology [accessed 26 August 2020].

Skripcak T, Belka C, Bosch W, et al. (2014) "Creating a Data Exchange Strategy for Radiotherapy Research: Towards Federated Databases and Anonymised Public Datasets". *Radiother Oncol.* Elsevier Ireland Ltd.

Vallières M, Freeman CR, Skamene SR and Naqa EL (2015) "A Radiomics Model from Joint FDG-PET and MR Texture Features for the Prediction of Lung Metastases in Soft-Tissue Sarcomas of the Extremities - IOPscience". *Phys Med Biol* **60**(14).

Part III

Radiation Safety

The hospital radiation safety team plays a crucial role in supporting the safe use of ionising radiation in all areas of the hospital, as well as in the local community, such as dentists and GP practices. Clinical Scientists specialising in both non-ionising and ionising radiation safety have specialist scientific skills in measuring and calculating radiation exposure and dose, but also need to be good at assessing risk and working with local teams to encourage a strong radiation safety culture and put measures in place to keep radiation exposures As Low As Reasonably Practicable (ALARP) and within limits set by strict regulations. Clinical Scientists and MR safety experts, laser protection advisers (LPAs) and radiation protection advisers (RPAs) work with colleagues to put required measures into place. Finally, if there is an incident, Clinical Scientists need to use their detective skills to identify what went wrong and how similar incidents can be prevented in the future. All these aspects of the role make radiation safety one of the most varied and interesting areas of Medical Physics, requiring attention to detail, a detailed knowledge of legislative requirements and excellent people skills.

Radiation safety in medicine covers the use of electromagnetic radiation including the use of magnetic fields, X-rays, lasers, and infrared, if visible, or ultraviolet (UV) light. The electromagnetic spectrum is summarised in Figure 3.1.1.

Radiation safety also covers the use of radioactive materials within hospitals (as described in Chapter 5) as well as energy imparted to the body by ultrasound. Radiation safety considerations for the use of magnetic resonance imaging (MRI) and ultrasound were previously described in Part I, as these usually fall under the remit of MRI and ultrasound experts. This chapter summarises hospital radiation safety associated with the use of ionising radiation, UV and lasers.

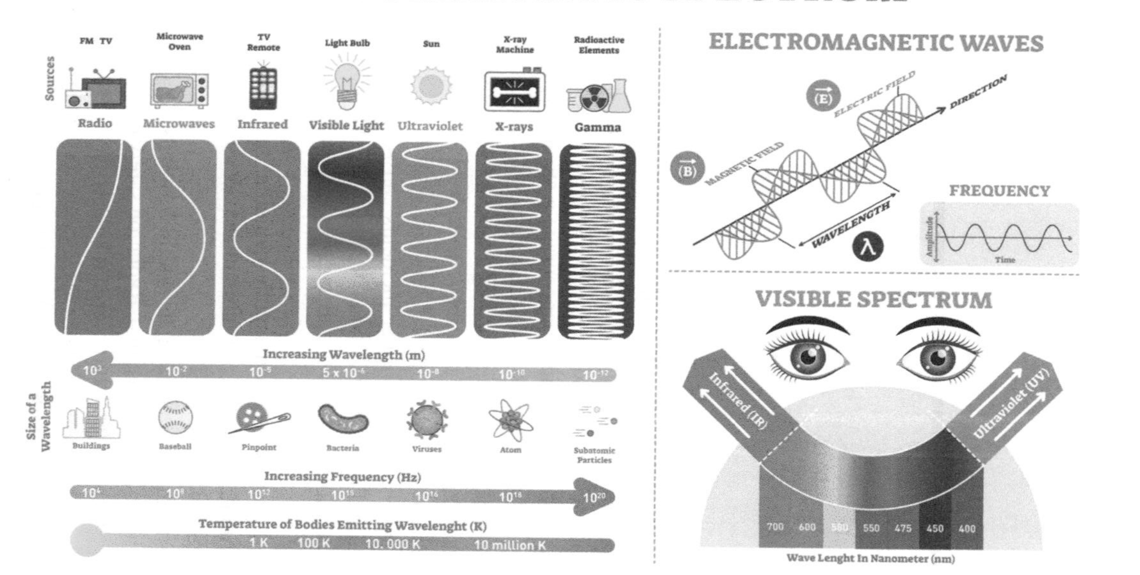

FIGURE 3.1.1 The electromagnetic spectrum. Ionising radiation includes short-wavelength X-rays and gamma rays. Non-ionising electromagnetic radiation used in hospitals covers the full range of the electromagnetic spectrum from UV and visible laser light to MRI radiofrequency (RF) signals.

7 Radiation Safety

Debbie Peet, Elizabeth Davies and Alimul Chowdhury

CONTENTS

7.1 THE ROLE OF THE CLINICAL SCIENTIST IN RADIATION SAFETY

Clinical Scientists working in radiation safety need to have a good understanding of the quantities and units used in radiation safety to describe radiation exposure. These include measures of dose and radiation exposure. Radiation safety experts also have a good understanding of how radiation passes through and interacts with human tissue, shielding and building materials, and how medical equipment is used. Another role of Clinical Scientists working in radiation safety is to provide hospitals with guidance on how to comply with legislation and best practice surrounding the uses of both ionising and non-ionising radiation in healthcare. It is a specialism where even the most experienced scientists must keep learning and developing their skills to support the implementation of solutions to problems and challenges.

The work of Radiation Safety Scientists may specifically involve the following tasks:

- Design of facilities
- Involvement in licensing/permitting premises to use radiation
- Commissioning new equipment and facilities
- Operational radiation protection during routine activities
- Providing expert advice in the event of a radiation leak or incident
- Supporting inspections by regulatory authorities
- Monitoring/measurement of radiation levels
- Assessment of patient and staff radiation doses
- Auditing practices involving radiation
- Teaching/training and research and innovation in all areas of radiation safety

Radiation Safety Scientists need to be aware of techniques, equipment and technologies in use within the hospital so that they can take appropriate measures to promote safe clinical practice and protect staff, patients and the public from harm. Radiation Safety Scientists also need to be able to react quickly to unforeseen and rapidly changing events.

A Clinical Scientist in radiation safety acts as a role model for their peers and colleagues in other departments. They need to know and understand the theory and background behind the principles of radiation safety and recognise the hazards and clinical requirements of examinations and procedures, so they can provide advice and help staff to understand why precautions are needed. Good communication and persuasive skills are essential. Many "users" of ionising radiation in healthcare concentrate on clinical outcomes and patient experience and measures designed to reduce radiation exposure can appear to add complexity and time to some procedures. Legislative radiation safety requirements are therefore sometimes viewed as additional restrictions that cause unnecessary delays. The capacity to listen to clinical colleagues' concerns and work with them to identify pragmatic solutions, rather than imposing restrictions, is a key skill.

7.2 NON-IONISING RADIATION SAFETY

7.2.1 PHOTOTHERAPY

As the name suggests, phototherapy ultraviolet (UV) electromagnetic wave radiation is used to treat skin conditions. Broadband UV (UVB) light uses wavelengths ranging from 280 nm to 314 nm to treat conditions such as psoriasis and eczema. Narrowband UVB uses a narrower part of the spectrum (311 nm to 312 nm), which is found to be more effective than broadband UVB for treating severe dermatological cases. PUVA therapy (psoralen + UVA) uses the UVA part of the spectrum (315 nm to 400 nm) with a photosensitising drug, called psoralen, added to treat psoriasis, vitiligo, and cutaneous T-cell lymphoma. The drug is administered as a cream to the skin or orally. Phototherapy is widely available, although audits have shown enormous variations in the quality of service and incidence of injury.

Service guidance and standards for phototherapy services written by the British Association of Dermatologists (BAD 2016) highlight the multidisciplinary nature of the work and specifically mention the value of the involvement of scientists from Medical Physics. The guidelines require that systems are in place for accurate UV dosimetry and state that "routine calibration and service checks ensure early detection of equipment problems, minimising risks to patients, and traceable records of safety and quality assurance systems. Regular ultraviolet risk assessments of the phototherapy unit will ensure the safety of patients, staff and visitors. Expertise and equipment to undertake such dosimetry is only available from Medical Physics. Staff occupational exposure to ultraviolet radiation must be assessed and kept below recommended limits".

The equipment used is a walk-in light box containing fluorescent light bulbs. The definition of a "dose" in the context of phototherapy is the irradiance of the radiation over the time the skin is exposed. Irradiance is measured in the quantity power per unit area with units of mW/cm^2 ($mW.cm^{-2}$). As power is the rate of the energy imparted in joules divided by time in seconds, phototherapy dose is often expressed in $J.cm^{-2}$ for a specified exposure time. Calibration using a radiometer, which determines treatment times, should involve a Clinical Scientist experienced in phototherapy. Doses are typically required to be accurate to within 15%. The actual measurement is not straightforward or without risk to staff, so training is vital. Detailed information is given in IPEM report 104 (IPEM 2010) and the BAD guidelines (BAD 2016). For measurement of UV levels, details can be found in Moseley et al. (2015). Scientists are involved in risk assessments and safety audits and are asked to provide general advice. There is a degree of responsibility involved with this role as the Clinical Scientist is responsible for ensuring that patients receive the prescribed dose to an acceptable level of accuracy.

7.2.2 Lasers

Lasers are widely used in many branches of medicine. They are widely used in ophthalmology, and as a cutting or ablating device in other forms of surgery in many specialisms such as urology and gynaecology and as a heating or curing device in other disciplines (e.g. in physiotherapy and dentistry). The wavelengths of laser light extend beyond visible light and clinical lasers can have wavelengths ranging from 160 to 10,600 nm. Laser light for medical applications is formed within several media (including solid state, liquid or gas [e.g. CO_2 lasers]).

There are several hazards associated with the use of lasers. Lasers and phototherapy equipment fall under the Control of Artificial Optical Radiation at Work Regulations (2010). There are many other sets of regulations and other requirements that also apply to the use of lasers which are described in the MHRA document "Lasers, intense light source systems and LEDs – guidance for safe use in medical surgical, dental and aesthetic practices" (MHRA 2015).

The most hazardous lasers, which cause damage to the eye or the skin, or can cause a fire hazard, are known as class 3B and class 4 lasers with class 4 being the most hazardous. The nominal ocular hazard distance (NOHD) is the distance from the laser device required to ensure that the laser beam does not exceed the maximum permissible exposure (MPE). Calculations of NOHD and MPE depend on several parameters. Henderson and Schulmeister (2004) explain how to perform these calculations from first principles, although most centres will use software for NOHD and MPE calculations. Protective glasses are usually required to be worn wherever surgical lasers are in use. These are specific to the power and wavelength of the laser, so staff need to be careful to wear the correct glasses when working with different types of laser. Scientists working with lasers should consult the relevant British Standards, describing PPE requirements for the eyes (BS EN 208 1999; BS EN 60825-8 2006; BS EN 60601-2-22 2013; BS EN 60825-1 2014; BS EN 207 2017).

Some senior Scientists working in either non-ionising or in Radiation Safety become experts in the use and control of lasers in hospitals and gain certification as a Laser Protection Adviser (LPA). Clinical Scientists might provide advice on the choice of room and room layout and minimisation of hazards, as well as carrying out measurements of the laser beam, risk assessments and safety audits. The LPA may hold a senior role within an organisation and sit on safety committees.

CASE STUDY

The LPA is called to the ophthalmology outpatient department to advise on the suitability of a room proposed for laser use. The room is very small; it has windows at ground floor level and a mirror above a sink located within 1 m of the proposed laser location. The room opens onto an outpatient waiting area and the room cannot be locked. Observation shows that clinical staff enter the room without knocking. It is not proposed to connect the laser to a warning light outside the room to show when the laser was in use. Discussions with the laser protection supervisor (LPS), clinician, and nurse in charge of the area flag this and a number of other concerns and it is agreed that another location should be found. The LPA writes a report that is sent through to those involved.

7.3 IONISING RADIATION SAFETY – KEY CONCEPTS

Soon after the discovery of X-rays and radioactivity described in Chapter 1, people working with radiation began to show signs of what we now recognise as radiation sickness and radiation injuries. Unsafe practices from the late 19th and early 20th century resulted in high exposures to patients, staff and the public. In 1928, the International X-ray and Radium Committee was established in response to an increasing body of evidence showing that the use of radiation in medicine generated harmful effects amongst health workers. It was recognised, even then, that doses to workers should be restricted as far as is practicable. This body was later renamed the International Commission of Radiological Protection (ICRP). The ICRP is now linked to the World Health Organisation (WHO) and provides a global framework for radiation safety in all sectors of industry including healthcare. Following the atomic bombing of Hiroshima and Nagasaki in 1945, the devastating and long-term effects of high levels of ionising radiation on humans also started to become apparent. Doctors noticed a marked increase in the incidence of leukaemia in populations living close to the epicentres of the bombs compared to the rest of the population. A United Nations committee was set up in 1955 to collate scientific evidence from epidemiological studies on the effects of radiation on humans (UNSCEAR 2020). Continual progress in obtaining a better understanding of the effects of radiation has led to a common framework for radiation safety, which is followed by almost every country of the world. The latest recommendations are provided in the International Commission on Radiological Protection, Report 103 (ICRP 2007).

A key role of the Clinical Scientist specialising in radiation safety involves training radiation users to adhere to current legislative and best practice guidelines. This is partly achieved by ensuring that equipment safety checks are regularly performed, but also through encouraging safe working practices and a culture of continual improvement to ensure compliance with regulatory standards.

7.3.1 Ionising Radiation Effects – Stochastic and Deterministic

Radiation effects can be divided into two distinct categories: stochastic – where the risk, or probability, of an effect (cancer) is related to the level of exposure, and non-stochastic effects (such as radiation burns or tissue damage) where direct injury can occur above a certain threshold dose. Stochastic risk estimates have been largely derived from studies involving exposure to high doses of radiation at high dose rates.

In medicine, and particularly in diagnostic radiology, limited data are available relating radiation exposures to stochastic outcomes. This means that risk estimates carry large uncertainties. Compared to an individual's natural risk of cancer, the additional (or excess) risk associated with medical exposure is relatively small (Table 7.1). Multiple risk factors and gene-environment processes are also at play that can impact the risk of developing cancer over the lifetime of a specific individual.

Currently, stochastic effects are described using a Linear No-Threshold (LNT) model of radiation risk. This model simply assumes that there is no threshold below which exposure to radiation is completely safe, but that the level of risk increases linearly with the level of exposure (Figure 7.1). In this model, stochastic risk is assumed to vary as a function of age, dose and dose rate.

TABLE 7.1
Lifetime Risk of Cancer Incidence in a Male Composite Euro-American Population (from Wall et al. 2011)

Age (years)	Lifetime Risk of Cancer Incidence (% per Gy)
0–9	10.0
10–19	8.0
20–29	6.2
30–39	5.1
40–49	4.2
50–59	3.3
60–69	2.2
70–79	1.3
80–89	0.6
90–99	0.0

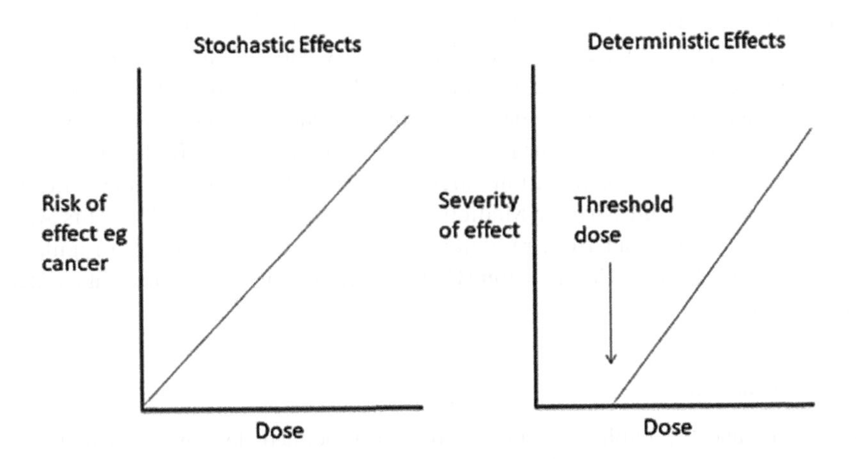

FIGURE 7.1 Linear threshold models describing (a) stochastic effects and (b) deterministic effects.

The second category of non-stochastic (deterministic) effects only occur above a certain threshold dose. Above this threshold, the severity of direct tissue damage, or other effects, increases linearly with dose. Non-stochastic effects are important in radiotherapy, and some forms of interventional radiology, where erythema (skin reddening similar to a burn) can be a consequence of exposure. Indeed in radiotherapy, specific doses of radiation are administered with the intention of damaging tumour cells, and Chapter 6 describes this and many techniques to minimise exposure to healthy tissues.

7.3.2 Legislative Requirements for the Use of Ionising Radiation

The International Atomic Energy Agency (IAEA) "Basic Safety Standards" document outlines standards for reducing unnecessary exposure to ionising radiation (IAEA 2014). These apply across all sectors, including healthcare, nuclear and industry. Radiation safety professionals are employed across these sectors to support the IAEA core principles of justifying, optimizing and limiting radiation exposure.

In healthcare, ionising radiation is used to treat and diagnose a variety of medical conditions. In some circumstances, staff could be accidentally exposed to relatively high levels of radiation; for example, an individual might unwittingly remain within a radiotherapy bunker during treatment leading to high levels of exposure. Existing legislation, and best practice guidelines from professional bodies, provide a framework for minimising the risk and impact of these situations. The first task of a trainee Clinical Scientist working in radiation safety is to familiarise themselves with relevant guidelines.

Clinical Scientists working in Radiation Safety need to understand how relevant legislation applies in a range of settings including Diagnostic Radiology, Radiotherapy, and Nuclear Medicine. A number of the most important regulations and guidance documents are summarised briefly in Table 7.2. These include a useful guidance document called the Medical and Dental Guidance Notes (MDGN) produced by the Institute of Physics and Engineering in Medicine (IPEM 2002) which provides a working interpretation of the legislation applied to different clinical areas. This is a good first place to start when trying to understand the legislation. An updated version is expected imminently. Other professional bodies that issue practical accessible guidance in Radiation Safety are the Society for Radiological Protection (SRP) (https://srp-uk.org/) and the British Institute of Radiology (BIR; https://www.bir.org.uk/; see e.g. BIR 2016).

There is also an Approved Code of Practice issued by the UK Health and Safety Executive (HSE 2018). This should be viewed as the "Highway Code" of radiation use relating to the exposure of workers and the public to radiation. While these recommendations are not a legislative requirement, if you opt to deviate from the code of practice you should be able to justify why and demonstrate that your processes are equivalent. The RCR guidance published in 2020 on the implications of radiation safety for clinical practice in diagnostic radiology, interventional radiology and Nuclear Medicine is also useful (RCR 2020). This was a multidisciplinary project with input from radiation safety experts. There is also regular guidance produced by the Care Quality Commission (CQC) on compliance with incidents under IR(ME) R 17 (CQC 2020b).

7.3.3 Dose Limitation

Members of staff and the public are subject to legally defined dose limits which must not be exceeded. These are set down in IRR17 and enforced by the Health and Safety Executive (Table 7.2). Many staff who work with ionising radiation wear personal dosimeters (Figure 7.2) to confirm that any radiation dose is well within expected safe limits. It is usually scientists in

TABLE 7.2
Legislation in England

Regulation	Main Points and Key Guidance Documents
Justification of Practices Involving Ionising Radiation Regulations 2004 (amended 2018)	To ensure that no practice is undertaken unless there is prior justification of the practice by a justifying authority. This covers most uses of radiation in healthcare.
IRR (2017): Ionising Radiations Regulations 2017	To ensure the safety of workers and the public when radiation is used. These regulations are enforced by the Health and Safety Executive (HSE) through: Notification, registration or consent for the use of radiation Prior risk assessment Local rules Information, instruction and training Consultation of a Radiation Protection Adviser Appointment of a Radiation Protection Supervisor Co-operation of employers when employees work over multiple sites Dose limitation Personal protective equipment use, maintenance and storage Contingency plan implementation and rehearsal Designation of controlled area Designation of classified staff Safe use of sealed sources The Approved Code of Practice (ACOP) provides practical methods of implementing the requirements. Guidance in the HSE document L121 (HSE 2018) and the Medical and Dental Guidance Notes (IPEM 2002).
IR(ME)R17: Ionising Radiation (Medical Exposures) Regulations 2017 (amended 2018)	The aim of this legislation is to ensure that governance is in place to optimise the use of radiation in healthcare. It is enforced by the Care Quality Commission (CQC). It does this by defining key roles of: **Referrer:** A doctor who can ask for an X-ray to be performed. This may also be a non-medical member of clinical staff if they are in a registered profession, have had appropriate training, and have been authorised to do so by their employer. **Practitioner:** The person who is responsible for ensuring that the benefit to the patient is greater than the risk associated with an exposure. This is the ARSAC (PHE 2020) license holder in Nuclear Medicine and for Molecular Radiotherapy (see Chapter 5 for more on this). **Operator:** Anyone who performs any practical task that could affect a patient's radiation dose. This includes Radiographers, Radiologists, Technicians and Scientists. **Medical Physics Expert:** An individual who has enough knowledge of Medical Physics to provide advice on the exposure aspects and application of IR(ME)R (2017).

(*Continued*)

TABLE 7.2 (Continued)

Regulation	Main Points and Key Guidance Documents
	IR(ME)R (2017) also sets out the governance requirements around: Training requirements Procedures that must be in place, for example how to identify the correct patient Audits that must be in place Quality control requirements The requirement to have diagnostic reference levels rather than dose limits for medical exposures Managing incidents Guidance is found in CQC (2017), Department of Health and Social Care (DHSC 2019) guidance documents and in RCR guidance.
EPR16: The Environmental Permitting Regulations (EPR 2016)	This legislation covers all types of waste production and embodies the principles of reduce, reuse and recycle. Schedule 23 covers radioactive substances to ensure: Control over keeping, use, and security of sources of radioactivity. Adequate systems are in place for the accumulation and disposal of radioactive waste to manage the radiological impact on both individuals and the environment. Adequate provisions are in place for the use of High Activity Sealed Sources (HASS). Guidance is issued by the Environment Agency on EPR16 and guidance for hospitals is in the MDGN.
Carriage of Dangerous Goods Act (2019)	This covers the transport of radioactive material to hospitals and between departments. Requirements for packaging, labelling, and transport documentation are described. Training requirements for drivers, including contingency plans in the event of road traffic accident or theft from a vehicle, are also covered. Hospitals involved in transport are required to consult a dangerous goods safety adviser who is required to keep up to date with the measures required. Guidance can be found in the MDGN (IPEM 2002).

radiation safety who advise their organisation as to who should be monitored. In a hospital setting, most staff receive low whole-body doses that are well within recommended limits.

For radiation workers, the dose limit for whole-body exposure is set at a dose where any additional risk of developing cancer above the 1 in 2 chance occurring naturally (CRUK 2020) is considered acceptable factoring in other occupational hazards. In addition to this whole-body dose limit, separate limits are set for individual organs to ensure that radiation levels are well below the threshold for deterministic effects (such as erythema). The most applicable limit will depend on the type of work being undertaken in each area. For example, in Nuclear Medicine, where staff

FIGURE 7.2 Personal whole-body radiation dosimeter.

may be manually handling radioactive sources, the most appropriate limit is the finger dose limit. In interventional radiology, it may be the eyes that are most likely to approach a dose limit. The dose limit for the lens of the eye is set well below the level where radiation-induced cataracts have been found to occur. While radiation-induced cataracts were originally thought to result from deterministic effects, these are now thought to be "quasi" stochastic (Hamada et al. 2014; ICRP 2012) which resulted in a significant reduction in eye dose limit in IRR17.

7.4 HOSPITAL IONISING RADIATION HAZARDS

A knowledge of how various sources of radiation are used throughout healthcare is needed to be able to quantify and mitigate the risks of exposure.

7.4.1 Diagnostic X-ray Imaging

An X-ray tube is a source of radiation with a well-collimated beam of X-rays directed towards the patient so that an image can be obtained. While the tube is encased in metal to reduce the intensity of X-rays in other directions, leakage from the tube itself can be hazardous to anyone standing close to the tube. See Figure 7.3, for example.

A 2-mm-thick layer of lead is often used to surround diagnostic radiology rooms to prevent exposure to staff, patients and the public in surrounding areas. Collimated sources can be used in fixed orientations, or moved more freely, which may impact on radiation protection requirements. To carry out the required work staff often need to remain within the room alongside the patient and X-ray unit. Systems with a free range of movement, such as that shown in Figures 7.3 and 7.4, can be a challenge for radiation safety. Eyes can be protected by shields in the room, or by requiring staff to wear lead glasses (Figure 7.4). At diagnostic energies, most X-ray scatter is directed back towards the X-ray tube. Therefore, the tube is usually placed under the operating table (to protect the eyes and radiosensitive abdominal organs), whereas the legs are protected by lead skirts placed around the table (Figure 7.3).

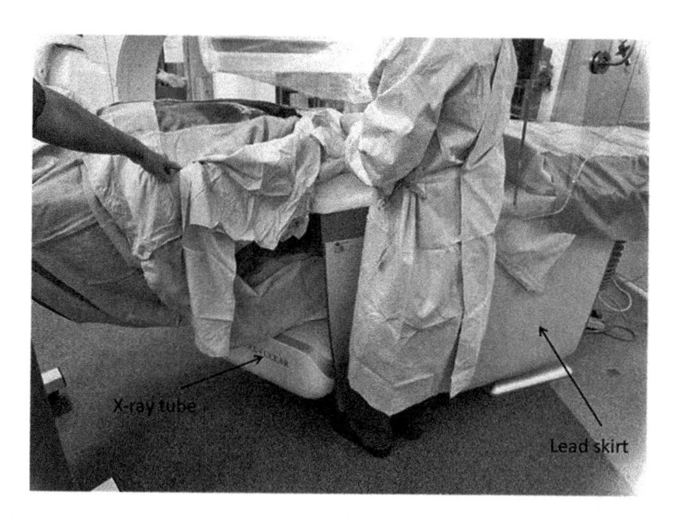

FIGURE 7.3 The surgeon's legs are protected from X-rays emitted by the X-ray tube by the lead skirt hanging from the table. However, her foot is close to the tube housing and should be retracted back to prevent unnecessary X-ray exposure.

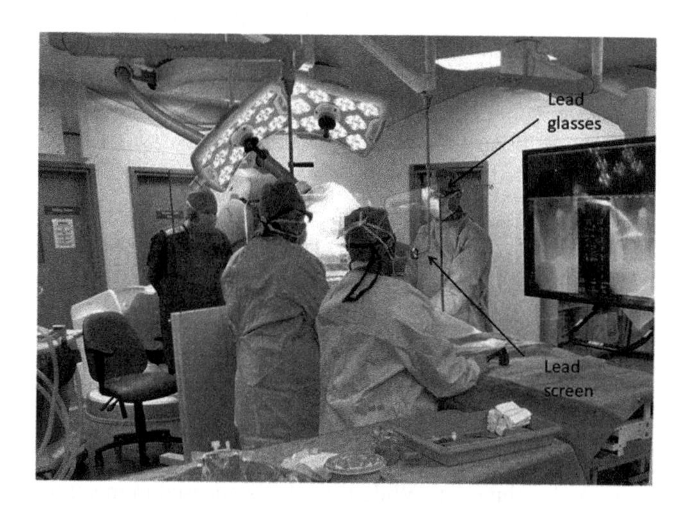

FIGURE 7.4 Measures to protect eyes from significant exposure. In this environment, staff without a lead screen between them and the source of the radiation will need to wear lead glasses. In this instance, the patient is the source.

7.4.2 Nuclear Medicine

In Nuclear Medicine, hazards are more varied and complex. In addition to the potential for contamination, inhalation, and ingestion of radioactive material, once patients are injected with a radionuclide, they effectively become a mobile radiation source with potential to take public transport, urinate, breastfeed and engage in a variety of activities that could irradiate or contaminate and thereby increase the radiation risk to those around them. More details of how the public are protected following nuclear medicine treatment of patients were described in Chapter 5.

Radioactive sources can be classified as either sealed sources (if they are completely encased in a solid material), or unsealed sources. Unsealed sources have potential to become an external dose-rate hazard and a contamination hazard. In the event of contamination, all reasonable precautions must be taken to reduce the risk of ingestion or inhalation of radioactive material. Once a source is inside the body, it becomes difficult to limit radiation exposure. Sealed sources are usually considered an external dose-rate hazard, although if the casing splits and the radioactive material is released, they could become an unsealed radioactive source and a contamination hazard. To reduce this risk, radioactive sources are usually stored in special containers, which need to be regularly tested for leaks.

7.4.3 Radiotherapy

As described in Chapter 6, the goal of radiotherapy is to directly destroy cancerous tumour cells by imparting a high, but targeted, dose of radiation to the patient. Localised doses can be imparted using focused beams of high-energy X-rays and other particles (e.g. electrons) generated by a linear accelerator.

In radiotherapy, levels of exposure are several orders of magnitude higher than for diagnostic imaging applications. Dose in radiotherapy is measured in units of Grays (Gy) to specify absorbed dose to a target organ, rather than the whole-body dose monitored in diagnostic radiology, which is measured in Sieverts (mSv). Additional radiation safety precautions are required to limit the level of exposure to staff and the public, and to parts of the patient that are not undergoing treatment. Most linear accelerators are surrounded by a 1–2.5 m concrete bunker to protect surrounding areas from high levels of radiation. In radiotherapy, it is imperative that no one other than the patient remains in the treatment room during exposure. Due to stricter engineering controls around radiotherapy, such as interlock and shielding, doses received by staff are often lower than for staff working in interventional radiology or Nuclear Medicine. This is because the extensive shielding, and ability to leave the patient in the treatment room, reduces the potential for radiation exposure. Figure 7.5 shows a radiotherapy treatment room.

Brachytherapy uses sealed sources to treat cancers directly. Some sources are implanted permanently, others are administered for short periods of time. Because the activity of these sources is relatively high, the potential for excessive staff exposure exists and additional controls will need to be implemented.

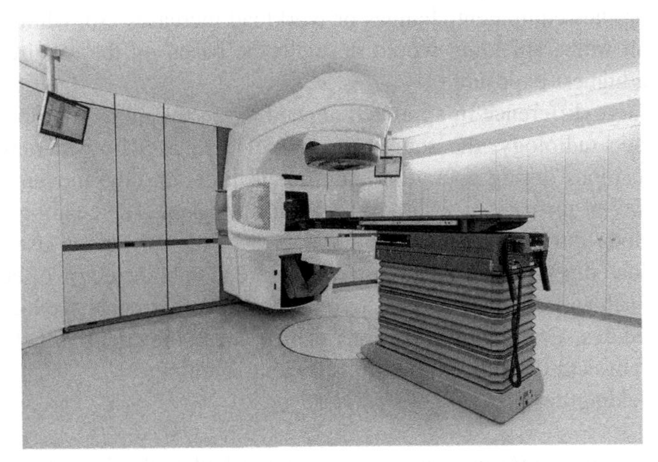

FIGURE 7.5 A radiotherapy treatment room.

7.5 ROOM DESIGN

Designing a facility to protect individuals from ionising radiation can be quite a complicated task and involves several steps from start to finish. It can be easy for the clincial scientist to have a narrow view and only consider any shielding required around a room. In reality, the process of room design involves multiple stages and large numbers of specialist staff. Room design must consolidate many, sometimes conflicting, requirements with the purpose of designing a room that is safe, meets the appropriate electrical and structural requirements, has suitable storage and access for clinical requirements, and can accommodate the necessary fixtures and fittings required.

From a radiation safety perspective, the position of equipment, the intended clinical use, and the work carried out in surrounding rooms should be considered. Placement of equipment is a major factor, as distance between the equipment and walls or screens can significantly reduce the amount of shielding required, but also determines the risks resulting from direct beam exposure and scattered radiation at key locations. Clinical use is also important to consider, as different projection angles and use of different techniques, or different isotopes in nuclear medicine, can impact the ideal positioning for shielding. In surrounding rooms, an office occupied full time may require more shielding than an external wall or roof. Liaison with the architect can be useful for identifying surrounding areas, distances involved, and entry and exit points for electrics, water, air and gas flow.

Specialist staff involved in the process can be unaware of the requirements outside of their expertise, particularly if they have not been involved in designing and building a radiation facility before. Poor communication at this stage can result in costly corrections and delays. Ideally, the design process should include discussions between the manufacturer, architects, clinical staff, construction/facilities staff and Clinical Scientists. During these conversations, key points can be raised and informed decisions can be reached regarding conflicting requirements. For example, a Clinical Scientist might suggest that an X-ray room door opens inward to form a barrier between individuals entering the room and the X-ray unit, but this may prove impractical for entering the room with a patient on a trolley. Face-to-face meetings provide a good opportunity to discuss the expectations for designating areas. This impacts on the location of warning lights and signs as these should be placed at the threshold of controlled areas. If the area is un-designated at times (e.g. with X-ray equipment electrically isolated or after a radioisotope room has been checked for contamination), then no permanent signs should indicate it as a controlled area. Reversible signs, or two-stage warning lights, can be installed to accommodate this.

It is common practice to install the same thickness of shielding to all walls to allow for the possibility of the equipment being moved in the future. This reduces the risk of errors during installation and simplifies equipment changes or workload increases. So, while calculations could be performed for all walls, shielding would normally be based on the requirements of walls or floors receiving the highest exposure.

There is also a large difference in approach depending on the modality. For example, in fixed installation diagnostic radiology, the source position will have a limited range of motion and specific standard positions during operation. Additional limits on tube movement can be used to prevent the possibility of primary exposure to the controlled area. The beam itself will only be on for very short periods during exposures and the energy of the beams used in clinical imaging are relatively low; due to this the main interaction are the photoelectric effect and high atomic number materials, such as lead, provide effective barriers. For high energy Nuclear Medicine and Radiotherapy procedures, Compton interactions can be more prevalent, which may require materials with a high physical density to provide sufficient shielding.

Principles of shielding design:

- Workload and equipment/source usage are used to calculate the dose at a critical point where a worker or member of the public may spend time.
- A constraint is set; often 0.3 mSv per annum for a member of the public. Higher doses, such as 1 mSv per annum might be set for staff who work with radiation.
- The ratio of these is called the attenuation factor.
- Tables/calculators can be used to calculate the thickness of material required to achieve the desired level of attenuation. The material might be concrete, or lead, or standard bricks or blocks.

7.5.1 Engineering Controls/Interlocks/Warning Signs

Because the consequence of exposure to anyone other than the patient receiving treatment is serious, several measures are routinely used to ensure no one other than the patient is in the bunker before the radiation beam is initiated. All rooms will have warning lights that illuminate when a source is exposed (Figure 7.6) as required under the Safety Signs Regulations 2015 (HSE 2015). Some rooms will have supplementary warning signs giving further details about the type of source and access restrictions.

When leaving the patient in the room before radiotherapy treatment, someone takes the role of last person out (LPO); after visually sweeping the room, they press the LPO button. This gives a short time period for that person to leave the room before setting an interlock in the form of a "light curtain" or stable door. The beam can only be initiated if this safety step has been completed. If the interlock is broken, the beam cannot start and the LPO system needs to be reset while checking that no one has inadvertently entered the room. Likewise, if the interlock is broken during treatment, the beam will stop. These failsafe measures are checked every day.

7.5.2 During Construction

If construction and installation of equipment is / are required, the scientist may need to visit the site to visually inspect the shielding fitted and check the positions of any warning lights and other engineering controls. They may also discuss the operation of the equipment with the installation engineer.

7.5.3 Attenuation/Transmission Measurements

When a room is being fitted within an existing building there may not be adequate records of how the wall was built to allow the transmission to be estimated. A transmission measurement may then be required to determine the amount of shielding already in place. There may be defects from poor construction or gaps for sinks, drainage or air conditioning, which may not have been noted on plans. It is also common to perform transmission measurements on new buildings prior to use to ensure that adequate shielding is in place if this cannot be verified during the build.

Measurements of transmission of the X-ray beam through a barrier can be made using suitable measurement devices.

For a diagnostic X-ray room, the air kerma from the X-ray tube can be measured at a distance close to the source (d_2) (Figure 7.7) using an ionisation chamber (Figure 7.8). Measurements can be extrapolated using the inverse square law to estimate the Kerma expected at a distance further from the source at d_1 in the absence of shielding. By obtaining a further measurement at d_1 using a

FIGURE 7.6 Warning light at the entrance to a linear accelerator bunker. The area is designed as a controlled area. When the beam is on, the red light is illuminated. During treatment, a stable door is closed. If someone were to enter, the interlock would terminate the exposure.

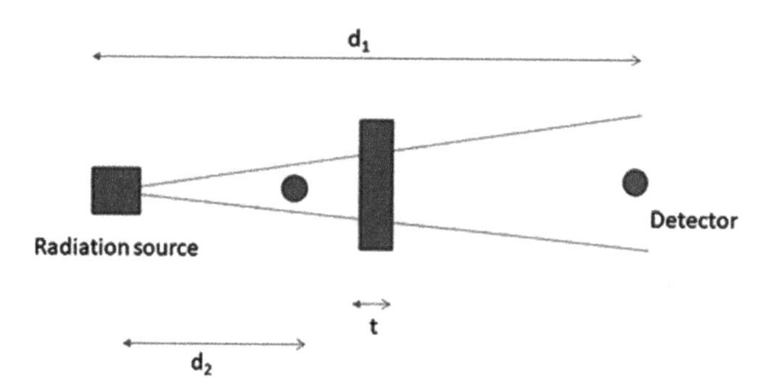

FIGURE 7.7 Attenuation of a photon beam through a medium of thickness, t.

detector, it is then possible to determine by how much the radiation has been attenuated by the shielding. The level of exposure may be quite low through the barrier, so a large ionisation chamber with greater sensitivity might be needed, or a specially designed and calibrated solid-state device or survey meter. The ratio of air kerma at this point gives the attenuation factor – independent of material in the walls. The Clinical Scientist should recognise that it can sometimes be difficult to accurately assess the thickness, t, of shielding and building materials, which can increase errors in assessing the attenuation of shielding materials.

For conventional radiography, exposure times are very short and so identification of defects in construction can be difficult unless multiple measurements are taken to identify gaps. Visual examination of the walls during construction is often the best way to identify such defects.

An alternative assessment method is the use of a portable radioactive source. This is passed along the barrier at the same time as a contamination monitor (Figure 7.9) on the other side to identify any inconsistencies in the barrier's attenuation. Ideally this should be performed with an americium-241 source, which closely replicates the diagnostic spectrum, however purchase and licencing of these sources can be prohibitive. A collimated technetium-99m source, absorbed into cotton wool to reduce contamination risks, can provide a practicable alternative although this provides a less accurate assessment due to the differences in x-ray energy between technetium and diagnostic X-rays. The lead equivalence of the wall can be estimated based on the known attenuation of the source that is used via a similar method indicated in Figure 7.7. For technetium sources, this measurement should be made under broad beam conditions. These estimates provide a convenient estimate but should be treated with caution due to errors involved in the measurement process.

7.5.4 Post Installation Equipment checks

Once the equipment is installed, the installer (often the equipment manufacturer) is required under the Ionising Radiations Regulations HSE(2018) to perform a critical examination as described in the diagnostic radiology chapter of all the safety features of the equipment. This may be carried out by a Clinical Scientist on behalf of the manufacturer. They will check that emergency off switches operate as intended, that lights and warning systems operate correctly, and that the equipment has been safely installed from a radiation safety perspective. They will operate the X-ray equipment using the systems of work agreed and will check that all procedures and processes are in place before clinical work can begin. New equipment will be also be commissioned by scientists and or clinical technologists to ensure it is performing as expected. These measurements might be carried out by the radiation safety team or by scientists in the clinical area where the equipment is used. Baseline measurements of performance measures such as dose and image quality should be made so that repeated measurements can be compared throughout the life cycle of the equipment in line with standards, as described in Chapter 4.

FIGURE 7.8 Measurement of transmission through a wall using a large volume ionisation chamber.

Several patient safety measures are also required under the IR(ME)R (2017) regulations. For example, general exposure protocols should be developed for the use of the unit. These "exposure charts" list the exposure settings used for individual examinations. There is increasing concern about the increased use of CT scans resulting in higher doses at a population level (as described by the "COMARE 16" report, DHSE (2014)). There are now efforts to reduce individual doses for CT scans, see, for example, the "Image Gently" programme (https://www.imagegently.org/). If children are scanned, different charts listing paediatric exposure levels should also be available. For individual exposures, a trained and experienced radiographer will be required to individually optimise the exposure through the appropriate use of automatic exposure control systems, exposure field size and set up. Scientists should be involved in the setup of scanning protocols, working closely with the Applications Specialist, Radiographer and Radiologist. This is important in ensuring equipment is optimised and that doses to patients are ALARP. The whole team must work together to ensure that all their different areas of expertise are fully utilised. A Senior Clinical Scientist will then be able to sign off that equipment as safe for clinical use. The Scientist will be involved throughout the whole life cycle of the equipment and may also provide advice on equipment disposal.

7.5.5 Environmental Monitoring

The next stage is to carry out long-term environmental monitoring. This is usually undertaken using monitors obtained from personal dosimetry suppliers. It is important that environmental monitoring is placed in a location likely to be exposed to a similar level of background radiation as users, without

FIGURE 7.9 Contamination monitor equipped with a scintillation detector and rate meter, showing counts per second (cps).

being artificially exposed. Elevated readings may be obtained where the building materials are slightly active; therefore, careful choice of the control position is important. Over time, the occupancy of surrounding areas may change, so a regular review of the doses in those areas should be conducted as part of the risk assessment (Section 7.6). It is also important to periodically re-monitor to ensure that nothing has changed, such as the workload increasing. Monitoring is usually repeated every 3–5 years but may be measured sooner if there may have been changes to workload or structure or processes.

7.5.6 Diagnostic Radiology Room Design

For diagnostic radiology, rooms require some shielding, typically around 2 mm of lead or 20 cm of blockwork. A detailed description of the calculations can be found in Sutton et al. (2012).

7.5.7 Radiotherapy Room Design

Linear accelerators, although much higher energy, are also relatively fixed, with motion usually limited to a circle with a fixed isocentre. Again, the beam "on time" is short, although longer than in diagnostic radiology applications. Linear accelerators are installed within purpose-designed "bunkers" (Figure 7.5). Patients and staff should be able to enter the bunker so that treatment can take place, but the patient entrance (or maze) can be the hardest part of the installation to design safely. Radiation is scattered from the X-ray beam interacting with the patient and every surface the beam hits (IPEM 2017). The walls, floor and ceiling around the treatment unit need to be designed in such a way to

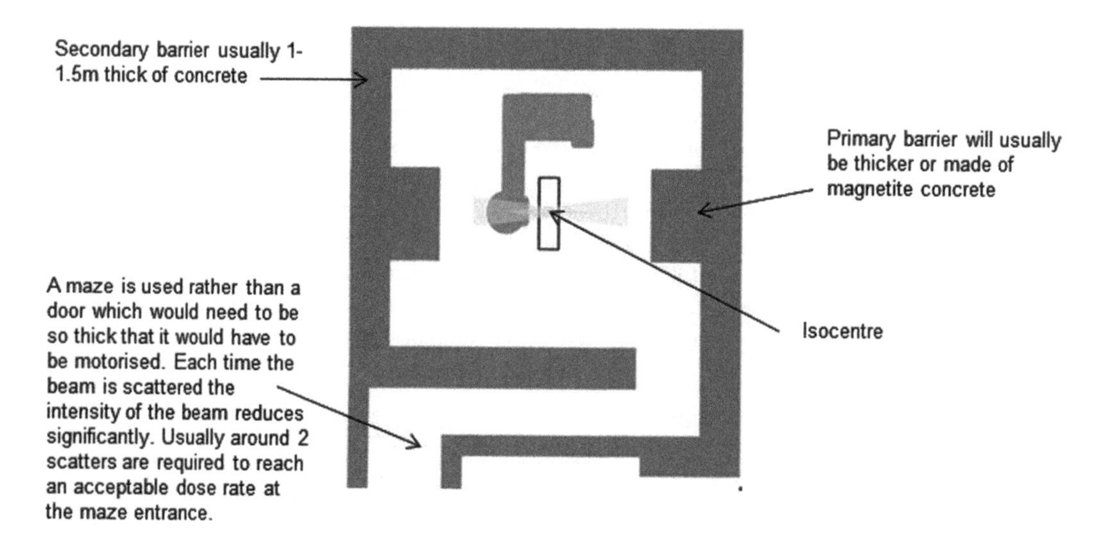

FIGURE 7.10 Typical layout of a linear accelerator.

ensure that scattered radiation is absorbed by the walls and other barriers to levels that are acceptable outside of the room. Radiotherapy bunkers are often installed in single-story buildings so that no additional shielding is required under the bunker floor, and less material is needed on the roof. The room itself will be designated as a controlled radiation area (Figure 7.10).

The photon beam is limited in size. The barriers that the beam can directly hit therefore only need to be wider than this divergent beam and are known as primary barriers. These barriers can include up to 2.5 m of concrete in the walls to reduce the dose rate, and doses outside the room, to acceptable levels that are well within required dose limits. The design of these bunkers can be very complicated, but the basic principles will need to be understood by all Clinical Scientists working in these areas. A longer description of the design and potential pitfalls is found in IPEM report 75 (IPEM 2017), NCRP report 151 (NCRP 2005) and Sutton's Practical Radiation Protection (Sutton et al. 2012).

CASE STUDY: PRIMARY BARRIER SHIELDING CALCULATION

A linear accelerator rotates around a single point in space (known as an isocentre) (Figure 7.10). The machine produces a beam of X-rays with a maximum dose rate at this point – D_0. Typical values might be 6 Gy/min at this point or as high as 24.5 Gy/min with modern machines.

Outside the room, there may be a corridor or an office. If this critical point is in an office and is located 5000 mm from the isocentre a simple inverse square law calculation shows that the dose rate at this point will be $D_0/(5 \times 5)$ or 14.4 Gy/hr for a beam of 6 Gy/min at the isocentre.

The 1 mSv dose limit for a member of the public would be exceeded in 0.25 seconds, so clearly some material is required to reduce this or attenuate the beam. Assuming the space is an office requires the highest level of shielding as the design constraint could be as low as 0.3 mSv/year.

Knowing the time that the beam is on, the proportion of time it points at an individual barrier (in this case assumed to be 30% of the time - an orientation factor of 0.3), and the occupancy of the surrounding areas, the dose to the critical point can be calculated as 8,640 Gy. The attenuation factor (AF) required is the dose (D) divided by the dose constraint at the critical point (DC) multiplied by the orientation factor and in this case is 2.8×10^6.

Converting this to an actual thickness of material (e.g. concrete) requires the use of Lambert's law:

$$I = I_0 e^{-\mu t}$$

where I_0 is the initial intensity, μ is the attenuation coefficient and t is the thickness of the material.

The transmission factor is given by:

$$\text{TF} = I/I_0 = \text{DC/D}$$

The thickness of material that reduces the intensity by a factor of two is called the half-value layer (HVL). More usefully, for shielding calculations, the concept of tenth-value layer (TVL) is often more helpful.

The number of TVLs, n, required can be calculated using:

$$n = -\log_{10}(1/\text{TF})$$

In this example, the required thickness, t, can then be calculated using:

$$t = n \times \text{TVL}$$

In this example, this would lead to $t = 6 \times \text{TVL}$. Given a limiting TVL for concrete is 370 mm for 10 MV (IPEM 2017); therefore, the material required to achieve this attenuation at 10 MV would be 2.3 m of concrete.

Experienced radiation shielding designers would normally perform these calculations and work with architects, users and other professionals to ensure the design is safe, yet works, both clinically and ergonomically.

This case study also demonstrates that it is vital that no one other than the patient is in the treatment room while the beam is on, as dose limits could be exceeded in less than a second.

7.5.8 Nuclear Medicine Room Design

In Nuclear Medicine, the sources are the patients themselves, once injected they remain radioactive for a period determined by the isotope. As patients can move around this can make shielding design much more complex.

CASE STUDY: NUCLEAR MEDICINE ROOM DESIGN

Taking the case of an uptake room in a PET scanner suite as an example. A radioactive source is injected, and the patient then rests in the room for up to an hour. Critical points in the room requiring calculation first need to be identified. These may include the closest point to the radiation source (patient) or an area where staff spend their time. Areas above and below the room might also need to be considered as described by Peet et al. (2012). For example, see the plan in Figure 7.11.

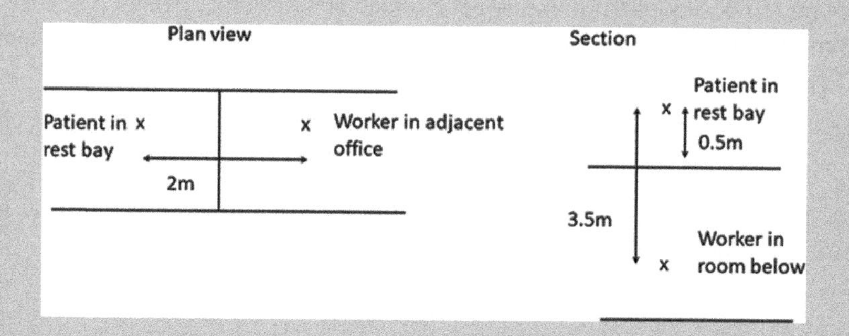

FIGURE 7.11 Plan and section through a building used for PET/CT scanning.

Consider the worker in the office adjacent to the treatment room. The distance from the patient is 2 m.

The air kerma rate is 37 μGy/hr at 1 m.

The kerma rate at 2 m is therefore 9.25 μGy/hr.

The dose per annum (assuming 2000 working hours) is 18.5 mSv compared to a constraint of 0.3 mSv.

Transmission required TF is 0.3/18.5 = 0.02

No of TVLs = $-\log_{10}$(TF)

1.79 TVLs required – 270 mm concrete/30 mm lead

The same calculation can be repeated for work in a room on the floor below.

This case study illustrates that a PET installation requires much more shielding than a standard diagnostic X-ray room (typically around 2 mm of lead) but is lower than for a linear accelerator. This is because the energy of the radiation source is much higher, and the photons are therefore more penetrating.

7.5.9 Dose Rate Measurements

For CT scanners and linear accelerators, which have much longer exposure times than the few milliseconds of diagnostic X-ray exposure, the dose rate outside the room might be measured

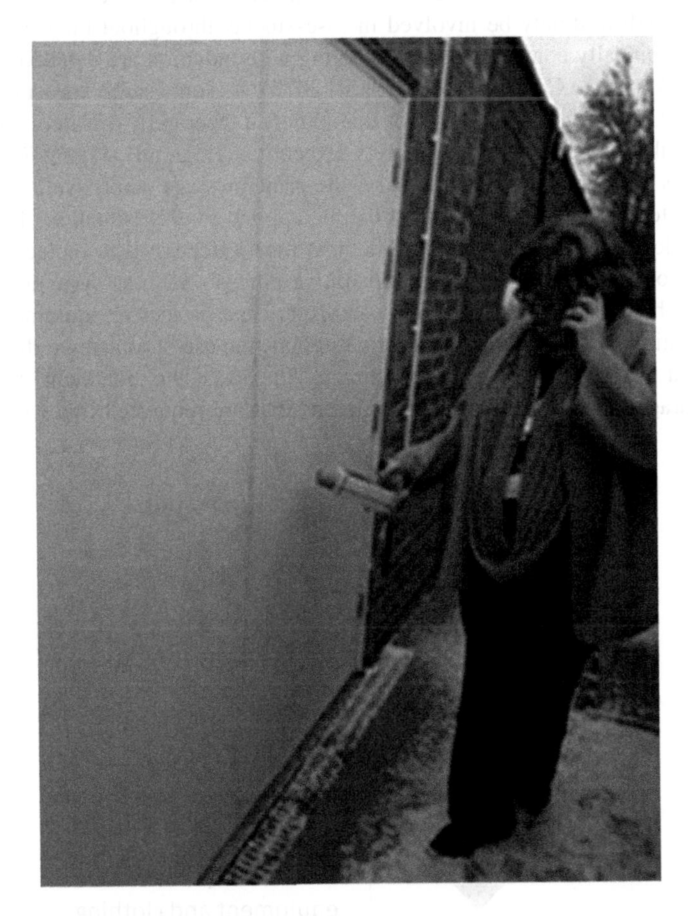

FIGURE 7.12 Survey of a CT scanner. Measurements can be taken in all weathers. Mobile phones and walkie-talkies are used to communicate when the radiation beam is on. A second person is useful to help record the readings.

directly using a dose rate meter (Figure 7.12). This might be a scintillation detector, an energy-compensated Geiger Muller tube, or a semiconductor detector. The choice of source and detector can depend on the environment, sensitivity required, and required accuracy of the measurement, as described by Knoll (2010). Measurements are taken along a barrier in a raster movement and recorded as spot measurements for locations along the barrier at waist, head and ankle height.

7.6 RADIATION SAFETY RISK ASSESSMENTS

The regulations require a Radiation Protection Adviser (RPA) to be involved with all radiation risk assessments. All work with radiation should be risk assessed before it is undertaken, with an assessment of risk being continually undertaken during the work. The basic principles of risk assessment were described in Chapter 1. A radiation safety risk assessment will specifically consider who might be exposed to radiation, the level of exposure, engineering controls such as shielding, and potential for accidents and incidents. The results of the risk assessment will inform the implementation of any additional measures required for safe, legal operation of the equipment to keep radiation exposures ALARP.

7.6.1 What Should a Radiation Safety Risk Assessment Consider?

Clinical Scientists are expected to undertake a risk assessment for a radiation facility as part of their training and will routinely be involved in assessments throughout their working life. Areas that need to be specifically considered when working with radiation are detailed in the Health and Safety Executive Approved Code of Practice (HSE 2018). As some safety measures are likely to be more effective and less prone to human error than others a "hierarchy of safety controls" has been proposed as described by the Health and Safety Executive (HSE, nd) (Figure 7.13). While this is described in terms of the construction industry, the principles also apply well to radiation safety. This can be used to consider what measures can be applied to each situation. The most effective control is to physically remove the hazard. The next most effective is to replace the hazard, or to use engineering controls to control it. Administrative controls such as systems of work are next most effective with least effective being the use of personal protective equipment (Figure 7.13).

In a clinical setting, it is often not possible to eliminate the use of ionising radiation as the use of radiation forms a crucial part of accurate clinical diagnosis and treatment. Clinicians are increasingly encouraged to consider whether the scans that are requested (known as referrals) will

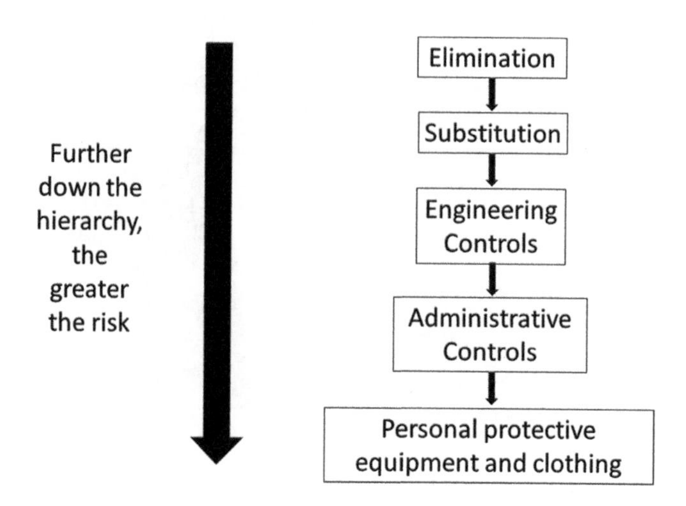

FIGURE 7.13 Hierarchy of controls.

really benefit the patient or alter the course of their treatment. A key strategy for reducing radiation exposure at a population level is to reduce the number of unnecessary scans performed, especially in children. It is also sometimes possible to substitute ionising with non-ionising techniques, such as MRI or ultrasound, without compromising patient care.

For a Clinical Scientist specialising in Radiation Safety controls that should be prioritised in order of effectiveness:

1. *Engineering Controls* such as protective shielding placed around the room and door interlocks. Engineering controls are favoured over other types of protection as these are less prone to human error and can be designed to be "fail safe".
2. *Administrative controls* aim to change the ways that people work. In a radiation safety setting, this will include introducing systems of work (training, local rules, instructions and standard operating procedures [SOPs]) to reduce radiation exposure to tolerable levels. Systems of work are often supported by the room design, signage and warning devices (such as warning signs on doors, warning lights, and audible exposure indicators). It may be necessary to designate areas of the room as controlled, or supervised, and consider access restrictions to control and restrict the risk of radiation exposure.
3. PPE, such as wearing lead aprons (and safety glasses and gloves in Nuclear Medicine), should be considered where protection is not practical by other means. The use of PPE requires a strong safety culture and is not always particularly well controlled.

A completed risk assessment should identify specific actions required to ensure doses are ALARP. Monitoring of staff dose through personal dosimetry (section 7.8.3) and regular audits will help to ensure the effectiveness of the above measures to identify potential for accidental exposure, or reasons for staff not following the safety measures (non-compliance). Risk assessments should be regularly reviewed.

Since the level of risks involved in the medical use of ionising radiation is well known, it is often possible to apply a number of assumptions based on standard practice; fixed and mobile X-ray units have dedicated rooms, or a controlled area around the unit. Standard systems of work can be expected within departments and PPE requirememnts fall into a number of standard categories. As part of your risk assessment, it is often reasonable to assume that standard training and safety practices are in place, especially for healthcare professionals, such as Radiographers and Radiologists, who have specialised in ionising radiation modalities. Often Clinical Scientists start from the absolute worst-case assumptions about workload and workflow consulting standard texts (Martin and Sutton 2015; Sutton et al. 2012). If the exposure is acceptable (usually defined as 0.3 mSv per year), no further measures are necessary. However, if it is found to be unacceptable, then the assumptions might be refined to more realistic worst cases. For example, more accurate exposure times or workloads could be used to better reflect expected practice. The constraint is often set at 0.3 mSv per year but other values might be more appropriate depending on the situation.

Radiation risks assessments are more detailed than the generic approaches described in Chapter 1 and there are a lot of details that should be considered as part of a radiation risk assessment. Many centres develop templates to aid this. In any case, the risk assessment must cover the applicable elements listed in the Approved Code of Practice regarding Regulation 8 (HSE 2018). Following the requirements of the ACOP the risk assessment should consider: training requirements; management responsibilities; maintenance and testing schedules; audit arrangements; contingency plans; and regular testing of radioactive sources for leaks. A particular consideration should be to identify any anticipated changes in work should a staff member declare they are pregnant or breastfeeding. In most situations the radiation risk should be negligible and changes are not anticipated. The one exception to this is likely to be Nuclear Medicine where contamination risks and doses can be more significant.

Whether the majority of the risk assessment is drafted by a Clinical Scientist, local Radiographers, or other staff, responsibility for managing radiation risk lies with the clinical team so it is essential to ensure their involvement from the outset and to develop good working relationships with staff. This might require observing standard clinical practice to better understand the requirements of the scan or treatment. Striking a balance between safety and clinical goals can sometimes be challenging.

A good place to start a radiation safety risk assessment is often by mapping out the layout of the room, marking any safety features. The intended uses of the space should be talked through with staff to understand exactly what they want to do and why. You may need to ask staff to draw diagrams, or to imagine where people will be standing during the procedure and what activities they will be undertaking. Once a layout and safety measures have been proposed it may then be beneficial to discuss these with a wider group to identify "reasonably foreseeable" problems; different people will have different perspectives, so the more people who are involved at the review stage the better. A full risk assessment is an iterative process and should be repeated or reviewed as new information becomes available.

Working on the assumption that there will be a controlled area involved (Section 7.7), it is important to consider which staff will be able to enter the area. If outside workers, visitors, or people under the age of 18 might enter the space, this is also important to know, as the legal restrictions and radiation safety requirements for these groups differ from those of adult patients.

Once there is a plan for the area, the effective dose likely to be received by staff under routine and accidental situations should be estimated. These should be compared to dose constraints which should be well below applicable dose limits defined under IRR (2017). Estimating doses will likely require several assumptions to be made using experience or reference data. It is good practice to detail all assumptions, including the source of any numeric data, so that estimates can be reviewed, and if necessary adjusted, as part of future risk assessments. It is advisable to consider realistic worst-case scenarios to ensure staff are not at greater risk than documented.

The risk assessment will determine the required personal protective equipment or personal monitoring and the systems of work in the local rules (Section 7.8). Systems of work will be designed to restrict exposure as much as practicable. The risk assessment would then be reviewed periodically to ensure that the assessment remains valid and doses are kept as low as reasonably practicable – ALARP.

When a new procedure or activity is implemented, it can be an exciting time for those involved. However, a key activity that Radiation Safety Scientists are involved with is risk assessing the activity. The aim is not to prevent that activity but to ensure that it is done in the safest way possible.

CASE STUDY: A NEW X-RAY PROCEDURE

A new X-ray procedure is being proposed for use in an operating theatre. Some staff need to be close to the patient during the exposure, including the surgeons, scrub nurse and the anaesthetist. Other staff are advised to leave the theatre during the exposure. A circulating nurse has to remain in the theatre but is advised to stand as far away from the patient as possible during the X-ray procedure (Figure 7.14).

Standing 2 m away from the patient reduces the circulating nurse's exposure by a factor of 16 compared with the surgeon who is standing 0.5 m from the radiation source (the X-ray tube/patient). Staff must be trained to ensure that they are aware of this. All staff in the theatre wear a lead apron giving some shielding to radiosensitive organs. The radiographer operating the machine only exposes the patient when necessary to keep the exposure time ALARP.

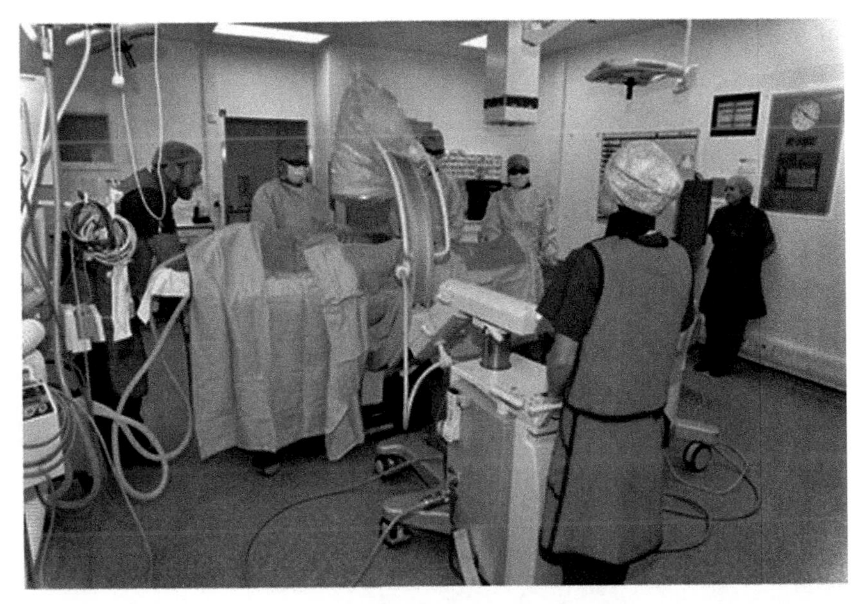

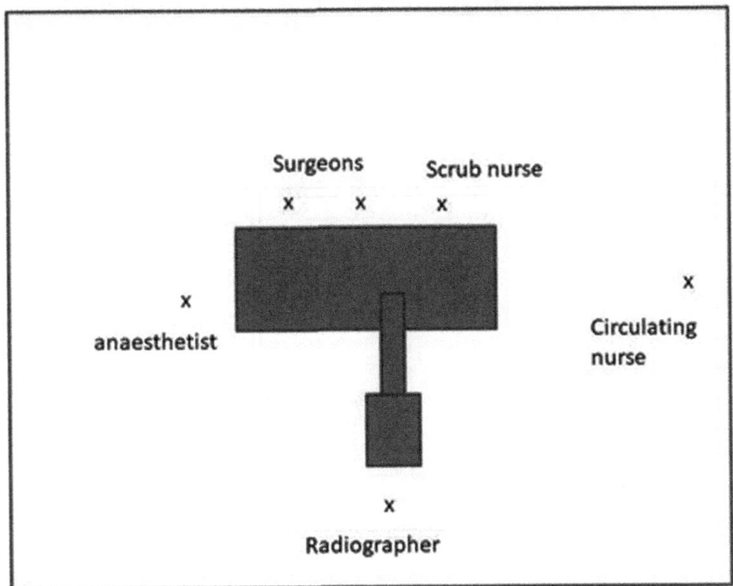

FIGURE 7.14 Staff positions within an operating theatre.

CASE STUDY: X-RAY SERVICE IN A GP SURGERY

As part of moving care closer to the patient, a GP surgery wishes to implement an "in house" plain film X-ray service. They do not know how many examinations they will do but want to offer a service 5 days per week between 9 am and 5 pm. The GP has obtained a basic plan for their X-ray room from an X-ray manufacturer, with the equipment laid out as shown in Figure 7.15.

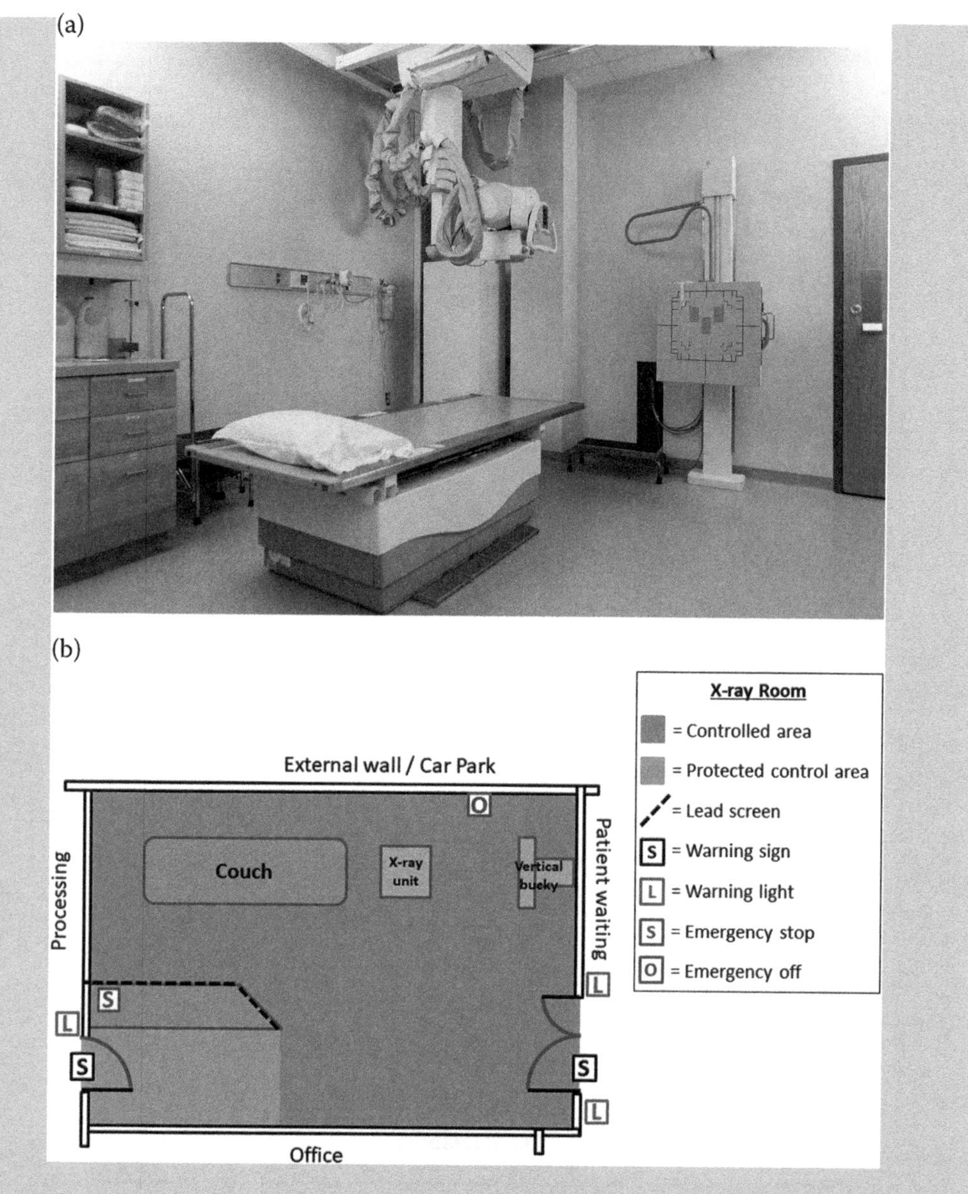

FIGURE 7.15 X-ray room layout: (a) photograph, (b) plan view.

The room is in a single-story building built from concrete blocks. It is on the ground floor with nothing below and there is a window opening out onto a car park. Areas adjacent to the room include the car park, an image processing area, a waiting area and an office. The doors into the room open into a waiting area. There is no information on the thickness or density of the concrete blocks.

If we simply place an X-ray system in the room, a risk assessment will demonstrate that there is an unjustifiably high risk to staff and the public. Initial discussions identified that the practice team want to be able to take X-rays of the chest, spine, abdomen, pelvis and extremities. The Scientist agreed dose levels with the GP that would be acceptable for staff and members of the public in surrounding areas. The Scientist then performed calculations to determine whether additional shielding would be required to improve the attenuation of X-rays by walls and doors, and to screen operators in the adjacent office.

For this example, the following set of simple instructions were issued:

- All doors must be closed before an exposure is made.
- The operator should check that no one other than the patient is in the room before making the exposure.
- If individuals need to remain within the room for some reason, a lead apron should be worn.
- Individuals that remain within the controlled area will require personal dosimetry.

The Scientist will support staff throughout this process to identify potential radiation hazards and ensure new processes are consistent with regulatory requirements. The Scientist can make suggestions to ensure doses are ALARP and will think about all the things that might go wrong, to ensure that everyone involved is aware of any risk and measures are in place to reduce the likelihood and severity of mistakes. Of course, there are always problems, or combinations of events, that could not have been reasonably foreseen, but we do our best to ensure accidental or over-exposure incidents are avoided.

7.6.2 Radon Risk Assessments

Radon is a gas released from the radioactive decay of uranium in rocks and soil. High levels of radon lead to an increased risk of lung cancer. Therefore, radon risk assessments are required for all premises, not just hospitals. In most areas, it is enough to check that the building is not in a radon area. However, in some areas, further investigation may be required. More details can be found on the UK radon website (https://www.ukradon.org/).

7.7 CONTROLLED AREAS

Areas where the risk assessment has shown that exposure has the potential to exceed thresholds lower than recommended dose limits are designated as "controlled areas", where exposure to radiation needs to be monitored and controlled. Any area that exceeds a dose rate of 7.5 μSv/hr should be subject to additional consideration to decide whether controlling the area is required (HSE 2018).

Controlled areas are ideally separated from other areas (preferably by solid structures, such as walls, floors and ceilings) so that it is easy to control access. Controlled areas should also be clearly demarcated with warning signs posted at each entrance. Requirements will depend on the type of equipment being used. For example, for planar X-ray equipment, it is customary to shield only to a height of 2.2 m above finished floor level, but for CT it is usually necessary to shield up to the underside of the soffit so that radiation cannot escape from the room and reflect back down from the ceiling to people in adjacent areas. Where practicable areas associated with higher exposures should be delineated by physical barriers.

In many healthcare settings, the situation may be more complicated. For example, a mobile X-ray unit might be taken to a ward to image a patient while another patient is in an adjacent bed. In this case, the Medical and Dental Guidance Notes MDGN (IPEM 2002) may provide enough information to set a fixed distance as the limit of the controlled area. Access will need to be controlled by the Radiographer taking the X-ray.

In more complicated situations, calculations of dose, dose rate and distances from the source to critical persons might be carried by a Clinical Scientist (Section 7.6). In Nuclear Medicine, injections may be taken up onto wards so that less mobile patients do not have to visit the Nuclear Medicine department. The dose rate from the source in this case could mean a controlled area needs to temporarily exist around the syringe. The syringe would normally be shielded and further placed in a lead-lined carrycase to mitigate this requirement during transport. Once the patient has been injected, although the dose rate emitting from them might be relatively high, there is no requirement to create a controlled area around the patient, although dose that might result from exposure to such patients will be considered in local protocols.

Exercise:

Risk assessment of the diagnostic X-ray room drawn below revealed several issues with potential to increase the risk of X-ray exposure to staff and the public. Before looking at the model answer, critically review the layout in Figure 7.16 and see how many issues you can spot? Can you propose a better layout?

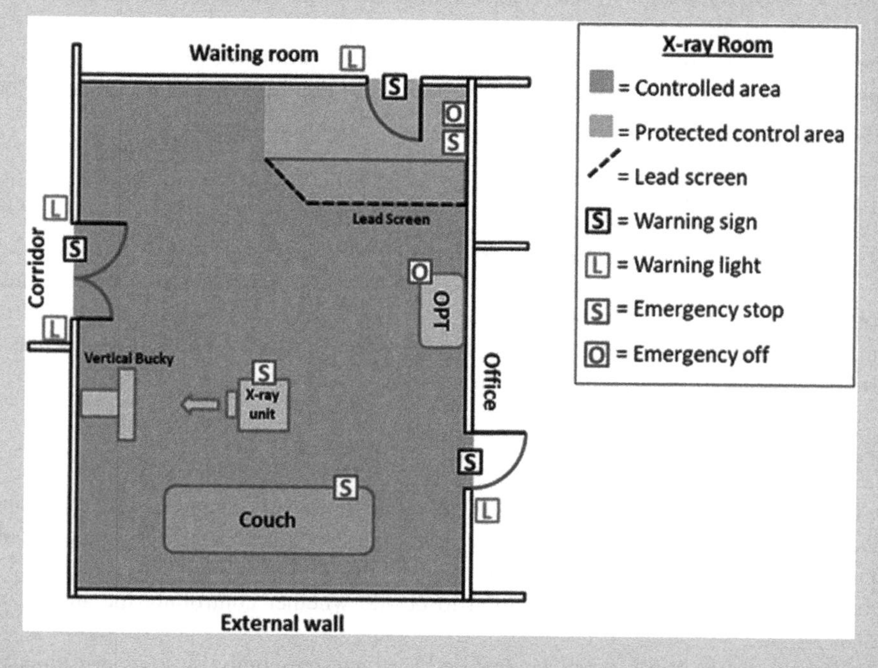

FIGURE 7.16 Plan of a diagnostic X-ray room.

Answer:

The office can be accessed through the X-ray room. It could be high occupancy and so should be shielded to prevent exposure of office staff. This is a very unusual and undesirable arrangement. Although the office features a light and a sign to show when an exposure is made there is nothing to prevent someone entering the X-ray room from the office while the X-ray is taken. Local rules should be in place to either keep the office door locked (if there is access from elsewhere) or check that the office is empty before starting an exposure.

The vertical "bucky", which holds the X-ray detector is very close to the door. Whilst it is the small leaf for the door, if anyone walks in through this door they are likely to be close to the beam. Any spill-over will also be incident on the door, which does not provide adequate shielding for the primary beam. There is not much to be done about this layout once it is in place but ensuring that the small leaf (least used) is closed will minimise the risk.

On the plan, the area behind the bucky is of unknown use. This should be established as part of the planning process.

Exercise:

A person ignores the warning light outside of the room and walks into the room when a chest X-ray is being taken. Assuming they are 1 m from the patient during the exposure, if the maximum scatter at 1 m (S_{max}) is [((0.031∗kV) + 2.5)] microGy/Gycm^2 (Sutton et al. 2012) what dose would that individual receive?

Answer:

The maximum dose is ((0.031 × 125) + 2.5) µGy/Gycm^2

The National Diagnostic Reference Level for a chest exam is 0.15 Gycm^2

Therefore, the scatter at 1 m = 0.15 ∗ 6.4 µGy

I.e. <1 µGy equivalent to a few hours of natural background radiation in the UK.

At a 0.5 m distance, the dose would be <4 µGy.

While the dose for a single occurrence is quite low, this is not ALARP. If the dose were higher interlocks might be fitted to doors to actively prevent this from happening. Clinical flow and freedom of movement are considered important in the healthcare setting, so this is managed through local rules and systems of work.

7.8 LOCAL RULES

For any areas designated as either controlled or supervised, IRR (2017) requires local rules, or "systems of work" to be in place. These usually take the form of written instructions describing how to work safely in the controlled/supervised area. As with conducting radiation risk assessments, the level of involvement of Clinical Scientists in drafting and reviewing local rules will vary between teams.

Local rules should provide a simplified summary of key findings from any risk assessments, a description of the areas and their designation, dose levels, and the name of a local Radiation Protection Supervisor. The local rules should clearly describe the working instructions required to restrict exposure and any contingency plans (these will be discussed later). It is often useful to include a diagram of the controlled area and the surrounding rooms, especially if one set of rules covers multiple rooms. If areas are designated as controlled areas only for specific uses, the local rules need to be clear about when the designation is in place.

Radiation Protection Supervisors are usually senior members of staff who have enough authority and knowledge of legislative requirements to ensure that local rules are followed. A good working relationship with hospital Radiation Protection Supervisors is important for ensuring that issues are picked up and communicated effectively. The appointment of a Radiation Protection Supervisor is required by IRR(2017) to ensure that the local rules are followed, and any issues are appropriately reported and followed-up.

Entry into a controlled area is limited to either classified radiation workers or staff working under the system of work in the local rules. Classified workers are highly trained and will normally in healthcare also follow the same systems of work. The local rules, therefore, need to indicate who may enter the room and under what conditions they may do so. Dose levels from working in the controlled area should be stated clearly to help ensure that doses remain ALARP. These can be based on dose estimates from the risk assessment, or from previous monitoring of personal dosimetry badges. Local rules should explain what to do if expected dose levels are exceeded. Effective local rules will be brief, truly local to that area, and focus on the main risks and working instructions required in practice, rather than in an ideal situation.

7.8.1 Classified Radiation Workers

Under IRR (2017) (HSE, 2018) employers must classify staff members where it is reasonably foreseeable that they could receive:

- An effective dose greater than 6 mSv per year, or
- An equivalent dose greater than 15 mSv per year to the lens of the eye, or
- A dose greater than 150 mSv per year for the skin or the extremities (hands forearms, feet or ankles).

This could apply, for example, to staff working in a radiopharmacy in Nuclear Medicine whose fingers may receive a high cumulative dose over time. Cardiologists and Interventional Radiologists can also receive high doses to their eyes and/or their fingers as they spend long periods close to the patient and X-ray tube as part of their routine work. Staff must also be classified if it is reasonably foreseeable that annual dose limits could be reached in an accident situation (for example, during a single accidental exposure to high dose-rate brachytherapy).

If you become a classified radiation worker, you will:

- Be informed that you are classified. Undergo an occupational health check and annual medical examinations to verify that you are fit to work with ionising radiation.
- Have your radiation dose assessed and monitored by an Approved Dosimetry Service (ADS). Dose records are submitted annually to the Central Index of Dose Information (CIDI).

7.8.2 Personal Dose Monitoring

Radiation workers wear personal dosimeters to demonstrate that the dose they receive is well within any dose limit and that the dose received is as low as reasonably practicable (Martin et al. 2018).

Most workers wear whole body dosimeters on their chest or waist to estimate their whole-body dose (personal dose equivalent). These are known as passive detectors and might be made from TLD (thermoluminescent material) or optically stimulated dose meters as described by Sutton et al. (2012). These badges are changed periodically at a frequency of between 2 weeks and 3 months. Where it is important to know the dose in real time, there are some active dose meters that can be worn and semi-passive devices that allow periodic Bluetooth reporting of dose.

Reviews of dose over time can be an important part of a Scientist's work. They might be involved in collating graphs with data such as that below (Figure 7.17), looking at trends year on year, reviewing individual results that might exceed a threshold and making recommendations on monitoring.

Such reviews require knowledge of the work carried out as well as the properties of the dosimeters in use. Although there are relatively few individuals recording higher doses it is important to ensure that these individuals are not recording significantly higher doses than colleagues doing the same work as this would indicate that best practice is not being followed.

It is also important to be aware of how dosimeters are worn. Currently in some hospitals, dosimeters are often worn under lead aprons. However, going forward there is a move towards wearing dosimeters above lead aprons, which will result in a step change in the historical dose. The reason for this change is that the dose limit that is most likely to be breached is for the eye. By

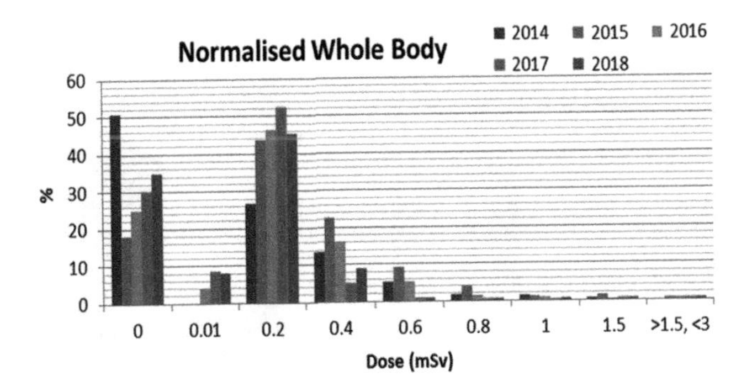

FIGURE 7.17 Whole-body doses received by all staff in a hospital over time.

monitoring outside the lead apron, an indication of eye dose can be obtained, with additional monitoring of the eye undertaken once the whole-body dose recorded reaches a certain level.

7.9 PATIENT SAFETY

From the moment that a patient is referred for a procedure involving radiation, under IR(ME)R (2017) their care is covered by several written procedures in place to protect the patient from exposure to unnecessary radiation. These procedures will be part of an organisational radiation safety policy, which defines governance structures, roles and responsibilities. To really understand how key these procedures are to the safety of patients, it is instructive to map the patient pathway and correlate this with the procedures (Figure 7.18).

There is also a requirement in Nuclear Medicine for a procedure for the patient to be issued with instructions on how to minimise dose to others. Non-medical imaging procedures cover any exposure to radiation that is not for a medical purpose. Examples may include X-rays for legal requirements, occupational health monitoring or to determine the age of an individual.

Further to this, there is a requirement to have procedures in place covering the governance of:

- How operators, practitioners, referrers and Medical Physics Experts are identified.
- How quality control is performed.
- How to minimise the likelihood and impact of accidents.

7.10 RADIATION SAFETY CULTURE

One of the most important tools for promoting radiation safety is a strong radiation safety culture. If a strong radiation safety culture exists, staff will automatically follow safety procedures (even when no-one is watching) as it is so embedded in their ways of working. Developing strong working relationships with staff in clinical areas (including Radiation Protection Supervisors) is an important first step in developing a strong culture of radiation safety. Good communication and influencing skills – spoken and written, are essential. Above all, Clinical Scientists working within Radiation Safety need to act as exemplary role models, sharing good practice with colleagues.

A safety culture maturity model that can be followed to help engender a strong radiation safety culture (Figure 7.19).

Radiation safety culture is a combination of attitudes, priorities, policies and practices. The first stage of developing a culture of safety is obtaining a commitment to safety from management. This

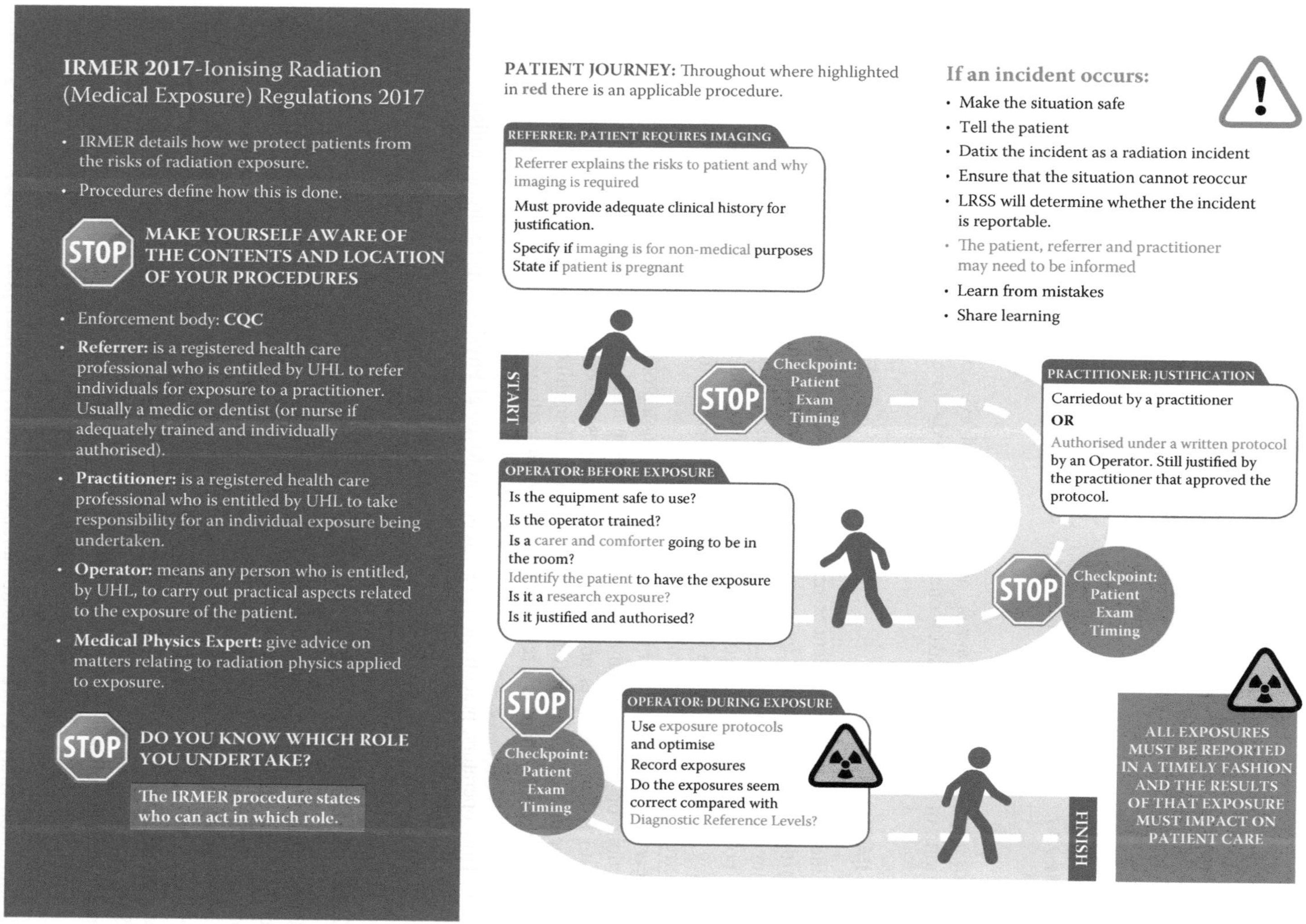

IRMER 2017-Ionising Radiation (Medical Exposure) Regulations 2017

- IRMER details how we protect patients from the risks of radiation exposure.
- Procedures define how this is done.

STOP **MAKE YOURSELF AWARE OF THE CONTENTS AND LOCATION OF YOUR PROCEDURES**

- Enforcement body: **CQC**
- **Referrer:** is a registered health care professional who is entitled by UHL to refer individuals for exposure to a practitioner. Usually a medic or dentist (or nurse if adequately trained and individually authorised).
- **Practitioner:** is a registered health care professional who is entitled by UHL to take responsibility for an individual exposure being undertaken.
- **Operator:** means any person who is entitled, by UHL, to carry out practical aspects related to the exposure of the patient.
- **Medical Physics Expert:** give advice on matters relating to radiation physics applied to exposure.

STOP **DO YOU KNOW WHICH ROLE YOU UNDERTAKE?**

The IRMER procedure states who can act in which role.

PATIENT JOURNEY: Throughout where highlighted in red there is an applicable procedure.

REFERRER: PATIENT REQUIRES IMAGING

Referrer explains the risks to patient and why imaging is required

Must provide adequate clinical history for justification.

Specify if imaging is for non-medical purposes
State if patient is pregnant

START

STOP Checkpoint: Patient Exam Timing

OPERATOR: BEFORE EXPOSURE

Is the equipment safe to use?
Is the operator trained?
Is a carer and comforter going to be in the room?
Identify the patient to have the exposure
Is it a research exposure?
Is it justified and authorised?

STOP Checkpoint: Patient Exam Timing

OPERATOR: DURING EXPOSURE

Use exposure protocols and optimise
Record exposures
Do the exposures seem correct compared with Diagnostic Reference Levels?

If an incident occurs:

- Make the situation safe
- Tell the patient
- Datix the incident as a radiation incident
- Ensure that the situation cannot reoccur
- LRSS will determine whether the incident is reportable.
- The patient, referrer and practitioner may need to be informed
- Learn from mistakes
- Share learning

PRACTITIONER: JUSTIFICATION

Carriedout by a practitioner
OR
Authorised under a written protocol by an Operator. Still justified by the practitioner that approved the protocol.

STOP Checkpoint: Patient Exam Timing

FINISH

ALL EXPOSURES MUST BE REPORTED IN A TIMELY FASHION AND THE RESULTS OF THAT EXPOSURE MUST IMPACT ON PATIENT CARE

FIGURE 7.18 Example of a leaflet designed by scientists to help with training staff to understand the IR(ME)R17 implications ado practical radiation safety controls in

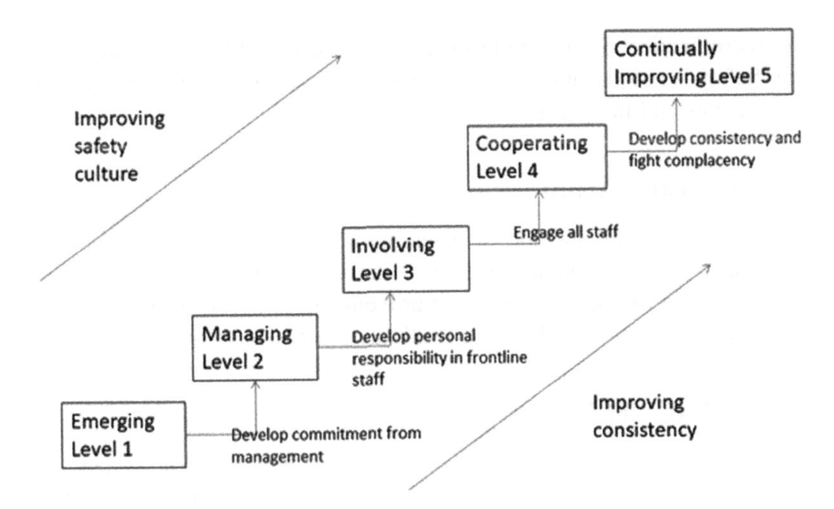

FIGURE 7.19 Safety culture maturity model.

is usually demonstrated initially by the development, ratification and implementation of a radiation safety policy within a hospital. Senior Clinical Scientists will usually contribute to these policies, but all staff are expected to read and abide by their principles.

Staff then need to be involved in ensuring that radiation safety becomes an integral part of clinical routine. It may be necessary to provide users with additional training and support to promote a safe working environment, especially if inspections by external bodies or regulators find the culture lacking. The successful establishment of a strong radiation safety culture requires active participation and collaboration of all stakeholders. Insufficient knowledge and lack of collaboration can be significant barriers.

In areas where a strong culture of radiation safety already exists, less oversight is usually required, however, it is still necessary to guard against complacency. Clinical Scientists may need to carry out regular audits to ensure high standards are consistently maintained by all staff. As radiation safety culture can be difficult to quantify, it is sometimes necessary to combine information from different sources (dosimetry, surveys or observations, incidents and audits) to evaluate the culture and highlight any issues.

7.11 MONITORING RADIATION SAFETY

Systems put in place to monitor radiation safety include, but are not limited to:

- Safety walkabouts by management to identify work practices within the area that need to be improved.
- Regular local audits of compliance with procedures and legislation.
- Monitoring of incidents and near-misses reported in the area, including feedback to staff to minimise the likelihood of reoccurrence.
- Regular meetings with frontline staff to discuss safety issues and support individuals in making positive changes.

It is important that there is a clear management structure and the method for escalation of issues via the Radiation Protection Supervisor, or other staff, is clearly defined. Staff must be empowered to "stop the line" if they believe that unsafe working practices are being undertaken and to do so without fear of recrimination. Staff at all levels must feel responsible for safe working within their area. Radiation Safety is not just about complying with relevant legislation, but about ensuring the safety of staff, patients and the public.

7.11.1 Radiation Safety Audits

A compliance audit provides a method of assessing safety culture and regulatory compliance. Audits provide scientists with an understanding of strengths and weaknesses within each area, to provide well-informed recommendations for improvement. There are many types of audits that could be completed by a Clinical Scientist and all have their own benefits and drawbacks.

7.12 CONTINGENCY PLANNING

No matter how safe the area and how well the people trained in that area no system can be assured to remain incident free forever. Therefore, individuals need to know how to respond when things go wrong. Some incidents identified by the risk assessment will be quite likely to occur but have minor repercussions, while others may be extremely unlikely but have serious consequences. Contingency planning requires radiation users to develop plans describing exactly what to do if an incident occurs that requires deviation from the normal work to ensure limitation of the consequences of that incident.

Contingency plans are required to be rehearsed so that, should something go wrong, the right actions are more likely to be followed. Clinical Scientists in the NHS are involved in drawing up these plans, in their rehearsal, and in carrying out investigations of incidents when things have gone wrong.

There are various pieces of radiation legislation relevant to contingency planning (Figure 7.20). The Radiation Emergency Preparedness Regulations are unlikely to apply to hospitals.

Local contingency plans, as required under IRR (2017), require involvement from all relevant local staff. Other levels of contingency planning are likely to be undertaken by specialist staff within Radiation Safety.

7.12.1 Local Contingency Planning

Following a risk assessment, as required by the Ionising Radiation Regulations (HSE 2018), any realistically possible accidents involving radiation must have contingency plans associated with them. These contingency plans must be detailed in the local rules. Typical incidents requiring a contingency plan are listed in the Medical and Dental Guidance Notes (IPEM 2002).

Development of contingency plans should involve:

- Representatives of the people that will undertake the actions detailed in the plan;
- An RPA or a Radiation Safety scientist;

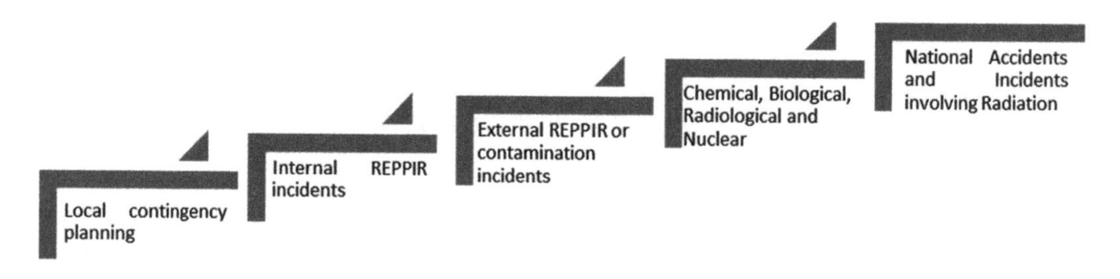

FIGURE 7.20 Levels of contingency planning.

- Managers of the area to which the plan applies;
- The Radiation Protection Supervisor.

If an accident occurs and the contingency plan is acted on, there is a requirement to record the outcome, including the details required by the Approved Code of Practice, HSE (2018), and retain this information for a period of 2 years. Therefore, it is important to ensure that contingency plans are limited to radiation incidents and do not cover regular activities such as low-level contamination within Nuclear Medicine. Therefore, there should be a clear definition within the contingency plan describing situations where the plan should be enacted.

The Approved Code of Practice (HSE 2018) gives the required contents of the contingency plan. The plan should contain enough detail for those working in the area to understand exactly what is required of them. Responsibilities and contact details should be clearly stated, and up to date. For detailed contingency plans, a flowchart may be useful as an aide-memoire. Training should be provided to relevant staff and it may be prudent to place the flowchart or details of the plan within areas where the contingency plan may need to be enacted. Finally, contingency plans need to be rehearsed at a frequency determined by the level of risk.

In diagnostic radiology applications, contingency plans rarely extend beyond the use of the emergency stop to terminate exposure. Therefore, extensive rehearsals are not usually undertaken. The position and use of the emergency stop should be part of staff training and this knowledge should be verified annually.

In Nuclear Medicine, where radioactive samples are frequently handled, more detailed rehearsals are required. This is commonly covered by training using mock spills. A typical rehearsal will include the following:

- Prior to the rehearsal taking place all staff involved with the rehearsal should sign to say that they have read the local rules.
- An individual will be made responsible for the planning of the exercise to be undertaken. They should determine the aims of the rehearsal, the aspects of the plan they wish to test and how best to do this. This may include a practical exercise, checking that contact details are correct, checking the availability of equipment, or holding a tabletop discussion.
- An observer should be appointed to make notes on the progress of the plan and any deviations from it. This is usually the planner.
- For practical exercises, the activity should be planned to minimise any risks. Where possible, record contemporaneous, timed, accounts of responses. Where this is not possible, record the actions taken as soon as possible after the event.
- After completion staff should be given time as part of the rehearsal to finish off their written accounts of what happened.

In brachytherapy, rehearsals are also needed in case a source is not returned to its shielded housing. In diagnostic radiology, emergency stops should be regularly tested if safe to do so. Some may force a hard shut down which might damage the equipment.

7.12.2 National Arrangements for Incidents Involving Radioactive Material

Scientists involved in radiation safety may be part of National Arrangements for Incidents Involving Radioactive Material (PHE, 2021). There is a call-out system through hospital switchboards and scientists can be asked to attend an incident at any time of the day or night.

Call-outs can be minor – for example, someone throws away a can with radioactive markings on the outside but there is no radioactive material involved. This might involve a simple measurement before closing the incident. If the presence of radioactive material is identified, the resources required to deal with an incident can be more significant. Training is given by Public Health England for incidents of this sort and individual scientists might be called a few times over their career. Any major incidents would usually also involve calling out a Health Physics team from the Nuclear power industry who have resources and instruments available to deal with large-scale incidents. Significant events in hospitals, such as the radiation poisoning of Alexander Litvinenko, although not falling under the NAIR scheme, stretch hospital resources requiring national support from Public Health England.

7.13 INVESTIGATING AN INCIDENT

In the event of something going wrong, such as a patient being inadvertently exposed to radiation, staff must report the incident (CQC 2017). An investigation then takes place to ensure the causes are understood and the results of the investigation are shared. Under IR(ME)R (2017), there is a duty to inform the referrer, practitioner and patient if radiation exposure due to the incident has the potential to be clinically significant. If there is a serious incident, it is likely that senior members of the safety team within the Trust will also be involved.

Radiation safety aims to cultivate a culture of reflective learning, openness and continuous improvement. There must be an understanding that most accidents are not caused solely by individual human error, but by systems that allow errors to occur. It is therefore important to ensure that a thorough root-cause analysis is applied to all incidents.

Root cause analysis is about identifying the underlying "“root-causes" of an incident. Basic skills in root cause analysis are a key component of the training of any Clinical Scientist. A simple root cause analysis might begin by using the "5 whys" method, as illustrated in the following example:

CASE STUDY: ACCIDENTAL EXPOSURE OF A PATIENT

A Clinical Scientist is asked to investigate the reasons leading to accidental over-exposure of a patient. Performing a root-cause analysis using the 5 whys method, they list the following reasons (or Whys):

1. **Why was the patient over-exposed?** The exposure was set to use the automatic exposure control (AEC) to terminate but the X-ray was not incident on the AEC chamber.
2. **Why was the exposure allowed?** Because the manual override key was in place.
3. **Why was the override key in place?** The previous Radiographer needed to perform an unusual exposure and forgot to remove the key after the previous patient had been scanned.
4. **Why did the first radiographer forget to remove the key?** They were dealing with a difficult patient.
5. **Why did the second radiographer fail to notice that the key was still in place?** The key is small and was covered by the handles on the X-ray unit.

It is not always possible to follow through to five whys; however, the process of continually questioning what happened gives you opportunities to think about potential solutions. In the aforementioned example, it might be useful to investigate:

- How common it is for Radiographers to override the automated exposure control system?

- Would it be possible to alter the equipment settings to bring more of these unusual exposures under standard protocols?
- Could the override key have been made more visible by adding a larger fob?
- Could a message come up on the screen to alert the users that the override key is activated before starting an exposure?

If human error was identified as the sole cause of the incident, and no further investigation undertaken, then the problem would be likely to reoccur. Radiographers will often be dealing with difficult patients, so systems need to be in place that take human error into account, making it as difficult as possible for these sorts of incidents to occur.

A further method, called Human factors analysis, has also been applied to radiation safety. This focuses on designing systems around the humans involved in a process, rather than asking staff to follow set rules. This theory is sometimes helpful when developing local rules and procedures.

Ishikawa (or fishbone) diagrams represent another form of cause and effect analysis. Ishikawa analysis invites you to consider all factors that may have contributed to an incident. Contributory factors are usually depicted in the form of a fishbone, but essentially the aim is to generate a mind-map listing all contributory factors (Figure 7.21).

After identifying contributory factors and potential preventative actions, the following steps should be undertaken:

- All process changes should be risk assessed to ensure that no new risks are introduced to other points in the system.

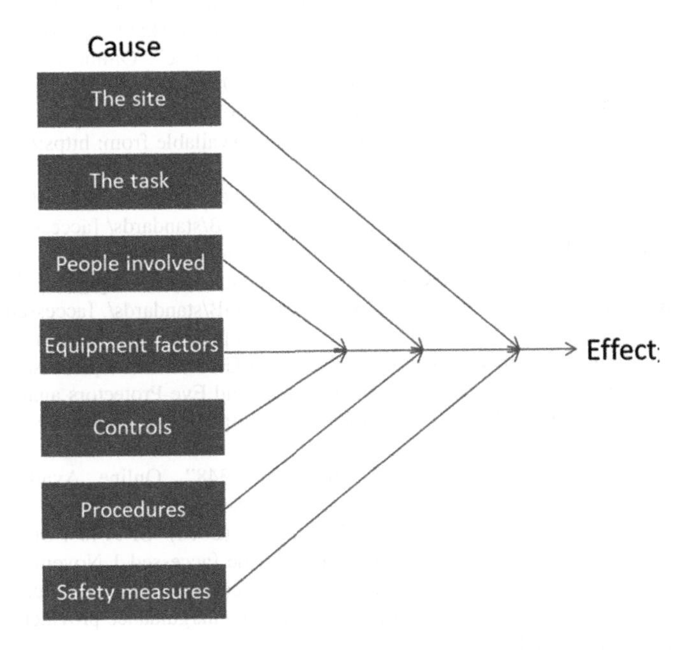

FIGURE 7.21 Ishikawa diagram example. Each cause may have multiple contributory factors.

- A SMART (specific, measurable, action-orientated, realistic and time-bound) action plan should be proposed. The action plan should be "owned" by the area with actions attributed to specific individuals.
- Action plans should include a plan for monitoring and completion (or closure).
- Any issues with closure should be escalated up the management chain.

7.14 THE FUTURE OF RADIATION SAFETY

Radiation safety is as varied as the work done in the hospital. As techniques and equipment develop, then so too must safety requirements. Molecular radiotherapy is a developing area which requires considerable radiation safety resource.

Radiation detectors are developing for personal dosimetry. Dynamic readout and artificial intelligence to look at the results of dosimetry measurements are active areas of research and development. Novel ways of training staff, involving other disciplines such as psychology, are beginning to develop.

Further to this, it is expected that as health and safety practices become more evidence-based, human factors and failure mode effects analysis will become more influential. The culture surrounding radiation safety is also always changing, during recent years there has been a move from "blame culture" to "no blame culture" and again to just "culture". Similarly, the introduction of checklists into safety culture within hospitals has been introduced to radiation.

REFERENCES

BAD (2016) "Service Guidance and Standards for Phototherapy Units". Online Available from: https://www.bad.org.uk/library-media/documents/Phototherapy%20Service%20Guidance%20and%20Standards%202018(4).pdf [accessed 30 October 2020].

BEIS (2004) "The Justification of Practices Involving Ionising Radiation Regulations". Online Available from: https://assets.publishing.service.gov.uk/government/uploads/system/uploads/attachment_data/file/804958/Justification-of-practices-involving-ionising-radiation-regulations-2004.pdf [accessed 1 November 2020].

BIR (2016) "Personal Protective Equipment for Diagnostic X-ray Use". Online Available from: https://www.birpublications.org/doi/book/10.1259/book.9780905749846 [accessed 1 November 2020].

BS EN 208 (1999) "Personal Eye-Protection. Eye-Protectors for Adjustment Work on Lasers and Laser Systems (Laser Adjustment Eye-Protectors)". BSI. Online Available from: https://www.bsigroup.com/en-GB/standards/ [accessed 15 March 2021].

BS EN 60825-8 (2006), "Safety of Laser Products – Part 8 Guidelines for the Safe Use of Lasers on Humans". BSI. Online Available from: https://www.bsigroup.com/en-GB/standards/ [accessed 15 March 2021].

BS EN 60601-2-22 (2013) "Medical Electrical Equipment – Particular Requirements for Basic Safety and Essential Performance of Surgical, Cosmetic, Therapeutic and Diagnostic Laser Equipment". BSI. Online Available from: https://www.bsigroup.com/en-GB/standards/. [accessed 15 March 2021].

BS EN 60825-1 (2014) "Safety of Laser Products – Part 1 Equipment Classification and Requirements". BSI. Online Available from: https://www.bsigroup.com/en-GB/standards/ [accessed 15 March 2021].

BS EN 207 (2017) "Personal Eye-Protection Equipment. Filters and Eye-Protectors against Laser Radiation (Laser Eye-Protectors)". BSI. Online Available from: https://www.bsigroup.com/en-GB/standards/ [accessed 15 March 2021].

Carriage of Dangerous Goods Act (2019) "SI 2009 No 1348". Online Available from: https://www.hse.gov.uk/cdg/regs.htm [accessed 1 November 2020].

Control of Artificial Optical Radiation (AOR) at Work Regulations (2010) "SI 1140". Online Available from: https://www.legislation.gov.uk/uksi/2010/1140/contents/made [accessed 1 November 2020].

CQC (2017) "Guidance on Investigation and Notification of Medical Exposures much greater than Intended, 16 January 2017". Online Available from: https://www.cqc.org.uk/guidance-providers/ionising-radiation/definition-significant-accidentalor-unintended-exposures

CQC (2020) "Care Quality Commission IR(ME)R Annual Report". Online Available from: https://www.cqc.org.uk/guidance-providers/ionising-radiation/ionising-radiation-medicalexposure-regulations-IR(ME)R [accessed 30 October 2020].

CRUK (2020) "Data and Statistics". Online Available from: https://www.cancerresearchuk.org/health-professional/cancer-statistics/risk [accessed 9th May 2021]

DHSC (2014) Department of Health and Social Care (DHSC) Committee on Medical Aspects of Radiation in the Environment (COMARE) "16th Report. Patient Radiation Dose Issues Resulting from the Use of CT in the UK". Online Available from: https://assets.publishing.service.gov.uk/government/uploads/system/uploads/attachment_data/file/343836/COMARE_16th_Report.pdf [accessed 1 November 2020].

DHSC (2019) *Guidance to the Ionising Radiation (Medical Exposure) Regulations*. Department of Health and Social Care.

EPR (2016) *The Environmental Permitting (England and Wales) Regulations*. SI 2010/67. The Stationery Office, London, England.

Hamada N, Fujimichi Y, Iwasaki T, et al. (2014) "Emerging Issues in Radiogenic Cataracts and Cardiovascular Disease". *J Radiat Res* **55**(5): 831–846.

Henderson R and Schulmeister K (2004) *Laser Safety*. CRC Press, Boca Raton.

HSE (nd) Online Available from: https://www.hse.gov.uk/construction/lwit/assets/downloads/hierarchy-risk-controls.pdf [accessed 23 March 2021].

HSE (2015) *The Health and Safety (Safety Signs and Signals) Regulations*, L64, 3rd edition. Online Available from: https://www.hse.gov.uk/pubns/priced/l64.pdf [accessed 30 October 2020].

HSE (2018) "Work with Ionising Radiation. Ionising Radiation Regulations 2017. Approved Code of Practice and Guidance". L121, 2nd edition. Online Available from: https://www.hse.gov.uk/pubns/priced/l121.pdf [accessed 30 October 2020].

HSE (2019) "Radiation (Emergency Preparedness and Public Information Regulations) (REPPIR)". Online Available from: https://www.hse.gov.uk/radiation/ionising/reppir.htm [accessed 30 October 2020].

IAEA (2014) *Radiation Protection and Safety of Radiation Sources: International Basic Safety Standards*. Series No. GSR Part 3. IAEA, Vienna.

ICRP (2007) "Publication 103: The 2007 Recommendations of the International Commission on Radiological Protection". *Ann ICRP* **37**(2–4). Online Available from: https://journals.sagepub.com/doi/pdf/10.1177/ANIB_37_2-4 [accessed 30 October 2020].

ICRP (2012) "Publication 118: ICRP Statement on Tissue Reactions and Early and Late Effects of Radiation in Normal Tissues and Organs – Threshold Doses for Tissue Reactions in a Radiation Protection Context', *Ann ICRP* **41** (1-2). Online Available from: https://journals.sagepub.com/doi/pdf/10.1177/ANIB_41_1-2 [accessed 30 October 2020].

IPEM (2002) "The Medical and Dental Guidance Notes; A Good Practice Guide to Implement Ionising Radiation Protection Legislation in the Clinical Environment". IPEM, York, England.

IPEM (2010) *Phototherapy Physics: Principles, Sources, Dosimetry and Safety*. IPEM Report 104. IPEM, York.

IPEM (2017) *Report 75 Design and Shielding of Radiotherapy Treatment Facilities*, 2nd edition. IOP eBooks, London.

IR(ME)R (2017) *The Ionising Radiations (Medical Exposure) Regulations*. SI 2000/1059. The Stationery Office, London, England. Online Available from: https://www.legislation.gov.uk/uksi/2017/1075/contents/made [accessed 30 October 2020].

Knoll G (2010) *Radiation Detection and Measurement*, 4th edition. Wiley.

Martin, CJ et al. (2018) *IPEM report Guidance on the Personal Monitoring Requirements for Personnel Working in Healthcare*. IOP eBooks, London.

Martin CJ and Sutton DG (2015) *Practical Radiation Protection in Healthcare*, 2nd edition. Oxford University Press, Oxford, England.

MHRA (2015) "Lasers, Intense Light Source Systems and LED's – Guidance for Safe Use in Medical, Surgical, Dental and Aesthetic Practices". Online Available from: https://assets.publishing.service.gov.uk/government/uploads/system/uploads/attachment_data/file/474136/Laser_guidance_Oct_2015.pdf [accessed 30 October 2020].

Moseley H, Allan D, Amatiello H, et al. (2015) "Guidelines on the Measurement of Ultraviolet Radiation Levels in Ultraviolet Phototherapy: Report Issued by the British Association of Dermatologists and British Photodermatology Group 2015". *Br J Dermatol* **173**: 333–350. Online Available from: https://onlinelibrary.wiley.com/doi/full/10.1111/bjd.13937 [accessed 30 October 2020].

NCRP (2005) *Structural Shielding Design and Evaluation for Megavoltage X- and Gamma-Ray Radiotherapy Facilities*. Report 151. NCRP, Bethesda, MD.

Peet DJ et al. (2012) "Radiation Protection in Fixed PET/CT Facilities – Design and Operation". *Br J Radiol* **85**: 643–646.

PHE (2020) Publc Health England (PHE) Administration of Radioactive Substances Advisory Committee (ARSAC) "Notes for Guidance on the Clinical Administration of Radiopharmaceuticals and Use of Sealed Radioactive Sources". Online Available from: https://assets.publishing.service.gov.uk/government/uploads/system/uploads/attachment_data/file/912979/ARSAC_NfG_Sept_2020_FINAL_DRAFT_280820.pdf [accessed 1 November 2020].

PHE (2021) Public Health England "National arrangements for incidents involving radioactivity (NAIR)". Online Available from: https://www.gov.uk/guidance/national-arrangements-for-incidents-involving-radioactivity-nair [accessed 9th May 2021].

Sutton DG, Martin CJ, Williams JR and Peet DJ (2012) *Radiation Shielding for Diagnostic X-rays*, 2nd edition. British Institute of Radiology (BIR), London, England.

UNSCEAR (2020) "United Nations Scientific Committee on the Effects of Atomic Radiation Website". Online Available from: http://www.unscear.org/ [accessed 30 October 2020].

Wall BF, Haylock R, Jansen JTM, Hillier MC, Hart D and Shrimpton PC (2011) *Radiation Risks from Medical X-ray Examinations as a Function of the Age and Sex of the Patient*. Report HPA-CRCE-028. Health Protection Agency, Chilton, England.

Index

Note: *Italicized* page numbers refer to figures, **bold** page numbers refer to tables